Evolution vs. Creationism

An Introduction

Second Edition

EUGENIE C. SCOTT

FOREWORD BY NILES ELDREDGE

FOREWORD TO SECOND EDITION BY JUDGE JOHN E. JONES III

GREENWOOD PRESS
Westport, Connecticut • London

Library of Congress Cataloging-in-Publication Data

Scott, Eugenie Carol, 1945–
    Evolution vs. creationism : an introduction / Eugenie C. Scott; foreword by Niles
    Eldredge ; foreword to second edition by Judge John E. Jones III. — 2nd ed.
        p. cm.
    Includes bibliographical references and index.
    ISBN 978–0–313–34427–5 (alk. paper)
    1. Evolution (Biology) 2. Creationism. I. Title.
    QH367.S395   2009
    576.8—dc22   2008033529

British Library Cataloguing in Publication Data is available.

Library of Congress Catalog Card Number: 2008033529
ISBN: 978–0–313–34427–5

First published in 2009

Greenwood Press, 88 Post Road West, Westport, CT 06881
An imprint of Greenwood Publishing Group, Inc.
www.greenwood.com

Printed in the United States of America

The paper used in this book complies with the
Permanent Paper Standard issued by the National
Information Standards Organization (Z39.48–1984).

10 9 8 7 6 5 4 3 2 1

To my family, Charlie and Carrie

Contents

FOREWORD:
The Unmetabolized Darwin

A few weeks ago, I saw Darwin's name invoked in two separate articles in a single edition of *The New York Times*. One dwelled on a creationism controversy raging in a Midwestern state, while the other used the expression *Darwinian* in an offhand manner to allude to the dog-eat-dog competitiveness of the business world. I found it striking that, in both instances, it was *Darwin*, and not *evolution*, that was the key word. For in the beginning of the twenty-first century, it is Charles Robert Darwin who still stands out as the towering nineteenth-century intellectual figure who still gives modern society fits. Both Sigmund Freud and Karl Marx (to choose two others whose work also shook up Western society), though far from forgotten, after a good run have begun to fade from the front pages. Darwin recently replaced Charles Dickens on the British ten-pound note—ostensibly because his beard looks better, but in reality because he remains out front in our collective consciousness, increasingly alone among the voices of the past.

Why? Why does Darwin still bother so many of us in the Western world? Is it because Darwin's ideas of evolution are so difficult to understand? Or is it the very *idea* of evolution that is causing the problem?

The answer, of course, is the latter: the evolution of life through natural processes—and especially the recognition that our own species, *Homo sapiens*, is as inextricably linked to the rest of the living world as are redwood trees, mushrooms, sponges, and bacteria—still does not sit well with an awful lot of the citizenry of the United States and other Western countries. It is not that such skeptics are stupid—or even, at least in terms of their spokespersons, ill-informed. It's not, in other words, that creationists don't understand evolution: it's that they don't like it. Indeed, they revile it.

The reason that Darwin's name is still invoked so routinely is that social discourse on the cosmic origins of human beings has been stuck in a rut since the publication of his *On the Origin of Species* (1859). Roughly half of modern society at large grasps his point and is thereby able to understand why we look so much like chimps and orangutans—similar to the way people look at the matching shorelines of South

America and Africa and have no problem with the idea of continental drift. It seems commonsensical to this 50 percent of society to see us as the product of natural evolutionary processes—and when new facts come along, such as the astonishing 98.4 percent genetic similarity between humans and chimps, they seem to fit right in. These people have absorbed the evolutionary lesson and have moved on with their lives.

Darwin would be troubled but not especially surprised that the other roughly 50 percent of Americans (perhaps fewer numbers in his native England and on the European continent) still intransigently reject evolution. He had fully realized that life had evolved through natural selection—and that humans had evolved along with everything else—by the late 1830s. Yet, as is well known, Darwin pretty much kept his views a secret until virtually forced to "come out of the closet" and publish his views in the late 1850s by Alfred Russel Wallace's disclosure in a letter to Darwin that he had developed the same set of ideas. Darwin didn't want his earth-shattering idea to be scooped, so he hurriedly wrote the *Origin*—a book that sold out its initial print run on its first day of publication.

Although Darwin sometimes said that he waited twenty years to publish his ideas because he wanted to hone his concepts and marshal all the evidence he could (in itself a not-unreasonable claim), it is clear that the real reason for the delay was his fear of the firestorm of anger that his ideas were sure to unleash. His own wife was unhappy with his ideas; indeed, the marriage was almost called off when Darwin told her, against his father's advice, of his increasing religious doubts occasioned by his work. If Darwin's own faith was challenged by his conviction that life, including human life, had evolved through natural causes, he knew full well that the religiously faithful—nearly 100 percent of the population of Great Britain—would see his ideas in the very same stark terms. They too would see evolution as a challenge to the basic tenets of the Christian faith, and they would be very, very upset.

I agree with those historians who point to Darwin's nearly daily bouts with gastrointestinal upset as a manifestation of anxiety rather than of any systemic physical illness. Darwin finally did tell his new friend Joseph Hooker in 1844 a little bit about his secret ideas on evolution—telling him at the same time, though, that "it was like confessing a murder." Darwin knew he had the equivalent of the recipe for an atomic bomb, so devastating an effect would his ideas have on British society when he finally announced them. No wonder he was so hesitant to speak out; no wonder he was so anxious.

And, of course, his fears were well grounded. If it is the case that the majority of practitioners of the mainstream Judeo-Christian religions have had little problem concluding that it is the job of science to explain the material contents of the universe and how it works, and the task of religion to explore the spiritual and moral side of human existence, it nonetheless remains as true today as it was in the nineteenth century that a literal reading of Genesis (with its two and a half nonidentical accounts of the origin of the earth, life, and human beings) does not readily match up with the scientific account. There was a conflict then, and there remains a conflict today, between the scientific account of the history of earth and the evolution of life, on the one hand, and received interpretations of the same in some of the more hard-core Judeo-Christian sects. Darwin remains unmetabolized—the very reason that his name is still so readily invoked so long after he died in 1882.

Thus, it is not an intellectual issue—try as creationists will to make it seem so. Science—as many of the writings in this book make clear—cannot deal with the supernatural. Its rules of evidence require any statement about the nature of the world to be testable—to be subjected to further testing by asking the following: If this statement is true about the world, what would I expect to observe? If the predictions are borne out by experimentation or further observation, the idea is confirmed or corroborated—but never in the final analysis actually "proved." If, on the other hand, our predictions are not realized, we must conclude that our statement is in fact wrong: we have falsified it.

What predictions arise from the notion of evolution—that is, the idea that all organisms presently on Earth are descended from a single common ancestor? There are two major predictions of what life should look like if evolution has happened. As Darwin first pointed out, new features appearing within a lineage would be passed along in the same or further modified form to all its descendants—but would *not* be present in other lineages that had diverged prior to the appearance of the evolutionary novelty (Darwin knew that the idea of evolution must also include the diversification of lineages, simply because there are so many different kinds of organisms on earth). Thus the prediction: more closely related organisms share more similarities with each other than with more remotely related kin; rats and mice are more similar to each other than they are to squirrels; but rats and mice and squirrels (united as rodents) share more similarities than any of them share with cats. In the end, there should be a single nested set of similarities linking up all of life.

This is exactly what systematic biologists and paleontologists find as they probe the patterns of similarities held among organisms—in effect testing over and over again this grand prediction of evolution. Rats, squirrels, and mice share many similarities, but with all other animals—plus fungi and many microscopic forms of life—they share a common organization of their (eukaryotic) cells. They share even with the simplest bacteria the presence of the molecule RNA, which, along with the slightly less ubiquitous DNA, is the feature that is shared by all of life—and the feature that should be there if all life has descended from a single common ancestor.

Does this "prove" evolution? No, we don't speak of absolute proof, but we have so consistently found these predicted patterns of similarity to be there after centuries of continual research that scientists are confident that life has evolved.

The second grand prediction of the very idea of evolution is that the spectrum of simple (bacteria) to complex (multicellular plant and animal life) should be ordered through time: the earliest forms of life should be the simple bacteria; single celled eukaryotic organisms should come next in the fossil record—and only later do the more complex forms of multicellular life arrive. That is indeed what we do find: bacteria going back at least as far as 3.5 billion years; more complex cells perhaps 2.2 billion years; and the great "explosion" of complex animal life between five and six hundred million years—a rapid diversification that nonetheless has simpler animals (e.g., sponges and cnidarians [relatives of corals and sea anemones]) preceding more complex forms (like arthropods and mollusks). Among vertebrates, fishes preceded amphibians, which in turn preceded reptiles, which came, as would be expected, before birds and mammals.

Again, evolution is not proven—but it certainly is fundamentally and overwhelmingly substantiated by the failure to falsify this prediction of increasing complexity through time.

What do creationists have to refute the very idea of evolution? They trot out a mishmash of objections to specific scientific claims; to the extent that they are testable, creationists' ideas have long been refuted. More recently, they have reverted to notions of irreducible complexity and intelligent design—ideas presented as new, but actually part of the creationist war chest before Darwin ever published the *Origin*. The fact that organisms frequently display intricate anatomies and behaviors to perform certain functions—such as flying—has inspired the claim that there must be some intelligent designer behind it all, that a natural process like natural selection would be inadequate to construct such exquisite complexity.

There is, of course, no scientific way to test for the existence of the intelligent designer; on the other hand, we can study natural selection in the wild, in the laboratory, and in mathematical simulations. We can, however, ask whether patterns of history in systems that we *know* are intelligently designed—like cars, computers, or musical instruments—resemble those of biological history. I have actually done some work along these lines—and the answer, predictably and unsurprisingly, is that the evolutionary trees of my trilobites (the fossils I study) do not resemble the trees generated by the same program for my favorite man-made objects—the musical instruments known as cornets. The reason in a nutshell is obvious: the information in biological systems is transferred almost entirely "vertically" from parent to offspring via the DNA in sperm and egg; in man-made systems, like cornets, the information is spread as much "horizontally" (as when people copy other people's ideas) as it is vertically from old master to young pupil. The details of the history of man-made objects is invariably many times more complex than what biologists find for their organisms. I think the hypothesis of intelligent design, in this sense, is indeed falsifiable—and I think we have falsified it already.

But pursuit of scientific and intellectually valid truth is not really what creationism is all about. Creationism is about maintaining particular, narrow forms of religious belief—beliefs that seem to their adherents to be threatened by the very idea of evolution. In general, it should not be anyone's business what anyone else's religious beliefs are. It is because creationism transcends religious belief and is openly and aggressively political that we need to sit up and pay attention. For in their zeal to blot evolution from the ledger books of Western civilization, creationists have tried repeatedly for well over a hundred years to have evolution either watered down, or preferably completely removed, from the curriculum of America's public schools. Creationists persistently and consistently threaten the integrity of science teaching in America—and this, of course, is of grave concern.

Perhaps someday schools in the United States will catch up to those in other developed countries and treat evolution as a normal scientific subject. Before that happens, though, people need to understand evolution, and also understand the creationism and evolution controversy. *Evolution vs. Creationism: An Introduction* is a step toward this goal, and readers will indeed learn a great deal about the scientific, religious, educational, political, and legal aspects of this controversy. Then those of us lucky enough to study evolution as a profession won't be the only ones to appreciate this fascinating field of study.

Niles Eldredge
Division of Paleontology
The American Museum of Natural History

Foreword to the Second Edition

In September 2005, I convened the bench trial in the now famous case of *Kitzmiller v. Dover Area School District*. As the lengthy and complex trial testimony unfolded, I occasionally glanced at the substantial gallery that each day assembled to watch the proceedings. Many faces became familiar to me, although in most instances I did not know the names of these frequent attendees. One such visitor to my courtroom was an attractive and somewhat professorial looking woman. I did not know either her name or affiliation, although because she sat in proximity to the plaintiffs I assumed that she was aligned with their cause. She appeared at all times to be totally and intently engaged in the trial testimony. It was only after the case concluded that I learned through watching media interviews that the person I had almost daily observed was Eugenie C. Scott of the National Center for Science Education. I also learned that she had been substantially responsible for coordinating the plaintiffs' expert testimony.

In October 2006, well after the *Kitzmiller* case had ended, I found myself in Chicago speaking to judges from around the country at a national conference dealing with scientific evidence. Scott, whom I've since come to know as Genie, was one of my fellow presenters. Serendipitously we were seated next to each other at a dinner organized by our hosts. Given our common experiences during the previous year, we had much to talk about. That evening I learned several things. First, Genie is a most pleasant conversationalist! But more than that, she is virtually encyclopedic regarding the myriad issues that attend the debate over evolution and creationism. While she undoubtedly favors the former, I was tremendously impressed by her ability to objectively relate the salient points raised by advocates of the latter. Moreover, she possesses a comprehensive grasp of the long history of the underlying controversy.

When speaking publicly about the *Kitzmiller* case, I have candidly admitted that prior to having the case appear on my docket, other than generalized knowledge gained from my liberal arts education at Dickinson College in Pennsylvania years ago, coupled with somewhat eclectic reading tastes and a basic understanding of what took place in the John Scopes trial in Dayton, Tennessee, in 1925, I had little exposure to this debate or to the science of evolution. However, deciding *Kitzmiller* and experiencing its aftermath have informed me that in this regard I was decidedly in the company of the majority of my fellow citizens. Simply put, evolution is poorly understood by most Americans, if indeed it is grasped at all. And too many Americans do not understand the constitutional reasons for not advocating religious views in the classroom.

After Genie asked me to write a foreword for this edition of *Evolution vs. Creationism: An Introduction*, I had considerable pause, and thus initially demurred. But on reflection, and after reading the updated version, I reconsidered this somewhat reflexive position. After all, in the years since *Kitzmiller* I have frequently found myself saying to audiences and individuals when describing what I saw, heard, and read: "You should have been there." By that I have meant that the testimony in support of evolution was both compelling and understandable. But this comment is also directed to the historical antecedents in the evolution versus creationism debate. Manifestly, this is not a simple area, and the passions brought to it by advocates on both sides tend at times to impede clear understanding. I had the advantage of a full year of litigation on the topic, including a six-week trial containing abundant expert testimony. Few others will be so fortunate. But Genie Scott has rendered a book that both educates the uninformed and enlightens those who possess a basic but not detailed knowledge of the debate. In effect then, the aptly nicknamed Genie has granted my wish that others experience what I did in 2005. To the extent that someone either could not witness the whole of the *Kitzmiller* trial or lacks the time to wade through thousands of pages of dry transcripts, here is a compendium that in my view accurately depicts both the historical and scientific facets of the controversy.

In the last several years, I have developed a passion for speaking in public about topics such as judicial independence, the rule of law, and a better understanding of our democracy and Constitution. Genie Scott quite obviously brings that same passion to bear as it relates to science. Any tool that facilitates better teaching of these subjects in our high schools and colleges is vital. Here, then, is a superior work that I believe is a "must read" relating to science education in the United States. I commend it not just to students, but to anyone who seeks a better understanding of one of the important and enduring issues for our time.

Judge John E. Jones III
U.S. District Judge
Middle District of Pennsylvania

Preface

The second edition of *Evolution vs. Creationism: An Introduction* has been expanded from the first edition and includes several new readings, but it has the same goal as the earlier version. My intent is to provide a single reference that examines the creationism and evolution controversy from a broad perspective that includes historical, legal, educational, political, scientific, and religious perspectives. Although more depth in any of these topics can be found in several specialized books, this book presents, as its subtitle implies, an introduction.

I have attempted to write at a level suitable to the abilities of bright high school students and college undergraduates (it's OK if others wish to read the book, too!). At the National Center for Science Education, where I work, we regularly get calls or e-mails from students (and their teachers or professors) looking for information to help in the writing of research papers on the creationism/evolution controversy; this book is a good place to begin (note to students—don't stop with just one source!). Students often flounder while attempting such assignments, lacking enough basic science (and philosophy of science) to understand why creationist critiques of evolution are resisted so strongly by scientists, and similarly lacking the theological background to understand why the claims of creationists are not uniformly accepted by religious people. The first few chapters (on science, evolution, creationism, and religion) are intended to provide the background information necessary to understand the controversy. The second section, on the history of the controversy, puts today's headlines in context; an understanding of history is essential to make sense of the current situation, which did not arise in a vacuum. The second edition includes a new chapter, "Testing Intelligent Design and Evidence against Evolution" (chapter 7) that brings history up to the present, targeting on recent court cases. These include *Kitzmiller v. Dover* and *Selman v. Cobb County*. The rise of intelligent design and the so-called evidence-against-evolution (or critical analysis of evolution, or strengths and weaknesses of evolution) approach presents some of the most interesting manifestations of the controversy in the first decade of the twenty-first century. The other new chapter is the final chapter of the book, chapter 14, which looks at media treatments of the creationism and evolution controversy and at public opinion polls.

Evolution vs. Creationism includes excerpts from the creationist literature as well as rebuttals. Much of the creationist literature is not readily available except in sectarian publications and Christian bookstores, and public school libraries are properly reluctant to carry such obviously devotional literature. I have made selections from the literature that are representative of the major themes found in the creationism/evolution controversy, and I have attempted to let antievolutionists speak in their own voices.

Unfortunately, most proponents of intelligent design (ID) creationism—Stephen Meyer, David DeWolf, Percival Davis, Dean Kenyon, Jonathan Wells, Walter Bradley, Charles Thaxton, and Roger Olsen—refused, en masse, to grant me permission to reproduce their works in the first edition of this book Through their representative at the Seattle-based ID think tank, the Discovery Institute, these authors refused permission to reprint readily available material on the grounds that these excerpts from popular books and articles (e.g., opinion-editorial articles and magazine articles) that I sought to reprint would not do justice to the complexity of ID "theory." This rationale does make one wonder why such apparently inadequate works were published in the first place and continue, in several cases, to be available on or linked to from the Discovery Institute's Web site. The exception was ID proponent Phillip Johnson, who cordially and promptly granted permission for me to use excerpts from his publication. I thank him for this courtesy.

When the current, second edition was being written, I again requested permission from these ID proponents to excerpt their works. My requests—mailed and e-mailed— were ignored. Consequently, as was necessary in the first edition, many of the selections from the ID literature presented in chapters 8, 9, 10, and 12 consist of summaries of the articles I was denied permission to reprint. References to the original articles are provided, and because most of these writings are readily available on the Internet, readers can judge for themselves whether my summaries are accurate. The exception to this second generation of stonewalling was Michael Behe, who in the current edition kindly permitted me to reprint his article from *Natural History*, for which I thank him.

However, in the years between the first and second editions of this book, a series of trials have produced a volume of witness statements, amicus (friend of the court) briefs, depositions, and other legal documents that, by virtue of being part of a court's record, are in the public domain. I have taken advantage of this to include some new selections from the ID literature in part 3. You thus will be able to read some views of ID supporters in their own words, rather than my summaries.

In contrast to the behavior of the ID supporters, the late Henry M. Morris, John Morris, and other personnel at the Institute for Creation Research treated my requests for permission to reprint materials from ICR authors with professionalism. They were aware that their works would be juxtaposed with the writings of individuals who disagree with them, but they did not consider this sufficient reason to deny an honest presentation of their views. I was pleased that Henry Morris reviewed the first edition of *Evolution vs. Creationism*, and although he clearly believed that the selections from the creationist literature trumped those from the anticreation side, he said, "I believe that she has conscientiously tried to be objective in discussing this inflammatory subject in her book" (Morris 2004: a). I also thank Don Batten of Answers in Genesis, who worked with me in a professional manner to resolve disagreements over selections from literature published by AIG.

The juxtaposition of articles by creationists and articles by anticreationists requires a caveat, lest students be misled. Students are ill served if in the name of fairness or critical thinking they are misled into believing that there is a controversy in the scientific world over whether evolution occurred. There is none. Although the teaching of evolution is often regarded as controversial at the K–12 level, the subject is taught matter-of-factly in every respected secular and sectarian university or college in this country, including the Baptist institution Baylor, the Mormon flagship university Brigham Young, and, of course, the Catholic Notre Dame. There is scientific controversy concerning the details of mechanisms and patterns of evolution, but not over whether the universe has had a history measured in billions of years, nor over whether living things share a common ancestry. It would be dishonest as well as unfair to students to pretend that a public controversy over the teaching of evolution is also a scientific controversy over whether evolution occurred.

But a public controversy there is, and its complex foundation in history, science, religion, and politics will, I hope, be interesting to readers.

REFERENCE

Morris, Henry M. 2004. Creation versus evolutionism: A book report. *Back to Genesis* (191): a–c.

Acknowledgments

The second edition of *Evolution vs. Creationism: An Introduction* builds on the first, and therefore the contributions of NCSE staff members past and present who helped in that effort is still very much in evidence, and very much appreciated. I have greatly benefited from working with and sharing ideas with David Almandsmith, Josephina Borgeson, Wesley Elsberry, Skip Evans, Alan Gishlick, Charles Hargrove, Peter Hess, Anne Holden, Abraham Kneisley, David Leitner, Molleen Matsumura, Nicholas Matzke, Louise Mead, Eric Meikle, Jessica Moran, Josh Rosenau, Carrie Sager, and Susan Spath. Of course, none of us would get any work done if Nina Hollenberg, Philip Spieth, and Tully Weberg weren't keeping track of the business side of NCSE. My indebtedness to many other students of the creation/evolution controversy will be clear upon reading the introductory chapters. I have learned much about pedagogical issues from Rodger Bybee and the rest of the Biological Sciences Curriculum Study crew, Brian Alters, Craig Nelson, and Judy Scotchmoor; about traditional creationism from John Cole, Tom McIver, the late Robert Schadewald, and William Thwaites; about the history of the controversy from Ronald Numbers, Edward Larson, and James Moore; about philosophical issues from Philip Kitcher, Michael Ruse, and Rob Pennock; about scientific aspects of the controversy from Brent Dalrymple, Niles Eldredge, Doug Futuyma, Ken Miller, Kevin Padian, the late Art Strahler, and many others. I have acquired an appreciation for the complexity of the science and religion aspects of the controversy from many, including, to name only a few, Jack Haught, Jim Miller, and Robert John Russell.

I want to give an extra thank you to my colleague, Alan Gishlick, for assistance with illustrations, and to NCSE member and artist, Janet Dreyer, for the fossil and other drawings in chapter 2. If you peruse issues of *Reports of NCSE*, you will see her whimsical and sometimes-barbed covers and other artwork, which we appreciate greatly. Another skilled artist, Sarina Bromberg, contributed some new artwork to the second edition, which readers should enjoy.

I thank the authors who kindly allowed me permission to reprint their essays. I have necessarily had to reduce a large number of potential topics to a smaller number treatable in a book like this, but of course there is much left unexplored. I have tried to select writings regarding these topics that honestly and clearly express the views of both antievolutionists and those who accept evolution. I especially appreciate the cooperation of authors whose views are opposed to mine, especially Henry and John Morris from the Institute for Creation Research, and Don Batten from Answers in Genesis. Phillip Johnson and Michael Denton likewise were cordial and helpful, and I appreciate Michael Behe's willingness to reprint his *Natural History* essay in the second edition.

Feedback from readers of the first edition was very helpful in shaping the second. Because of reviewer suggestions, a section on cladistics has been added to chapter 2, and several small errors (which out of embarrassment, I won't iterate!) have been corrected. Most of these errors were called to account by the sharp editorial eyes of NCSE board member Frank Sonleitner and my good friend Larry Lerner, and I thank them. Dave Chapman's considerable advice greatly improved my understanding of cosmological evolution in the second edition, and I sincerely thank him.

A very special thanks to my colleague, NCSE Deputy Director Glenn Branch, who has contributed substantially to this book from its planning to its completion. Glenn provided valuable suggestions on the organization of chapters as well as their content and skillfully edited the whole first edition making my prose much clearer. The usefulness of this book owes much to his efforts. Glenn also assembled the References for Further Exploration section, which benefited greatly from his encyclopedic appetite for books and resources and his phenomenal recall of just about everything he has ever read. Anyone at NCSE who is looking for a reference knows whose desk to camp out at.

My husband Charlie put up with a lot during both the first and second editions of this book. He and I know how much, and I'm not tellin'.

There is no way to thank everyone to whom I am indebted for whatever useful information this book will have. Similarly, I have no one to blame but myself for any errors, which I hope are few. With luck, the contents of this book may inspire some reader to in turn contribute to a further understanding of this vexing problem of antievolutionism, and dare we hope, contribute thereby to a solution to it.

INTRODUCTION:
The Pillars of Creationism

This book examines the creationism/evolution controversy from a broad perspective. You will read about science, religion, education, law, history, and even some current events, because all of these topics are relevant to an understanding of this controversy. In this introduction, I will examine three antievolutionist contentions that provide a framework for thinking about this complex controversy. These "pillars of creationism" include scientific, religious, and educational arguments, respectively, and have been central to the antievolution movement since at least the Scopes trial in 1925. As you read the following chapters and selections, it may be helpful to keep the pillars of creationism in mind.

EVOLUTION IS A THEORY IN CRISIS

In 1986, the New Zealand physician Michael Denton wrote a book titled *Evolution: A Theory in Crisis*, which became and remains very popular in creationist circles. Denton claimed that there were major scientific flaws in the theory of evolution. This idea is not new: throughout the nineteenth and twentieth centuries, there was no shortage of claims that evolution scientifically was on its last legs, as documented delightfully by Glenn Morton (http://home.entouch.net/dmd/moreandmore.htm). Of course, such claims continue to be made in the twenty-first century as well. Ironically, Denton has rejected the antievolutionary claims of some of his readers, and describes his 1986 book as opposing Darwinism (i.e., evolution through natural selection) rather than rejecting evolution itself (Denton 1999).

Through constant reiteration in creationist literature and in letters to the editor in newspapers around the country, the idea that evolution is shaky science is constantly spread to the general public, which by and large is unaware of the theoretical and evidentiary strength of evolution. Evolution as a science is discussed in chapter 2.

EVOLUTION AND RELIGION ARE INCOMPATIBLE

Darwin made two major points in *On the Origin of Species*: that living things had evolved, or descended with modification, from common ancestors, and that the mechanism of natural selection was evolution's major cause. These two components of his book often are jumbled together by antievolutionists, who argue that if natural selection can be shown to be inadequate as an evolutionary mechanism, then the idea of common descent necessarily fails. But the two constituents of Darwin's argument are conceptually and historically distinct. Common descent was accepted by both the scientific and the religious communities more quickly than was the mechanism of natural selection. Further separating the two components of Darwinism is the fact that the religious objections to each are quite distinct. For these reasons, I will separate these two theoretical concepts in discussing religious objections to evolution.

Common Ancestry

Biblical literalists are strongly opposed to the idea of common ancestry—especially common ancestry of humans with other creatures. According to some literal interpretations of the Bible, God created living things as separate "kinds." If living things instead have descended with modification from common ancestors, the Bible would be untrue. Many biblical literalists (Young Earth Creationists, or YECs) also believe that Earth's age is measured in thousands rather than billions of years.

Yet even before Darwin published *On the Origin of Species*, there was compelling evidence for an ancient Earth and the existence of species of living things before the advent of humans. Fossils of creatures similar to but different from living forms were known, which implied that Genesis was an incomplete record of creation. More troubling was the existence of fossils of creatures not known to be alive today, raising the possibility that God allowed some creatures to become extinct. Did the evidence of extinction mean that God's Creation was somehow not perfect? If Earth was ancient and populated by creatures that lived before humans, death must have preceded Adam's fall—which has obvious implications for the Christian doctrine of original sin. These theological issues were addressed in a variety of ways by clergy in the nineteenth and early twentieth centuries (see chapters 3, 4, and 12, and references).

Unquestionably, evolution has consequences for traditional Christian religion. Equally unquestionably, Christian theologians and thoughtful laymen have pondered these issues and attempted to resolve the potential contradictions between traditional religion and modern science. Some of these approaches are discussed in chapter 12.

Natural Selection

Natural selection refers to Darwin's principal mechanism of evolution, which you will learn about in more detail in chapter 2. Those individuals in a population that (genetically) are better able to survive and reproduce in a particular environment leave more offspring, which in turn carry a higher frequency of genes promoting adaptation to that environment. Though effective in producing adaptation, natural selection is a wasteful mechanism: many individuals fall by the wayside, poorly adapted, and fail to survive and/or reproduce.

Even Christians who accept common descent may be uneasy about Darwin's mechanism of natural selection as the major engine of evolutionary change. Common ancestry itself may not be a stumbling block, but if the variety of living things we see today is primarily the result of the incredibly wasteful and painful process of natural selection, can this really be the result of actions of a benevolent God? The theodicy issue (the theological term for the problem raised by the existence of evil in a world created by a benevolent God) is a concern for both biblical literalist and nonliteralist Christians and, as discussed in chapter 6, is a major stumbling block to the acceptance of evolution by intelligent design creationists (IDCs). Yet the evidence for the operation of natural selection is so overwhelming that both IDCs and YECs now accept that it is responsible for such phenomena as pesticide resistance in insects or antibiotic resistance in bacteria. YECs interpret the wastefulness of natural selection as further evidence of the deterioration of creation since the fall of Adam. Both YECs and IDCs deny that natural selection has the ability to transform living things into different kinds or to produce major changes in body plans, such as the differences between a bird and a reptile.

Thus, religious objections to evolution are not simple; they span a range of concerns. Religious objections to evolution are far more important in motivating antievolutionism than are scientific objections to evolution as a weak or unsupported theory.

"BALANCING" EVOLUTION (FAIRNESS)

A third antievolution theme present as far back as the 1925 Scopes trial and continuing today is the idea that if evolution is taught, then creationism in some form should also be taught, as a matter of fairness. The fairness theme has, however, had many manifestations through time, largely evolving in response to court decisions (see chapters 6, 10, and 11).

The fairness pillar reflects American cultural values of allowing all sides to be heard, and also a long-standing American democratic cultural tradition that assumes an individual citizen can come to a sound conclusion after hearing all the facts—and has the right to inform elected officials of his or her opinion. Indeed, for many local and even national issues, Americans do not defer to elected and appointed officials but vigorously debate decisions in town meetings, city council meetings, and school board meetings.

As a result, in the United States there are disputes at the local school board level over who—scientists, teachers, or members of the general public—should decide educational content. In the 1920s, the populist orator, politician, and lawyer William Jennings Bryan raged at the audacity of "experts" who would come to tell parents what to teach their children, when (as he thought) the proposed subject matter (evolution) was diametrically opposed to parental values (see chapter 4).

Many modern-day antievolutionists make this same point, arguing that conservative Christian students should not even be exposed to evolution if their religious beliefs disagree with evolution's implications. Educators and scientists counter that a student must understand evolution to be scientifically literate and insist that the science curriculum would be deficient if evolution were omitted. Efforts to ban the teaching of evolution failed, as a result of both rulings by the Supreme Court and the growth of evolution as a science (see chapters 2, 4, 5, and 10). Antievolutionists shifted their emphasis from banning evolution to having it "balanced" with the teaching of a form of creationism called creation science (see chapters 3 and 5). When this effort also failed, antievolutionists began to lobby school boards and state legislatures to balance evolution with the teaching of evidence against evolution, which in content proved identical to creation science.

The perceived incompatibility of evolution with religion (especially conservative Christian theology) is the most powerful motivator of antievolutionism for individuals. However, the fairness concept, because of its cultural appeal, may be even more effective, for it appeals broadly across many diverse religious orientations. Even those who are not creationists may see value in being fair to all sides, whether or not they believe that there is scientific validity to creationist views. Scientists and teachers argue, however, that to apply fairness to the science classroom is a misapplication of an otherwise worthy cultural value (see chapters 9, 11, and 12).

A LOOK FORWARD

Consider these three themes, then, as you read the following chapters. Reflect on how these pillars of creationism have influenced the history of this controversy and continue to be reflected in creationism/evolution disputes you read about in the news or see on television. Should you encounter such a local or state-level controversy, you will, I predict, easily be able to place creationist arguments into one (or more) of these categories. The following chapters will provide context for understanding these three themes as well as the creationism/evolution controversy itself.

REFERENCE

Denton, Michael. 1999. Comments on special creationism. In *Darwinism defeated? The Johnson-Lamoureux debate on biological origins*, ed. P. E. Johnson and D. O. Lamoureux. Vancouver, BC: Regent College Publishing.

PART I

· · · · · · · · · · ·

Science, Evolution, Religion, and Creationism

The creationism/evolution controversy has been of long duration in American society and shows no sign of disappearing. To understand it requires some background in the two subject areas most closely concerned with the controversy, science and religion. Within science and religion, the subareas of evolution and creationism are clearly central to the dispute.

Most people will recognize that religion and creationism are related concepts, as are science and evolution, but there also is something called creation science, and there is even a form of religion called scientism. In this introductory section, then, you will read about science, evolution, religion, creationism, and scientism.

These and other subjects constitute part 1 of *Evolution vs. Creationism: An Introduction*. I assume that readers of this book will vary greatly in their understanding of these subjects, so I have tried to present material at a level that does not leave behind the beginner but has enough detail to interest a reader with a more-than-average background in philosophy of science, evolution, or religious studies. At a minimum, readers will at least know how I define and use the terms that will recur throughout the book.

In the first chapter, "Science," I consider different ways of knowing and how the way of knowing called science is especially appropriate to knowing about the natural world. Testing is the most important component in science, and I discuss different kinds of testing. In the second chapter, "Evolution," I discuss some of the basic ideas in this broad scientific discipline. The third chapter, "Beliefs," discusses religion as a universal set of beliefs, with particular attention to origin stories and creationisms. It also discusses naturalism as a belief. Because of the importance of the Christian religion to the creationism/evolution controversy, most of this chapter deals with Christian creationism.

CHAPTER 1

• • • • • • • • • • • • •

Science:
Truth without Certainty

We live in a universe made up of matter and energy, a material universe. To understand and explain this material universe is the goal of science, which is a methodology as well as a body of knowledge obtained through that methodology. Science is limited to matter and energy, but as will become clear when we discuss religion, most individuals believe that reality includes something other than matter and energy. The methodology of science is a topic on which any college library has dozens of feet of shelves of books and journals, so obviously just one chapter won't go much beyond sketching out the bare essentials. Still, I will try to show how science differs from many other ways of knowing and how it is particularly well suited to explaining our material universe.

WAYS OF KNOWING

Science requires the testing of explanations of the natural world against nature itself and the discarding of those explanations that do not work. What distinguishes science from other ways of knowing is its reliance upon the natural world as the arbiter of truth. There are many things that people are interested in, are concerned about, or want to know about that science does not address. Whether the music of Madonna or Mozart is superior may be of interest (especially to parents of teenagers), but it is not something that science addresses. Aesthetics is clearly something outside of science. Similarly, literature or music might generate or help to understand or cope with emotions and feelings in a way that science is not equipped to do. But if one wishes to know about the natural world and how it works, science is superior to other ways of knowing. Let's consider some other ways of knowing about the natural world.

Authority

Dr. Jones says, "Male lions taking over a pride will kill young cubs." Should you believe her? You might know that Dr. Jones is a famous specialist in lion behavior who has studied lions for twenty years in the field. Authority leads one to believe that Dr. Jones's statement is true. In a public bathroom, I once saw a little girl of perhaps four or five years old marvel at faucets that automatically turned on when hands were placed below the spigot. She asked her mother, "Why does the water come out, Mommy?" Her mother answered brightly, if unhelpfully, "It's magic, dear!" When we are small, we rely on the authority of our parents and other older people, but authority clearly can mislead us, as in the case of the magic spigots. And Dr. Jones might be wrong about lion infanticide, even if in the past she has made statements about animal behavior that have been reliable. Yet it is not "wrong" to take some things on authority. In northern California, a popular bumper sticker reads Question Authority. Whenever I see one of these, I am tempted to pencil in "but stop at stop signs." We all accept some things on authority, but we should do so critically.

Revelation

Sometimes people believe a statement because they are told it comes from a source that is unquestionable: from God, or the gods, or some other supernatural power. Seekers of advice from the Greek oracle at Delphi believed what they were told because they believed that the oracle received information directly from Apollo; similarly, Muslims believe the contents of the Koran were revealed to Muhammad by God; and Christians believe the New Testament is true because the authors were directly inspired by God. A problem with revealed truth, however, is that one must accept the worldview of the speaker in order to accept the statement; there is no outside referent. If you don't believe in Apollo, you're not going to trust the Delphic oracle's pronouncements; if you're not a Mormon or a Catholic, you are not likely to believe that God speaks directly to the Mormon president or the pope. Information obtained through revelation is difficult to verify because there is not an outside referent that all parties are likely to agree upon.

Logic

A way of knowing that is highly reliable is logic, which is the foundation for mathematics. Among other things, logic presents rules for how to tell whether something is true or false, and it is extremely useful. However, logic in and of itself, with no reference to the real world, is not complete. It is logically correct to say, "All cows are brown. Bossy is not brown. Therefore Bossy is not a cow." The problem with the statement is the truth of the premise that all cows are brown, when many are not. To know that the proposition about cows is empirically wrong even if logically true requires reference to the real world outside the logical structure of the three sentences. To say, "All wood has carbon atoms. My computer chip has no carbon atoms. Therefore my computer chip is not made of wood" is both logically and empirically true.

Science

Science does include logic—statements that are not logically true cannot be scientifically true—but what distinguishes the scientific way of knowing is the requirement of going to nature to verify claims. Statements about the natural world are tested against the natural world, which is the final arbiter. Of course, this approach is not perfect: one's information about the natural world comes from experiencing the natural world through the senses (touch, smell, taste, vision, hearing) and instrumental extensions of these senses (e.g., microscopes, telescopes, telemetry, chemical analysis), any of which can be faulty or incomplete. As a result, science, more than any of the other ways of knowing described here, is more tentative in its claims. Ironically, the tentativeness of science ultimately leads to more confidence in scientific understanding: the willingness to change one's explanation with more or better data, or a different way of looking at the same data, is one of the great strengths of the scientific method. The anthropologist Ashley Montagu summarized science rather nicely when he wrote, "The scientist believes in proof without certainty, the bigot in certainty without proof" (Montagu 1984: 9).

Thus science requires deciding among alternative explanations of the natural world by going to the natural world itself to test them. There are many ways of testing an explanation, but virtually all of them involve the idea of holding constant some factors that might influence the explanation so that some alternative explanations can be eliminated. The most familiar kind of test is the direct experiment, which is so familiar that it is even used to sell us products on television.

DIRECT EXPERIMENTATION

Does RealClean detergent make your clothes cleaner? The smiling company representative in the television commercial takes two identical shirts, pours something messy on each one, and drops them into identical washing machines. RealClean brand detergent goes into one machine and the recommended amount of a rival brand into the other. Each washing machine is set to the same cycle, for the same period of time, and the ad fast-forwards to show the continuously smiling representative taking the two shirts out. Guess which one is cleaner.

Now, it would be very easy to rig the demonstration so that RealClean does a better job: the representative could use less of the other detergent, use an inferior-performing washing machine, put the RealClean shirt on a soak cycle forty-five minutes longer than for the other brand, employ different temperatures, wash the competitor's shirt on the delicate rather than regular cycle—I'm sure you can think of a lot of ways that RealClean's manufacturer could ensure that its product comes out ahead. It would be a bad sales technique, however, because we're familiar with the direct experimental type of test, and someone would very quickly call, "Foul!" To convince you that they have a better product, the makers of the commercial have to remove every factor that might possibly explain why the shirt came out cleaner when washed in their product. They have to hold constant or control all these other factors—type of machine, length of cycle, temperature of the water, and so on—so that the only reasonable explanation for the cleaner shirt is that RealClean is a better product. The experimental method— performed fairly—is a very good way to persuade people that your explanation is correct. In science, too, someone will call, "Foul!" (or at least, "You blew it!") if a test doesn't consider other relevant factors.

Direct experimentation is a very powerful—as well as familiar—research design. As a result, some people think that this is the only way that science works. Actually, what matters in science is that explanations be tested, and direct experimentation is only one kind of testing. The key element to testing an explanation is to hold variables constant, and one can hold variables constant in many ways other than being able to directly manipulate them (as one can manipulate water temperature in a washing machine). In fact, the more complicated the science, the less likely an experimenter is to use direct experimentation.

In some tests, variables are controlled statistically; in others, especially in biological field research or in social sciences, one can find circumstances in which important variables are controlled by the nature of the experimental situation itself. These observational research designs are another type of direct experimentation.

Noticing that male guppies are brightly colored and smaller than the drab females, you might wonder whether having bright colors makes male guppies easier prey. How would you test this idea? If conditions allowed, you might be able to perform a direct experiment by moving brightly colored guppies to a high-predation environment and monitoring them over several generations to see how they do. If not, though, you could still perform an observational experiment by looking for natural populations of the same or related species of guppies in environments where predation was high and in other environments where predation was low. You would also want to pick environments where the amount of food was roughly the same—can you explain why? What other environmental factors would you want to hold constant at both sites?

When you find guppy habitats that naturally vary only in the amount of predation and not in other ways, then you're ready to compare the brightness of color in the males. Does the color of male guppies differ in the two environments? If males were less brightly colored in environments with high predation, this would support the idea that brighter guppy color makes males easier prey. (What if in the two kinds of environments, male guppy color is the same?)

Indirect experimentation is used for scientific problems where the phenomena being studied—unlike color in guppies—cannot be directly observed.

INDIRECT EXPERIMENTATION

In some fields, not only is it impossible to directly control variables but also the phenomena themselves may not be directly observable. A research design known as indirect experimentation is often used in such fields. Explanations can be tested even if the phenomena being studied are too far away, too small, or too far back in time to be observed directly. For example, giant planets recently have been discovered orbiting distant stars—though we cannot directly observe them. Their presence is indicated by the gravitational effects they have on the suns around which they revolve: because of what we know about how the theory of gravitation works, we can infer that the passage of a big planet around a sun will make the sun wobble. Through the application of principles and laws in which we have confidence, it is possible to infer that these planetary giants do exist and to make estimates of their size and speed of revolution.

Similarly, the subatomic particles that physicists study are too small to be observed directly, but particle physicists certainly are able to test their explanations. By applying knowledge about how particles behave, they are able to create indirect experiments to test claims about the nature of particles. Let's say that a physicist wants to ascertain properties of a particle—its mass, charge, or speed. On the basis of observations of similar particles, he makes an informed estimate of the speed. To test the estimate, he might bombard it with another particle of known mass, because if the unknown particle has a mass of m, it will cause the known particle to ricochet at velocity v. If the known particle does ricochet as predicted, this would support the hypothesis about the mass of the unknown particle. Thus, theory is built piece by piece, through inference based on accepted principles.

In truth, most scientific problems are of this if-then type, whether or not the phenomena investigated are directly observable. If male guppy color is related to predation, then we should see duller males in high-predation environments. If a new drug stimulates the immune system, then individuals taking it should have fewer colds than the controls do. If human hunters were involved in the destruction of large Australian land mammals, we should see extinction events that correlate with the appearance of the first Aborigines. We test by consequence in science all the time. Of course—because scientific problems are never solved so simply—if we get the consequence we predict, this does not mean we have proved our explanation. If you found that guppy color does vary in environments where predation differs, this does not mean you've proved yourself right about the relationship between color and predation. To understand why, we need to consider what we mean by proof and disproof in science.

PROOF AND DISPROOF

Proof

Scientists don't usually talk about proving themselves right, because *proof* suggests certainty (remember Ashley Montagu's truth without certainty!). The testing of explanations is in reality a lot messier than the simplistic descriptions given previously. One can rarely be sure that all the possible factors that might explain why a test produced a positive result have been considered. In the guppy case, for example, let's say that you found two habitats that differed in the number of predators but were the same in terms of amount of food, water temperature, and number and type of hiding places—you tried to hold constant as many factors as you could think of. If you find that guppies are less colorful in the high-predation environment, you might think you have made the link, but some other scientist may come along and discover that your two environments differ in water turbidity. If turbidity affects predation—or the ability of female guppies to select the more colorful males—this scientist can claim that you were premature to conclude that color is associated with predation. In science we rarely claim to prove a theory—but positive results allow us to claim that we are likely to be on the right track. And then you or some other scientist can go out and test some more. Eventually we may achieve a consensus about guppy color being related to predation, but we wouldn't conclude this after one or a few tests. This back-and-forth testing of explanations provides a reliable understanding of nature, but the procedure is neither formulaic nor especially tidy over the short run. Sometimes it's a matter of two steps forward, a step to the side (maybe down a blind alley), half a step back—but gradually the procedure, and with it human knowledge, lurches forward, leaving us with a clearer knowledge of the natural world and how it works.

In addition, most tests of anything other than the most trivial of scientific claims result not in slam-dunk, now-I've-nailed-it, put-it-on-the-T-shirt conclusions, but rather in more or less tentative statements: a statement is weakly, moderately, or strongly supported, depending on the quality and completeness of the test. Scientific claims become accepted or rejected depending on how confident the scientific community is about whether the experimental results could have occurred that way just by chance—which is why statistical analysis is such an important part of most scientific tests. Animal behaviorists note that some social species share care of their offspring. Does this make a difference in the survival of the young? Some female African silver-backed jackals, for example, don't breed in a given season but help to feed and guard the offspring of a breeding adult. If the helper phenomenon is directly related to pup survival, then more pups should survive in families with a helper.

One study tested this claim by comparing the reproductive success of jackal packs with and without helpers, and found that for every extra helper a mother jackal had, she successfully raised one extra pup per litter over the average survival rate (Hrdy 2001). These results might encourage you to accept the claim that helpers contribute to the survival of young, but only one test on one population is not going to be convincing. Other tests on other groups of jackals would have to be conducted to confirm the results, and to be able to generalize to other species the principle that reproductive success is improved by having a helper would require conducting tests on other social species. Such studies in fact have been performed across a wide range of birds and mammals, and a consensus is emerging about the basic idea of helpers increasing survivability of the young. But there are many remaining questions, such as whether a genetic relationship always exists between the helper and either the offspring or the helped mother.

Science is quintessentially an open-ended procedure in which ideas are constantly tested and rejected or modified. Dogma—an idea held by belief or faith—is anathema to science. A friend of mine once was asked to explain how he ended up a scientist. His tongue-in-cheek answer illustrates rather nicely the nondogmatic nature of science: "As an adolescent I aspired to lasting fame, I craved factual certainty, and I thirsted for a meaningful vision of human life—so I became a scientist. This is like becoming an archbishop so you can meet girls" (Cartmill 1988: 452).

In principle, all scientific ideas may change, though in reality there are some scientific claims that are held with confidence, even if details may be modified. The physicist James Trefil (1978) suggested that scientific claims can be conceived of as arranged in a series of three concentric circles (see Figure 1.1). In the center circle are the core ideas of science: the theories and facts in which we have great confidence because they work so well to explain nature. Heliocentrism, gravitation, atomic theory, and evolution are examples. The next concentric circle outward is the frontier area of science, where research and debate are actively taking place on new theories or modifications and additions to core theories. Clearly no one is arguing with the basic principle of heliocentrism, but on the frontier, planetary astronomers still are learning things and testing ideas about the solar system. That matter is composed of atoms is not being challenged, but the discoveries of quantum physics are adding to and modifying atomic theory.

Figure 1.1
Scientific concepts and theories can be arranged as a set of nested categories with core ideas at the center, frontier ideas surrounding them, and fringe ideas at the edge (after Trefil 1978). Courtesy of Alan Gishlick.

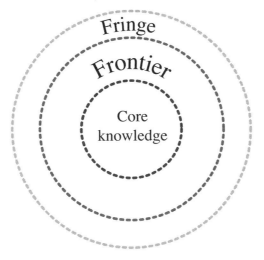

The outermost circle is the fringe, a breeding ground for ideas that very few professional scientists are spending time on: unidentified flying objects, telepathy and the like, perpetual motion machines, and so on. Generally the fringe is not a source of new ideas for the frontier, but occasionally (very occasionally!) ideas on the fringe will muster enough support to warrant a closer look and will move into the frontier. They may well be rejected and end up back in the fringe or be discarded completely, but occasionally they may become accepted and perhaps eventually become core ideas of science. That the continents move began as a fringe idea, then it moved to the frontier as data began to accumulate in its favor, and finally it became a core idea of geology when seafloor spreading was discovered and the theory of plate tectonics was developed.

Indeed, we must be prepared to realize that even core ideas may be wrong, and that somewhere, sometime, there may be a set of circumstances that could refute even our most confidently held theory. But for practical purposes, one needn't fall into a slough of despond over the relative tentativeness of scientific explanation. That the theory of gravitation may be modified or supplemented sometime in the future is no reason to give up riding elevators (or, even less advisedly, to jump off the roof). Science gives us reliable, dependable, and workable explanations of the natural world—even if it is good philosophy of science to keep in mind that in principle anything can change.

On the other hand, even if it is usually not possible absolutely to prove a scientific explanation correct—there might always be some set of circumstances or observations somewhere in the universe that would show your explanation wrong—to disprove a scientific explanation is possible. If you hypothesize that it is raining outside, and walk out the door to find the sun is shining and the ground is dry, you have indeed disproved your hypothesis (assuming you are not hallucinating). So disproving an explanation is easier than proving one true, and, in fact, progress in scientific explanation has largely come by rejecting alternative explanations. The ones that haven't been disconfirmed yet are the ones we work with—and some of those we feel very confident about.

Disproof

Now, if you are a scientist, obviously you will collect observations that support your explanation, but others are not likely to be persuaded just by a list of confirmations. Like proving RealClean detergent washes clothes best, it's easy to find—or concoct—circumstances that favor your view, which is why you have to bend over backward in setting up your test so that it is fair. So you set the temperature on both washing machines to be the same, you use the same volume of water, you use the recommended amount of detergent, and so forth. In the guppy case, you want to hold constant the amount of food in high-predation environments and low-predation environments, and so on. If you are wrong about the ability of RealClean to get the stains out, there won't be any difference between the two loads of clothes, because you have controlled or held constant all the other factors that might explain why one load of clothes emerged with fewer stains. You will have disproved your hypothesis about the allegedly superior stain-cleaning qualities of RealClean. You are conducting a fair test of your hypothesis if you set up the test so that everything that might give your hypothesis an advantage has been excluded. If you don't, another scientist will very quickly point out your error, so it's better to do it yourself and save yourself the embarrassment!

What makes science challenging—and sometimes the most difficult part of a scientific investigation—is coming up with a testable statement. Is the African AIDS epidemic the result of tainted oral polio vaccine (OPV) administered to Congolese in the 1950s? Chimpanzees carry simian immunodeficiency virus, which researchers believe is the source of the AIDS-causing virus HIV (human immunodeficiency virus). Poliovirus is grown on chimp kidney culture or monkey kidney culture. Was a batch of OPV grown on kidneys from chimps infected with simian immunodeficiency virus the source of African AIDS? If chimpanzee DNA could be found in the fifty-year-old vaccine, that would strongly support the hypothesis. If careful analysis did not find chimpanzee DNA, that would fail to support the hypothesis, and you would have less confidence in it. Such a test was conducted, and after very careful analysis, no chimp DNA was found in samples of the old vaccine. Instead, macaque monkey DNA was found (Poinar, Kuch, and Pääbo 2001).

The study by Poinar and colleagues did not disprove the hypothesis that African AIDS was caused by tainted OPV (perhaps some unknown batch of OPV is the culprit), but it is strong evidence against it. Again, as in most science, we are dealing with probabilities: if all four batches of OPV sent to Africa in the 1950s were prepared in the same manner, at the same time, and in the same laboratory, what is the probability that one would be completely free of chimp DNA and one or more other samples would be tainted? Low, presumably, but because the probability is not 0 percent, we cannot say for certain that the OPV-AIDS link is out of the question. However, we have research from other laboratories on other samples, and they also were unable to find any chimpanzee genes in the vaccine (Weiss 2001). Part of science is to repeat tests of the hypothesis, and when such repeated tests confirm the conclusions of early tests, it greatly increases confidence in the answers. Because the positive evidence for this hypothesis for the origin of AIDS was thin to begin with, few people now are taking the hypothesis seriously. Both disproof of hypotheses and failure to confirm are critical means by which we eliminate explanations and therefore increase our understanding of the natural world.

Now, you might notice that although I have not defined them, I already have used two scientific terms in this discussion: *theory* and *hypothesis*. You may already know what these terms mean—probably everyone has heard that evolution is "just a theory," and many times you have probably said to someone with whom you disagree, "Well, that's just a hypothesis." You might be surprised to hear that scientists don't use these terms in these ways.

FACTS, HYPOTHESES, LAWS, AND THEORIES

How do you think scientists would rank the terms *fact, hypothesis, law,* and *theory?* How would you list these four from most important to least? Most people list facts on top, as the most important, followed by laws, then theories, and then hypotheses as least important at the bottom:

Most important
Facts
Laws
Theories
Hypotheses
Least important

You may be surprised that scientists rearrange this list, as follows:

Most important
Theories
Laws
Hypotheses
Facts
Least important

Why is there this difference? Clearly, scientists must have different definitions of these terms compared to how we use them on the street. Let's start with facts.

Facts

If someone said to you, "List five scientific facts," you could probably do so with little difficulty. Living things are composed of cells. Gravity causes things to fall. The speed of light is about 186,000 miles/second. Continents move across the surface of

Earth. Earth revolves around the sun—and so on. Scientific facts, most people think, are claims that are rock solid, about which scientists will never change their minds. Most people think that facts are just about the most important part of science, and that the job of the scientist is to collect more and more facts.

Actually, facts are useful and important, but they are far from being the most important elements of a scientific explanation. In science, facts are confirmed observations. When the same result is obtained after numerous observations, scientists will accept something as a fact and no longer continue to test it. If you hold up a pencil between your thumb and forefinger, and then stop supporting it, it will fall to the floor. All of us have experienced unsupported objects falling; we've leaped to catch the table lamp as a toddler accidentally pulls the lamp cord. We consider it a fact that unsupported objects fall. It is always possible, however, that some circumstance may arise when a fact is shown not to be correct. If you were holding that pencil while orbiting Earth on the space shuttle and then let it go, it would not fall (it would float). It also would not fall if you were on an elevator with a broken cable that was hurtling at 9.8 meters/second^2 toward the bottom of a skyscraper—but let's not dwell on that scenario. So technically, unsupported objects don't always fall, but the rule holds well enough for ordinary use. One is not frequently on either the space shuttle or a runaway elevator, or in other circumstances in which the confirmed observation of unsupported items falling will not hold. It would in fact be perverse for one to reject the conclusion that unsupported objects fall just because of the existence of helium balloons.

Other scientific facts (i.e., confirmed observations) have been shown not to be true. Before better cell-staining techniques revealed that humans have twenty-three pairs of chromosomes, it was thought that we had twenty-four pairs. A fact has changed, in this case with more accurate means of measurement. At one point, we had confirmed observations of twenty-four chromosome pairs, but now there are more confirmations of twenty-three pairs, so we accept the latter—although at different times, both were considered facts. Another example of something considered a fact—an observation—was that the continents of Earth were stationary, which anyone can see! With better measurement techniques, including using observations from satellites, it is clear that continents do move, albeit very slowly (only a few inches each year).

So facts are important but not immutable; they can change. An observation, though, doesn't tell you very much about how something works. It's a first step toward knowledge, but by itself it doesn't get you very far, which is why scientists put it at the bottom of the hierarchy of explanation.

Hypotheses

Hypotheses are statements of the relationships among things, often taking the form of if-then statements. If brightly colored male guppies are more likely to attract predators, then in environments with high predation, guppies will be less brightly colored. If levels of lead in the bloodstream of children is inversely associated with IQ scores, then children in environments with greater amounts of lead should have lower IQ scores. Elephant groups are led by matriarchs, the eldest females. If the age (and thus experience) of the matriarch is important for the survival of the group, then groups with younger matriarchs will have higher infant mortality than those led by older ones. Each of these hypotheses is directly testable and can be either disconfirmed or confirmed (note that hypotheses are not proved "right"—any more than any scientific explanation is proved). Hypotheses are very important in the development of scientific explanations. Whether rejected or confirmed, tested hypotheses help to build explanations by removing incorrect approaches and encouraging the further testing of fruitful ones. Much hypothesis testing in science depends on demonstrating that a result found in a comparison occurs more or less frequently than would be the case if only chance were operating; statistics and probability are important components of scientific hypothesis testing.

Laws

There are many laws in science (e.g., the laws of thermodynamics, Mendel's laws of heredity, Newton's inverse square law, the Hardy-Weinberg law). Laws are extremely useful empirical generalizations: they state what will happen under certain conditions. During cell division, under Mendel's law of independent assortment, we expect genes to act like particles and separate independently of one another. Under conditions found in most places on Earth's surface, masses will attract one another in inverse proportion to the square of the distance between them, following the inverse square law. If a population of organisms is larger than a certain size, is not undergoing natural selection, and has random mating, the frequency of genotypes of a two-gene system will be in the proportion $p2 + 2pq + q2$. This relationship is called the Hardy-Weinberg law.

Outside of science, we also use the term *law*. It is the law that everyone must stop for a stoplight. Laws are uniform and, in that they apply to everyone in the society, universal. We don't usually think of laws changing, but of course they do: the legal system has a history, and we can see that the legal code used in the United States has evolved over several centuries primarily from legal codes in England. Still, laws must be relatively stable or people would not be able to conduct business or know which practices or behaviors will get them in trouble. One will not anticipate that if today everyone drives on the right side of the street, tomorrow everyone will begin driving on the left. Perhaps because of the stability of societal laws, we tend to think of scientific laws as also stable and unchanging.

However, scientific laws can change or not hold under some conditions. Mendel's law of independent assortment tells us that the hereditary particles will behave independently as they are passed down from generation to generation. For example, the color of a pea flower is passed on independently from the trait for stem length. But after more study, geneticists found that the law of independent assortment can be "broken" if the genes are very closely associated on the same chromosome. So minimally, this law had to be modified in terms of new information—which is standard behavior in science. Some laws will not hold if certain conditions are changed. Laws, then, can change just as facts can.

Laws are important, but as descriptive generalizations, they rarely explain natural phenomena. That is the role of the final stage in the hierarchy of explanation: theory. Theories explain laws and facts. Theories therefore are more important than laws and facts, and thus scientists place them at the top of the hierarchy of explanation.

Theories

The word *theory* is perhaps the most misunderstood word in science. In everyday usage, the synonym of theory is *guess* or *hunch*. Yet according to the National Academy of Sciences (2008: 11), "The formal scientific definition of theory is quite different from the everyday meaning of the word. It refers to a comprehensive explanation of some aspect of nature that is supported by a vast body of evidence." A theory, then, is an explanation rather than a guess. Many high school (and even, unfortunately, some college) textbooks describe theories as tested hypotheses, as if a hypothesis that is confirmed is somehow promoted to a theory, and a really, really good theory gets crowned as a law. But rather than being inferior to facts and laws, a scientific theory incorporates "facts, laws, inferences, and tested hypotheses" (National Academy of Sciences 1998: 7). Theories explain laws! To explain something scientifically requires an interconnected combination of laws, tested hypotheses, and other theories.

EVOLUTION AND TESTING

What about the theory of evolution? Is it scientific? Some have claimed that because no one was present millions of years ago to see evolution occur, evolution is not a scientific field. Yet we can study evolution in a laboratory even if no one was present to see zebras and horses emerge from a common ancestor. A theory can be scientific even if its phenomena are not directly observable. Evolutionary theory is built in the same way that theory is built in particle physics or any other field that uses indirect testing—and some aspects of evolutionary theory can be directly tested. I will devote chapter 2 to discussing evolution in detail, but let me concentrate here on the question of whether it is testable—and especially whether evolution is falsifiable.

The big idea of biological evolution (as will be discussed more fully in the next chapter) is descent with modification. Evolution is a statement about history and refers to something that happened, to the branching of species through time from common ancestors. The pattern that this branching takes and the mechanisms that bring it about are other components of evolution. We can therefore look at the testing of evolution in three senses: Can the big idea of evolution (descent with modification, common ancestry) be tested? Can the pattern of evolution be tested? Can the mechanisms of evolution be tested?

Testing the Big Idea

Hypotheses about evolutionary phenomena are tested just like hypotheses about other scientific topics: the trick (as in most science!) is to figure out how to formulate your question so it can be tested. The big idea of evolution, that living things have shared common ancestors, can be tested using the if-then approach—testing by consequences—that all scientists use. The biologist John A. Moore suggested a number of these if-then statements that could be used to test whether evolution occurred:

1. If living things descended with modification from common ancestors, then we would expect that "species that lived in the remote past must be different from the species alive today" (Moore 1984: 486). When we look at the geological record, this is indeed what we see.

There are a few standout species that seem to have changed very little over hundreds of millions of years, but the rule is that the farther back in time one looks, the more creatures differ from present forms.

2. If evolution occurred, we "would expect to find only the simplest organisms in the very oldest fossiliferous [fossil-containing] strata and the more complex ones to appear in more recent strata" (Moore 1984: 486). Again going to the fossil record, we find that this is true. In the oldest strata, we find single-celled organisms, then simple multicelled organisms, and then simple versions of more complex invertebrate multicelled organisms (during the early Cambrian period). In later strata, we see the invasion of the land by simple plants, and then the evolution of complex seed-bearing plants, and then the development of the land vertebrates.

3. If evolution occurred, then "there should have been connecting forms between the major groups (phyla, classes, orders)" (Moore 1984: 489). To test this requires going again to the fossil record, but matters are complicated by the fact that not all connecting forms have the same probability of being preserved. For example, connecting forms between the very earliest invertebrate groups are less likely to be found because of their soft bodies, which do not preserve as well as hard body parts such as shells and bones, which can be fossilized. These early invertebrates also lived in shallow marine environments, where the probability of a creature's preservation is different depending on whether it lived under or on the surface of the seafloor: surface-living forms have a better record of fossilization due to surface sediments being glued together by bacteria. Fossilized burrowing forms haven't been found—although their burrows have. It might be expected to find connections between vertebrate groups because vertebrates are large animals with large calcium-rich bones and teeth that have a higher probability of fossilization than do the soft body parts of the earliest invertebrates. There are, in fact, good transitions that have been found between fish and amphibians, and there are especially good transitions between reptiles and mammals. More and more fossils are being found that show structural transitions between reptiles (dinosaurs) and birds. Within a vertebrate lineage, there are often fossils showing good transitional structures. We have good evidence of transitional structures showing the evolution of whales from land mammals, and modern, large, single-hoofed horses from small, three-toed ancestors. Other examples can be found in reference books on vertebrate evolution such as those by Carroll (1998) or Prothero (2007).

In addition to the if-then statements predicting what one would find if evolution occurred, one can also make predictions about what one would not find. If evolution occurred and living things have branched off the tree of life as lineages split from common ancestors, one would not find a major branch of the tree totally out of place. That is, if evolution occurred, paleontologists would not find mammals in the Devonian age of fishes or seed-bearing plants back in the Cambrian. Geologists are daily examining strata around the world as they search for minerals, or oil, or other resources, and at no time has a major branch of the tree of life been found seriously out of place. Reports of "man tracks" being found with dinosaur footprints have been shown to be carvings, or eroded dinosaur tracks, or natural erosional features. If indeed there had not been an evolutionary, gradual emergence of branches of the tree of life, then there is no scientific reason why all strata would not show remains of living things all jumbled together.

In fact, one of the strongest sources of evidence for evolution is the consistency of the fossil record around the world. Another piece of evidence is the fact that when we look at the relationships among living things we see that it is possible to group organisms in gradually broader classifications. There is a naturally occurring hierarchy of organisms that has been recognized since the seventeenth century: species can be grouped into genera, genera can be grouped into families, and on and on into higher categories. The branching process of evolution generates hierarchy; the fact that animals and plants can be arranged in a tree of life is predicted and explained by the inference of common descent.

We can test not only the big idea of evolution but also more specific claims within that big idea. Such claims concern pattern and process, which require explanations of their own.

Pattern and Process

Pattern. Consider that if evolution is fundamentally an aspect of history, then certain things happened and other things didn't. It is the job of evolutionary biologists and geologists to reconstruct the past as best they can and to try to ascertain what actually happened as the tree of life developed and branched. This is the pattern of evolution, and indeed, along with the general agreement about the gradual appearance of modern forms over the past 3.8 billion years, the scientific literature is replete with disputes among scientists about specific details of the tree of life, about which structures represent transitions between groups and how different groups are related. Morphologically, most Neanderthal physical traits can be placed within the range of variation of living humans, but there are tests on fossil mitochondrial DNA that suggest that modern humans and Neanderthals shared a common ancestor very, very long ago—no more recently than 300,000 years ago (Ovchinnikov et al. 2000). So are Neanderthals ancestral to modern humans or not? There is plenty of room for argument about exactly what happened in evolution. But how do you test such statements?

Tests of hypotheses of relationships commonly use the fossil record. Unfortunately, sometimes one has to wait a long time before hypotheses can be tested. The fossil evidence has to exist (i.e., be capable of being preserved and actually be preserved), be discovered, and be painstakingly (and expensively) extracted. Only then can the analysis begin. Fortunately, we can test hypotheses about the pattern of evolution— and the idea of descent with modification itself—by using types of data other than the fossil record: anatomical, embryological, or biochemical evidence from living groups. One reason why evolution—the inference of common descent—is such a robust scientific idea is that so many different sources of information lead to the same conclusions.

We can use different sources of information to test a hypothesis about the evolution of the first primitive amphibians that colonized land. There are two main types of bony fish: the very large group of familiar ray-finned fish (e.g., trout, salmon, sunfish) and the lobe-finned fish, represented today by only three species of lungfish and one species of coelacanth. In the Devonian, though, there were nineteen families of lungfish and three families of coelacanths. Because of their many anatomical specializations, we know that ray-finned fish are not part of tetrapod (four-legged land vertebrate) ancestry; we and all other land vertebrates are descended from the lobe-fin line. Early tetrapods and lobe-fins both had teeth with wrinkly enamel and shared characteristics of the shoulder girdle and jaws, plus a sac off the gut used for breathing (Prothero 1998:

Figure 1.2
Are tetrapods more closely related to lungfish or to coela-
canths? Courtesy of Alan Gishlick.

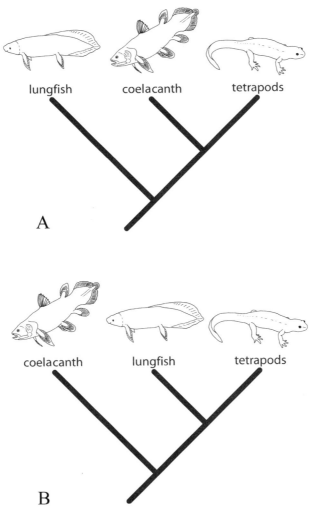

358). But are we tetrapods more closely related to lungfish or to coelacanths? Is the relationship among these three groups more like Figure 1.2A or Figure 1.2B? We can treat the two diagrams as hypotheses and examine data from comparative anatomy, the fossil record, biochemistry, and embryology to confirm or disconfirm A or B.

Anatomical and fossil data support hypothesis B (Thomson 1994). Studies on the embryological development of tetrapod and fish limbs also support hypothesis B. Now, when contemplating Figure 1.2, remember that these two diagrams omit the many known fossil forms and show only living groups. It isn't that tetrapods evolved from lungfish, of course, but that lungfish and tetrapods shared a common ancestor, and they shared that common ancestor with each other more recently than they shared a common ancestor with coelacanths. There is a large series of fossils filling the morphological gaps between ancestors of lungfish and tetrapods (Carroll 1998) and more are being discovered (Shubin, Daeschler, and Jenkins 2006).

Another interesting puzzle about the pattern of evolution is ascertaining the relationships among the phyla, which are very large groupings of kinds of animals. All the many kinds of fish, amphibians, reptiles, birds, and mammals are lumped together in one phylum (Chordata) with some invertebrate animals such as sea squirts and the wormlike lancelet (amphioxus). Another phylum (Arthropoda) consists of a very diverse group of invertebrates that includes insects, crustaceans, spiders, millipedes, horseshoe crabs, and the extinct trilobites. So you can see that phyla contain a lot of diversity. Figuring out how such large groups might be related to one another is a challenging undertaking.

Phyla are diagnosed on the basis of basic anatomical body plans—the presence of such features as segmentation, possession of shells, possession of jointed appendages, and so forth. Fossil evidence for most of these transitions is not presently available, so scientists have looked for other ways to ascertain relationships among these large groups. The recent explosions of knowledge in molecular biology and of developmental biology are opening up new avenues to test hypotheses of relationships—including those generated from anatomical and fossil data. Chordates for a long time have been thought to be related to echinoderms on the basis of anatomical comparisons (larvae of some echinoderms are very similar to primitive chordates) and this relationship is being confirmed through biochemical comparisons (e.g., ribosomal RNA) (Runnegar 1992). Ideas about the pattern of evolution can be and are being tested.

Process. Scientists studying evolution want to know not only the pattern of evolution but also the processes behind it: the mechanisms that cause cumulative biological change through time. The most important is natural selection (discussed in chapter 2), but there are other mechanisms (mostly operating in small populations, like genetic drift) that also are thought to bring about change. One interesting current debate, for example, is over the role of genetic factors operating early in embryological development. How important are they in determining differences among—and the evolution of—the basic body plans of living things? Are the similarities of early-acting developmental genes in annelid worms and in primitive chordates like amphioxus indicative of common ancestry? Another debate has to do with the rate and pace of evolution: do changes in most lineages proceed slowly and gradually, or do most lineages remain much the same for long periods that once in a while are punctuated with periods of rapid evolution? We know that individuals in a population compete with one another, and that populations of a species may outbreed one another, but can there be natural selection between lineages of species through time? Are there rules that govern the branching of a lineage through time? Members of many vertebrate lineages have tended to increase in size through time; is there a general rule governing size or other trends? All of these issues and many more constitute the processes or mechanisms of evolution. Researchers are attempting to understand these processes by testing hypotheses against the fossil and geological records as well as other sources of information from molecular biology and developmental biology (embryology).

Natural selection and other genetically based mechanisms are regularly tested and are regularly shown to work. By now there are copious examples of natural selection operating in our modern world, and it is not unreasonable to extend its operation into the past. Farmers and agricultural experts are very aware of natural selection as insects, fungi, and other crop pests become resistant to chemical controls. Physicians similarly are very aware of natural selection as they try to counter antibiotic-resistant microbes. The operation of natural selection is not disputed in the creationism/evolution controversy: both supporters and detractors of evolution accept that natural selection works. Creationists, however, claim that natural selection cannot bring about differences from one "kind" to another.

Pattern and process are both of interest in evolutionary biology, and each can be evaluated independently. Disputes about the pattern of evolutionary change are largely independent of disputes about the process. That is, arguments among specialists about how fast evolution can operate, or whether it is gradual or punctuated, are irrelevant to arguments over whether Neanderthals are ancestral to modern Europeans and vice versa. Similarly, arguments about either process or pattern are irrelevant to whether evolution took place (i.e., the big idea of descent with modification). This is relevant to the creationism/evolution controversy because some of the arguments about pattern or process are erroneously used to support the claim that descent with modification did not occur. Such arguments confuse different levels of understanding.

CREATIONISM AND TESTING

The topic of religion constitutes chapter 3, and creationism is a religious concept. Religion will be defined as a set of ideas concerning a nonmaterial reality; thus, it would appear that—given science's concern for material explanations—science and creationism have little in common. Yet the creationism/evolution controversy includes the claim made by some that creationism is scientific, or can be made scientific, or has scientific elements. The question naturally arises, then, Is creationism testable?

As discussed, science operates by testing explanations of natural phenomena against the natural world. Explanations that are disproved are rejected; explanations that are not disproved—that are corroborated—are provisionally accepted (though at a later time they may be rejected or modified with new information). An important element of testing is being able to hold constant some of the conditions of the test, so that a causative effect can be correctly assigned.

The ultimate statement of creationism—that the present universe came about as the result of the action or actions of a divine creator—is thus outside the abilities of science to test. If there is an omnipotent force in the universe, it would by definition be impossible to hold constant (to control) its effects. A scientist could control for the effects of temperature, light, humidity, or predators—but it would be impossible to control for the actions of God!

The question of whether God created cannot be evaluated by science. Most believers conceive of God as omnipotent, so God could have created everything just as we see it today, a theological position known as *special creationism*, or God could have created through a natural process such as evolution, a theological position known as *theistic evolution*. An omnipotent being could create the universe to appear as if it had evolved but actually have created everything five minutes ago. The reason that the ultimate statement of creationism cannot be tested is simple: the actions of an omnipotent creator are compatible with any and all observations of the natural world. The methods of science cannot choose among the possible actions of an omnipotent creator because by definition God is unconstrained.

Science is thus powerless to test the ultimate claim of creationism and must be agnostic about whether God did or did not create the material world. However, some types of creationism go beyond the basic statement "God created" to make claims of fact about the natural world. Many times these fact claims, such as those concerning the age of Earth, are greatly at variance with observations of science, and creationists sometimes invoke scientific support to support these fact claims. One creationist claim, for example, is that the Grand Canyon was laid down by the receding waters of Noah's flood. In cases like this, scientific methods can be used to test creationist claims, because the claims are claims of fact. Of course, it is always possible to claim that the creator performed miracles (that the layers of rocks in Grand Canyon were specially created by an omnipotent creator), but at this point one passes from science to some other way of knowing. If fact claims are made—assuming the claimer argues scientific support for such claims—then such claims can be tested by the methods of science; some scientific views are better supported than others, and some will be rejected as a result of comparing data and methodology. But if miracles are invoked, such occasions leave the realm of science for that of religion.

CONCLUSION

First, a caveat: the presentation of the nature of science and even the definitions of facts, hypotheses, laws, and theories I presented is very, very simplified and unnuanced, for which I apologize to philosophers of science. I encourage readers to consult some of the literature in philosophy of science; I think you'll find it a very interesting topic.

Science is an especially good way of knowing about the natural world. It involves testing explanations against the natural world, discarding the ones that don't work, and provisionally accepting the ones that do.

Theory building is the goal of science. Theories explain natural phenomena and are logically constructed of facts, laws, and confirmed hypotheses. Knowledge in science, whether expressed in theories, laws, tested hypotheses, or facts, is provisional, though reliable. Although any scientific explanation may be modified, there are core ideas of science that have been tested so many times that we are very confident about them and believe that there is an extremely low probability of their being discarded. The willingness of scientists to modify their explanations (theories) is one of the strengths of the method of science, and it is the major reason that knowledge of the natural world has increased exponentially over the past couple of hundred years.

Evolution, like other sciences, requires that natural explanations be tested against the natural world. Indirect observation and experimentation, involving if-then structuring of questions and testing by consequence, are the normal mode of testing in sciences such as particle physics and evolution, where phenomena cannot be directly observed.

The three elements of biological evolution—descent with modification, the pattern of evolution, and the process or mechanisms of evolution—can all be tested through the methods of science. The heart of creationism—that an omnipotent being created—is not testable by science, but fact claims about the natural world made by creationists can be.

In the next chapter, I will turn to the science of evolution itself.

REFERENCES

Carroll, Robert L. 1998. *Vertebrate paleontology and evolution.* New York: W. H. Freeman.

Cartmill, Matt. 1988. Seventy-five reasons to become a scientist: *American Scientist* celebrates its seventy-fifth anniversary. *American Scientist* 76: 450–463.

Hrdy, Sarah Blaffer. 2001. Mothers and others. *Natural History* (May): 50–62.

Montagu, M. F. Ashley. 1984. *Science and creationism.* New York: Oxford University Press.

Moore, John A. 1984. Science as a way of knowing—Evolutionary biology. *American Zoologist* 24 (2): 467–534.

National Academy of Sciences. 1998. *Teaching about evolution and the nature of science.* Washington, DC: National Academies Press.

National Academy of Sciences and Institute of Medicine. 2008. *Science, Evolution, and Creationism.* Washington, DC: National Academies Press.

Ovchinnikov, I. V., A. Gotherstrom, G. P. Romanova, V. M. Kharitonov, K. Liden, and W. Goodwin. 2000. Molecular analysis of Neanderthal DNA from the northern Caucasus. *Nature* 404: 490–493.

Poinar, Hendrik, Melanie Kuch, and Svante Pääbo. 2001. Molecular analysis of oral polio vaccine samples. *Science* 292 (5517): 743–744.

Prothero, Donald R. 1998. *Bringing fossils to life: An introduction to paleontology.* Boston: WCB.

Prothero, Donald R. 2007. *Evolution: What the fossils say and why it matters.* New York: Columbia University Press.

Runnegar, Bruce. 1992. Evolution of the earliest animals. In *Major events in the history of life,* ed. J. W. Schopf. Boston: Jones and Bartlett. pp. 64–93.

Shubin, Neil H., Edward B. Daeschler, and Farish A. Jenkins Jr. 2006. A Devonian tetrapod-like fish and the evolution of the tetrapod body plan. *Nature* 440: 757–763.

Thomson, Keith Stewart. 1994. The origin of the tetrapods. In *Major features of vertebrate evolution,* ed. D. R. Prothero and R. M. Schoch. Pittsburgh, PA: Paleontological Society. pp. 85–107.

Trefil, James. 1978. A consumer's guide to pseudoscience. *Saturday Review,* April 29, 16–21.

Weiss, Robin A. 2001. Polio vaccines exonerated. *Nature* 410: 1035–1036.

CHAPTER 2
• • • • • • • • • • • • •
Evolution

EVOLUTION BROAD AND NARROW

It has been my experience as both a college professor and a longtime observer of the creationism/evolution controversy that most people define evolution rather differently than do scientists. To the question, What does evolution mean? most people will answer, "Man evolved from monkeys," or invoke a slogan like "molecules to man." Setting aside the sex-specific language (surely no one believes that only males evolved; reproduction is challenging enough without trying to do it using only one sex), both definitions are much too narrow. Evolution involves far more than just human beings and, for that matter, far more than just living things.

The broad definition of evolution is a cumulative change through time. Not just any change counts as evolution, however. The Earth changes in position around the sun, but this is not evolution; an insect changes from egg to larva to adult during metamorphosis, but this is not evolution. An individual person (or a star) is born, matures, and dies but does not evolve. Evolution in this broad sense refers to the cumulative, or additive, changes that take place in phenomena like galaxies, planets, or species of animals and plants. It refers to changes that take place in groups rather than in individuals and to changes that accumulate over time.

Think of evolution as a statement about history. If we were able to go back in time, we would find different galaxies and planets, and different forms of life on Earth. Galaxies, planets, and living things have changed through time. There is astronomical evolution, geological evolution, and biological evolution. Evolution, far from the mere "man evolved from monkeys," is thus integral to astronomy, geology, and biology. As we will see, it is relevant to physics and chemistry as well.

Evolution needs to be defined more narrowly within each scientific discipline because both the phenomena studied and the processes and mechanisms of cosmological, geological, and biological evolution are different. Astronomical evolution deals with cosmology: the origin of elements, stars, galaxies, and planets. Geological evolution is concerned with the evolution of our own planet: its origin and its cumulative changes through time. Mechanisms of astronomical and geological evolution involve the laws and principles of physics and chemistry: thermodynamics, heat, cold, expansion, contraction, erosion, sedimentation, and the like. In biology, evolution is the inference that living things share common ancestors and have, in Darwin's words, "descended with modification" from these ancestors. The main—but not the only—mechanism of biological evolution is natural selection. Although biological evolution is the most contentious aspect of the teaching of evolution in public schools, some creationists raise objections to astronomical and geological evolution as well.

ASTRONOMICAL AND CHEMICAL EVOLUTION

Cosmologists conclude that the universe as we know it today originated from an explosion that erupted from an extremely dense mass, known as the Big Bang. Very soon thereafter, the universe inflated—it expanded at an inconceivably rapid rate. Within the first second after the Big Bang this rapid inflation had ceased, but the universe has continued expanding at a much slower pace ever since. Astronomers have found evidence that galaxies evolved from gravitational effects on swirling gases left over from the Big Bang. The total number of galaxies is estimated to be in the hundreds of billions. Stars formed within galaxies, and in the cores of the stars, helium and hydrogen fused into heavier elements. Additional elements are produced when stars explode. As stars die, many eject the heavy elements, enriching the gas and dust from which new generations of stars (and planets) will be born. Thus, the elements have evolved over the 13 billion years since the first galaxies began to form.

Cosmologists and geologists tell us that between 4 billion and 5 billion years ago the planet Earth formed from the accumulation of matter that was encircling the sun. In earliest times, Earth looked far different from what we see today: it was an inhospitable place scorched by radiation, bombarded by meteorites and comets, and belching noxious chemicals from volcanoes and massive cracks in the planet's crust. Yet it is hypothesized that Earth's atmosphere evolved from those gases emitted, and water might well have been brought to the planet's surface by those comets that were crashing into it.

Meteors and comets bombarded Earth until about 3.8 billion years ago. In such an environment, life could not have survived. After the bombardment ceased, however, primitive replicating structures evolved. Currently, there is not yet a consensus about how these first living things originated, and there are several directions of active research. Before there were living creatures, of course, there had to be organic (i.e., carbon-containing) molecules. Fortunately, such organic chemistry is common throughout space, so the raw material for life was probably abundant. Answering the question of chemical prebiotic evolution involves developing plausible scenarios for the emergence of organic molecules such as sugars, purines, and pyrimidines, as well as the building blocks of life, amino acids.

To explore this question, in the 1950s, scientists began experimenting to determine whether organic compounds could be formed from methane, ammonia, water vapor, and hydrogen—gases that were likely to have been present in Earth's early atmosphere. By introducing electrical sparks to combinations of gases, researchers were able to produce most of the amino acids that occur in proteins—which are the same amino acids found in meteorites—as well as other organic molecules (Miller 1992: 19). Because the actual composition of Earth's early atmosphere is not known, investigators have tried introducing sparks to various combinations of gases other than the original hypothesized blend. These also produce amino acids (Rode 1999: 774). Apparently, organic molecules form spontaneously on Earth and elsewhere, which has led one investigator to conclude, "There appears to be a universal organic chemistry, one that is manifest in interstellar space, occurs in the atmospheres of the major planets of the solar system, and must also have occurred in the reducing atmosphere of the primitive Earth" (Miller 1992: 20).

For life to emerge, some organic molecules had to be formed and then combined into amino acids and proteins, while other organic molecules had to be combined into something that could replicate: a material that could pass information from generation to generation. Modern living things are composed of cells, which consist of a variety of functioning components that are enclosed by a membrane; membranes set cells off from their environments and make them recognizable entities. As a result, origin-of-life research focuses on explaining the origin of proteins, the origin of heredity material, and the formation of membranes.

Origin-of-life researchers joke about their models falling into two camps: heaven and hell. Hell theories point to the present-day existence of some of the simplest known forms of life in severe environments, both hot and cold. Some primitive forms of life live in hot deep-sea vents where sulfur compounds and heat provide the energy to carry on metabolism and reproduction. Could such an environment have been the breeding ground of the first primitive forms of life? Other scientists have discovered primitive bacteria in permanently or nearly permanently frozen environments in the Arctic and the Antarctic. Perhaps deep in ice or deep in the sea, protected from harmful ultraviolet radiation, organic molecules assembled into primitive replicating structures.

The heaven theories note that organic molecules occur spontaneously in dust clouds of space and that amino acids have been found in meteorites. Perhaps these rocky visitors from outer space brought these basic components of life, which combined in Earth's waters to form replicating structures.

ORIGIN OF LIFE

Whether the proponents of hell or heaven theories finally convince their rivals of the most plausible scenario for the origin of the first replicating structures, it is clear that the origin of life is not a simple issue. One problem is the definition of life itself. From the ancient Greeks up through the early nineteenth century, people from European cultures believed that living things possessed an élan vital, or vital spirit—a quality that sets them apart from dead things and nonliving things such as minerals or water. Organic molecules, in fact, were thought to differ from other molecules because of the presence of this spirit. This view was gradually abandoned in science when more detailed study on the structure and functioning of living things repeatedly failed to discover any evidence for such an élan vital, and when it was realized that organic molecules could be synthesized from inorganic chemicals. Vitalistic ways of thinking persist in some East Asian philosophies, such as in the concept of chi, but they have been abandoned in Western science for lack of evidence and because they do not lead to a better understanding of nature.

How, then, can we define life? According to one commonly used scientific definition, if something is living, it is able to acquire and use energy, and to reproduce. The simplest living things today are primitive bacteria, enclosed by a membrane and not containing very many moving parts. But they can take in and use energy, and they can reproduce by division. Even this definition is fuzzy, though: what about viruses? Viruses, microscopic entities dwarfed by tiny bacteria, are hardly more than hereditary material in a packet—a protein shell. Are they alive? Well, they reproduce. They sort of use energy, in the sense that they take over a cell's machinery to duplicate their own hereditary material. But they can also form crystals, which no living thing can do, so biologists are divided over whether viruses are living or not. They tend to be treated as a separate special category.

If life itself is difficult to define, you can see why explaining its origin is also going to be difficult. Different researchers stress different components of the definition of life: some stress replication and others stress energy capture. Regardless, the first cell would have been more primitive than the most primitive bacterium known today, which itself is the end result of a long series of events: no scientist thinks that something like a modern bacterium popped into being with all its components present and functioning! Something simpler would have preceded it that would not have had all of its characteristics. A simple bacterium is alive: it takes in energy that enables it to function, and it reproduces (in particular, it duplicates itself through division). We recognize that a bacterium can do these things because the components that process the energy and allow the bacterium to divide are enclosed within a membrane; we can recognize a bacterium as an entity, as a cell that has several components that, in a sense, cooperate. But what if there were a single structure that was not enclosed by a membrane but that nonetheless could conduct a primitive metabolism? Would we consider it alive? It is beginning very much to look like the origin of life was not a sudden event, but a continuum of events producing structures that, early in the sequence, we would agree are not alive, and at the end of the sequence, we would agree are alive, with a lot of iffy stuff in the middle.

We know that virtually all life on Earth today is based on DNA, or deoxyribonucleic acid. This is a chainlike molecule that directs the construction of proteins and enzymes, which in turn direct the assembly of creatures composed of one cell or of trillions. A DNA molecule instructs cellular structures to link amino acids in a particular order to form a particular protein or enzyme. It also is the material of heredity, as it is passed from generation to generation. The structure of DNA is rather simple, considering all it does. A DNA molecule that codes for amino acids uses a "language" of four letters— A (adenine), T (thymine), C (cytosine), and G (guanine)—which, combined three at a time, determine the amino-acid order of a particular protein. For example, CCA codes for the amino acid proline and AGU for the amino acid serine. The exception to the generalization that all life is based on DNA is viruses, which can be composed of strands of RNA, another chainlike molecule that is quite similar to DNA. Like DNA, RNA is based on A, C, and G, but it uses uracil (U) rather than thymine.

The origin of DNA and proteins is thus of considerable interest to origin-of-life researchers, and many researchers approach the origin of life from the position that the replication function of life came first. How did the components of RNA and DNA assemble into these structures? One theory is that clay or calcium carbonate—both latticelike structures—could have provided a foundation upon which primitive chainlike molecules formed (Hazen, Filley, and Goodfriend 2001). Because RNA has one strand rather than two strands like DNA, some scientists are building theory around the possibility of a simpler RNA-based organic world that preceded our current DNA world (Joyce 1991; Lewis 1997), and very recently there has been speculation that an even simpler but related chainlike molecule, peptide nucleic acid (PNA), preceded the evolution of RNA (Nelson, Levy, and Miller 2000). Where did RNA or PNA come from? In a series of experiments combining chemicals available on early Earth, scientists have been able to synthesize purines and pyrimidines, which form the backbones of DNA and RNA (Miller 1992), but synthesizing complete RNA or DNA is extraordinarily difficult.

After a replicating structure evolved (whether it started out as PNA or RNA or DNA or something else), the structure had to acquire other bits of machinery to process energy and perform other tasks. Some researchers, the so-called metabolism-first investigators (Shapiro 2007), are looking at the generation of energy as the key element in the origin of life. In this scenario, replication is secondary to the ability to acquire energy.

Finally, this replicating and energy-using structure had to be enclosed in a membrane, and the origin of membranes is another area of research into the origin of life. A major component of membranes are lipids, which are arranged in layers. Precursors of lipids, layered structures themselves, apparently form spontaneously, and models are being developed to link some of these primitive compounds to simple membranes capable of enclosing the metabolizing and reproducing structures that characterize a cell (Deamer, Dworkin, Sandford, Bernstein, and Allamandola 2002). The origin of life is a complex but active research area with many interesting avenues of investigation, though there is not yet consensus among researchers on the sequence of events that led to the emergence of living things. But at some point in Earth's early history, perhaps as early as 3.8 billion years ago but definitely by 3.5 billion years ago, life in the form of simple single-celled organisms appeared. Once life originated, biological evolution became possible.

This is a point worth elaborating on. Although some people confuse the origin of life with evolution, the two are conceptually separate. Biological evolution is defined as the descent of living things from ancestors from which they differ. Evolution kicks in after there is something, like a replicating structure, to evolve. So the origin of life preceded evolution, and is conceptually distinct from it. Regardless of how the first replicating molecule appeared, we see in the subsequent historical record the gradual appearance of more complex living things, and many variations on the many themes of life. Predictably, we know much more about biological evolution than about the origin of life.

BIOLOGICAL EVOLUTION

Biological evolution is a subset of the general idea that the universe has changed through time. In the nineteenth century, Charles Darwin spoke of "descent with modification," and that phrase still nicely communicates the essence of biological evolution. *Descent* connotes heredity, and indeed, members of species pass genes from generation to generation. *Modification* connotes change, and indeed, the composition of species may change through time. Descent with modification refers to a genealogical relationship of species through time. Just as an individual's genealogy can be traced back through time, so too can the genealogy of a species. And just as an individual's genealogy has missing links—ancestors whose names or other details are uncertain—so too the history of a species is understandably incomplete. Evolutionary biologists are concerned both with the history of life—the tracing of life's genealogy—and with the processes and mechanisms that produced the tree of life. This distinction between the patterns of evolution and the processes of evolution is relevant to the evaluation of some of the criticisms of evolution that will emerge later in this book. First, let's look briefly at the history of life.

The History of Life

Deep Time. The story of life unfurls against a backdrop of time, of deep time: the length of time the universe has existed, the length of time that Earth has been a planet, the length of time that life has been on Earth. We are better at understanding things that we can have some experience of, but it is impossible to experience deep time. Most of us can relate to a period of one hundred years; a person in his fifties might reflect that one hundred years ago, his grandmother was a young woman. A person in her twenties might be able to imagine what life was like for a great-grandparent one hundred years ago. Thinking back to the time of Jesus, two thousand years ago, is more difficult; although we have written descriptions of people's houses, clothes, and how they made their living in those times, there is much we do not know of official as well as everyday life. The ancient Egyptians were building pyramids five thousand years ago, and their way of life is known in only the sketchiest outlines.

And yet the biological world of five thousand years ago was virtually identical to ours today. The geological world five thousand years ago would be quite recognizable: the continents would be in the same places, the Appalachian and Rocky mountains would look pretty much as they do today, and major features of coastlines would be identifiable. Except for some minor remodeling of Earth's surface due to volcanoes and earthquakes, the filling in of some deltas due to the deposition of sediments by rivers, and some other small changes, little has changed geologically. But our planet and life on it are far, far older than five thousand years. We need to measure the age of Earth and the time spans important to the history of life in billions of years, a number that we can grasp only in the abstract.

One second is a short period of time. Sixty seconds make up a minute, and sixty minutes make up an hour. There are therefore 3,600 seconds in an hour, 86,400 in a day, 604,800 in a week, and 31,536,000 in a year. But to count to 1 billion seconds at the rate of one per second, you would have to count night and day for approximately thirty-one years and eight months. The age of Earth is 4.5 billion years, not seconds. That is an enormous amount of time. As Stephen Jay Gould remarked, "An abstract, intellectual understanding of deep time comes easily enough—I know how many zeros to place after the 10 when I mean billions. Getting it into the gut is quite another matter. Deep time is so alien that we can really only comprehend it as metaphor" (Gould 1987: 3).

Figure 2.1 presents divisions of geological time used to understand geological and biological evolution. The solar system formed approximately 4.6 billion years ago; Earth formed about 4.5 billion years ago. The emergence of life was probably impeded by the bombardment of the Earth and moon by comets and meteorites until about 3.8 billion years ago, because only after the bombardment stopped do we find the first evidence of life. As discussed in the section "Astronomical and Chemical Evolution," there was a period of hundreds of millions of years of chemical evolution before the first structures that we might consider alive appeared on Earth: primitive one-celled organisms, less complex than any known bacterium.

After these first living things appeared between 3.5 billion and 4 billion years ago, life continued to remain outwardly simple for more than 2 billion years. Single-celled living things bumped around in water, absorbed energy, and divided—if some other organism didn't absorb them first. Reproduction was asexual: when a cell divided, the result was almost always two identical cells. Very slow changes occur with asexual reproduction, and this is probably an important reason that the evolution of life moved so slowly during life's first few billion years. Yet some very important evolutionary changes were taking place on the inside of these simple cells: the earliest living things gave rise to organisms that developed a variety of basic metabolic systems and various forms of photosynthesis.

Nucleated Cells. The first cells on Earth got along fine without a nucleus or a membrane around their DNA; in fact, bacteria today generate energy, carry out other cell functions, and reproduce new daughter bacteria without having nucleated DNA. Nucleated (eukaryotic) cells didn't evolve until about 1.5 billion years ago. Around 2 billion years ago, great changes in Earth's surface were taking place: continents were moving, and the amount of oxygen had increased in the atmosphere. Where did this oxygen come from? Oxygen is a by-product of photosynthesis, and indeed, oxygen produced by photosynthesizing bacteria built up in the atmosphere over hundreds of millions of years. This would explain the appearance of large red-colored geological deposits dating from this time: dissolved iron oxidized in the presence of free oxygen. In the words of researcher William Schopf, "The Earth's oceans had been swept free of dissolved iron; lowly cyanobacteria—pond scum—had rusted the world!" (Schopf 1992: 48). The increase of oxygen in the atmosphere resulted in a severe change in the environment: many organisms could not live in the new "poisonous" oxygenated environment. Others managed to survive and adapt.

The surface of Earth had been inhospitable for life: deadly radiation would have prevented life as we know it from existing at the planet's surface. The increase in oxygen as a result of photosynthesis resulted in the establishment of an ozone layer in the stratosphere. Oxygen is O_2; when ultraviolet radiation in the stratosphere strikes oxygen, ozone, which is O_3, is formed. The ozone shield protects living things from ultraviolet radiation, which permitted the evolution of life at the surface of the planet and eventually of the evolution of organisms composed of more than one cell (multicellular organisms, or metazoans).

Figure 2.1
Timescale of Earth's history. Courtesy of Alan Gishlick.

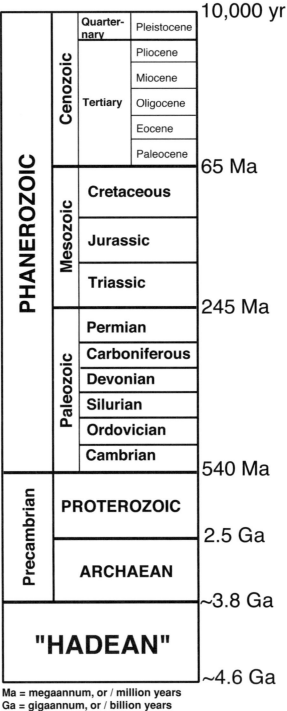

Ma = megaannum, or / million years
Ga = gigaannum, or / billion years

Eukaryotic cells may have evolved from unnucleated cells that were able to enclose their DNA in an interior membrane (forming the nucleus), and that incorporated other cells within their cell membranes. Nucleated cells have structures called organelles within their cytoplasm that perform a variety of functions having to do with energy capture and use, cell division, predation, and other activities. Some of these structures, such as mitochondria and chloroplasts, have their own DNA. Similarities between the DNA of such organelles and that of some simple bacteria have supported the theory that, early in evolution, the ancestors of eukaryotes absorbed certain bacteria and formed a cooperative, or symbiotic, relationship with them, whereby the newcomers functioned to enhance performance of metabolism, cell division, or some other task (Margulis 1993). The nucleus itself may have been acquired in a similar fashion, from "recycled" parts obtained after the absorption of other bacteria. Evidence for these theories comes, of course, not from the fossil record but from inferences based on biochemical comparisons of living forms.

Once nucleated cells developed, sexual reproduction was not far behind. Sexual reproduction has the advantage of combining genetic information from more than one individual, thus providing more variation to the population. Having more variation allows both the individual organism and the population of organisms to adjust to environmental change or challenge. Some researchers theorize that geological and atmospheric changes, together with the evolution of sexual reproduction, stimulated a burst of evolutionary activity during the late Precambrian period, about 900 million years ago, when the first metazoans (organisms composed of many cells) appear in the fossil record.

The Precambrian and the Cambrian Explosion. The first evidence we have of multicelled organisms comes from the Precambrian period, about 900 million years ago, and consists of fossils of sponges and jellyfish. Sponges are hardly more than agglomerations of individual cells; jellyfish are composed of two layers of cells that form tissues. Jellyfish, then, have a more consistent shape from organism to organism than do sponges, yet they lack a head and digestive, respiratory, circulatory, or other organs. Early echinoderms, represented today by starfish, sea urchins, and sea cucumbers, also occur in the Precambrian. Like the other Precambrian groups, early echinoderms have a rather simple body plan, but they do have a mouth and an anus, three tissue layers, and organs for digestion.

In the Cambrian, about 500 million years ago, there was rapid divergent evolution of invertebrate groups. New body plans appeared: "inventions" like body segmentation and segmented appendages characterized new forms of animal life, some of which died out but many of which continue to the present day. These new body plans appear over a geologically sudden—if not biologically sudden—period of about 10 million to 20 million years. Crustaceans, brachiopods, mollusks, and annelid worms, as well as representatives of other groups, appear during the Cambrian.

Evolutionary biologists are studying how these groups are related to one another and investigating whether they indeed have roots in the Precambrian period. In evolutionary biology, as in the other sciences, theory building depends on cross-checking ideas against different types of data. There are three basic types of data used to investigate the evolutionary relationships among the invertebrate groups: size and shape (morphological) comparisons among modern representatives of these groups, biochemical comparisons among the groups' modern descendants, and the fossil record. Largely because of problems in the preservation of key fossils at key times—and the fact that the evolution of these basic body plans might have taken "only" tens of millions of years, an eye blink from the perspective of deep time—the fossil evidence currently does not illuminate links among most of the basic invertebrate groups. Nonetheless, much nonfossil research is being conducted to understand similarities and differences of living members of these groups, from which we may infer evolutionary relationships.

One particularly active area of research has to do with understanding the evolution and developmental biology (embryology) of organisms, a new field referred to as "evo-devo."

Evo-Devo. Advances in molecular biology have permitted developmental biologists to study the genetics behind the early stages of embryological development in many groups of animals. What they are discovering is astounding. It is apparent that very small changes in genes affecting early, basic structural development can cause major changes in body plans. For example, there is a group of genes operating very early in animal development that is responsible for determining the basic front-to-back, top-to-bottom, and side-to-side orientations of the body. Other early-acting genes control such bodily components as segments and their number, and the production of structures such as legs, antennae, and wings. Major changes in body plan can come about through rather small changes in these early acting genes. What is perhaps the most intriguing result of this research is the discovery of identical or virtually identical early genes in groups as different as insects, worms, and vertebrates. Could some of the body plan differences of invertebrate groups be the result of changes in genes that act early in embryological development?

Probable evolutionary relationships among the invertebrate groups are being established through anatomy, molecular biology, and genetics, even if they have not been established through the fossil record. One tantalizing connection is between chordates, the group to which vertebrates belong (see the subsequent section), and echinoderms, the group to which starfish and sea cucumbers belong. On the basis of embryology, RNA, and morphology, it appears that the group to which humans and other vertebrates belong shared a common ancestor with these primitive invertebrates hundreds of millions of years ago. Although adult echinoderms don't look anything like chordates, their larval forms are intriguingly similar to primitive chordates. There are also biochemical similarities in the way they use phosphates—but read on to find out more about chordates!

Vertebrate Evolution. Our species belongs to the vertebrates, creatures with a bony structure encircling the nerve cord that runs along the back. Vertebrates are included in a larger set of organisms called chordates. Although all vertebrates are chordates, not all chordates are vertebrates. The most primitive chordates look like stiff worms. Characteristically, chordates have a notochord, or rod, running along the back of the organism with a nerve cord running above it. At some time in a chordate's life, it also has slits in the neck region (which become gills in many forms) and a tail. An example of a living chordate is a marine filter-feeding creature an inch or so long called

Figure 2.2
Amphioxus shows the basic body plan of chordates in having a mouth, an anus, a tail, a notochord, and a dorsal nerve chord. Courtesy of Janet Dreyer.

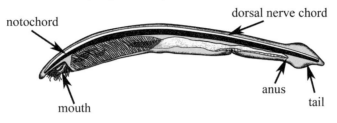

notochord

dorsal nerve chord

mouth

anus

tail

amphioxus. To look at it, you wouldn't think it was very closely related to vertebrates, but it is. Amphioxus lacks vertebrae, but like vertebrates, it has a notochord, a dorsal nerve cord, a mouth, an anus, and a tail. Like vertebrates, it is the same on the right side of the body as it is on the left (i.e., it is bilaterally symmetrical), and it has some other similarities in the circulatory system and muscle system that are structurally similar to vertebrates. It is probably fairly similar to an early chordate, but because it has been around the planet for a long time, it has evolved as well. Still, it preserves the diagnostic features of chordates in a relatively simple form (Figure 2.2).

Amphioxus is iconic in biological circles. There aren't very many evolution songs (there are far more antievolution songs!), but one that many biologists learn is sung to the tune of "It's a Long Way to Tipperary":

The Amphioxus Song
by Philip H. Pope

A fish-like thing appeared among the annelids one day.
It hadn't any parapods nor setae to display.
It hadn't any eyes nor jaws, nor ventral nervous cord,
But it had a lot of gill slits and it had a notochord.

Chorus:
It's a long way from Amphioxus. It's a long way to us.
It's a long way from Amphioxus to the meanest human cuss.
Well, it's goodbye to fins and gill slits, and it's welcome lungs and hair!
It's a long, long way from Amphioxus, but we all came from there.

It wasn't much to look at and it scarce knew how to swim,
And Nereis was very sure it hadn't come from him.
The mollusks wouldn't own it and the arthropods got sore,
So the poor thing had to burrow in the sand along the shore.

He burrowed in the sand before a crab could nip his tail,
And he said "Gill slits and myotomes are all to no avail.
I've grown some metapleural folds and sport an oral hood,
But all these fine new characters don't do me any good.

(chorus)

It sulked awhile down in the sand without a bit of pep,
Then he stiffened up his notochord and said, "I'll beat 'em yet!
Let 'em laugh and show their ignorance. I don't mind their jeers.
Just wait until they see me in a hundred million years.

My notochord shall turn into a chain of vertebrae
And as fins my metapleural folds will agitate the sea.
My tiny dorsal nervous cord will be a mighty brain
And the vertebrates shall dominate the animal domain.

(chorus)

Now that you have some idea of what a primitive chordate was like, let's return to my earlier comment that larval forms of echinoderms have similarities to primitive chordates. Unlike adult echinoderms, which are radially symmetrical (think of a starfish, where body parts radiate around a central axis), echinoderm larval forms are bilaterally symmetrical like chordates. In terms of embryology, echinoderms and chordates have a number of developmental similarities that set them apart from other bilaterally symmetrical animals. One hypothesis for chordate origins is that the larval form of an early echinoderm may have become sexually mature without growing up— that is, without going through the full metamorphosis to an adult. This phenomenon is uncommon, but it is not unknown. It occurs in salamanders such as the axolotl, for example.

In the Middle Cambrian is a small fossil called *Pikaia*, which is thought to be a primitive chordate because it looks rather like amphioxus (Figure 2.3). A new marine fossil discovered in the Late Cambrian Chengjiang beds of China might even be a primitive vertebrate. Although *Haikouella* swam, it certainly didn't look much like a fish as we think of fish today; it more resembled a glorified amphioxus (Figure 2.4). From such primitive aquatic chordates as these eventually arose primitive jawless

Figure 2.3
Pikaia, a Middle Cambrian fossil, shows some characteristics of primitive chordates. Courtesy of Janet Dreyer.

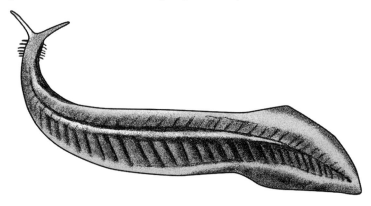

Figure 2.4
Haikouella, a Late Cambrian marine fossil, may be a primitive vertebrate. Courtesy of
Janet Dreyer.

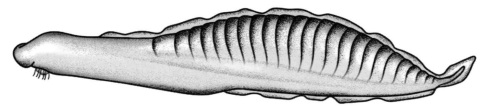

fish, then sharks and modern fish, and eventually the first land vertebrates: tetrapods
(which means "four footed"). These in turn became the ancestors of the other great
groups of land animals, reptiles and mammals. Later, more detail will be provided
about the evolution of many of these groups. But it is worthwhile to first present four
basic principles of biological evolution to keep in mind as you read the rest of the
book: natural selection, adaptation, adaptive radiation, and speciation.

MAJOR PRINCIPLES OF BIOLOGICAL EVOLUTION

Natural Selection and Adaptation

Natural selection is the term Charles Darwin gave to what he considered the most
powerful force of evolutionary change, and virtually all modern evolutionary biologists agree. In fact, the thesis that evolution is primarily driven by natural selection
is sometimes called Darwinism. Unfortunately, many people misapply the term to
refer to the concept of descent with modification itself, which is erroneous. Natural
selection is not the same as evolution. As discussed in chapter 1, there is a conceptual difference between a phenomenon and the mechanisms or processes that bring it
about.

When Darwin's friend T. H. Huxley learned of the concept of natural selection,
he said, "How extremely stupid not to have thought of that!" (Huxley 1888), so
obvious did the principle seem to him—after it was formulated. And indeed, it is a
very basic, very powerful idea. The philosopher Daniel Dennett has called natural
selection "the single best idea anyone has ever had" (Dennett 1995: 21). Because of
its generality, natural selection is widely found not only in nature but also increasingly
in engineering, computer programming, the design of new drugs, and other applications.

The principle is simple: generate a variety of possible solutions, and then pick the
one that works best for the problem at hand. The first solution is not necessarily the
best one—in fact, natural selection rarely results in even a good solution to a problem
in one pass. But repeated iterations of randomly generated solutions combined with
selection of the characteristics that meet (or come close to meeting) the necessary
criteria result in a series of solutions that more closely approximate a good solution.
Engineers attempting to design new airplane wings have used natural selection approaches; molecular biologists trying to develop new drugs have also used the approach

(Felton 2000). In living things, the problem at hand, most broadly conceived, is sur-vival and reproduction—passing on genes to the next generation. More narrowly, the problem at hand might be withstanding a parasite, finding a nesting site, being able to attract a mate, or being able to eat bigger seeds than usual when a drought reduces the number of small seeds. What is selected for depends on what, in the organism's par-ticular circumstances, will be conducive to its survival and reproduction. The variety of possible solutions consists of genetically based variations that allow the organism to solve the problem.

Variation among members of a species is essential to natural selection, and it is common in sexually reproducing organisms. Some of these variations are obvious to us, such as differences in size, shape, or color. Other variations are invisible, such as genetically based biochemical and molecular differences that may be related to disease or parasite resistance, or the ability to digest certain foods. If the environment of a group of plants or animals presents a challenge—say, heat, aridity, a shortage of hiding places, or a new predator—the individuals that just happen to have the genetic characteristics allowing them to survive longer and reproduce in that environment are the ones most likely to pass on their genes to the next generation. The genes of these individuals increase in proportion to those of other individuals as the population reproduces itself generation after generation. The environment naturally selects those individuals with the characteristics that provide for a higher probability of survival, and thus those characteristics tend to increase in the population over time.

So, the essence of natural selection is genetic variation within a population, an environmental condition that favors some of these variations more than others, and differential reproduction (some have more offspring than others) of the individuals that happen to have the favored variations.

A classic example of natural selection followed the introduction of rabbits into Australia—an island continent where rabbits were not native. In 1859, an English immigrant, Thomas Austin, released twelve pairs of rabbits so that he could go rabbit hunting. Unfortunately, except for the wedge-tailed eagle, a few large hawks, and din-goes (wild dogs)—and human hunters—rabbits have no natural enemies in Australia, and they reproduced like, well, rabbits. Within a few years, the rabbit population had expanded to such a large number that rabbits became a major pest, competing for grass with cattle, other domestic animals, and native Australian wildlife. Regions of the Australian outback that were infested with rabbits became virtual dust bowls as the little herbivores nibbled down anything that was green. How could rabbit numbers be controlled?

Officials in Australia decided to import a virus from Great Britain that was fatal to rabbits but that was not known to be hazardous to native Australian mammals. The virus produced myxomatosis, or rabbit fever, which causes death fairly rapidly. It is spread from rabbit to rabbit by fleas or other blood-sucking insects. The virus first was applied to a test population of rabbits in 1950. Results were extremely gratifying: in some areas the count of rabbits decreased from five thousand to fifty within six weeks. However, not all the rabbits were killed; some survived to reproduce. When the rabbit population rebounded, myxomatosis virus was reintroduced, but the positive effects of the first application were not repeated: many rabbits were killed, of course, but a larger percentage survived this time than had survived the first treatment. Eventually, myxomatosis virus no longer proved effective in reducing the rabbit population. Subsequently, Australians have resorted to putting up thousands of miles of rabbit-proof fencing to try to keep the rabbits out of at least some parts of the country.

How is this an example of natural selection? Consider how the three requirements outlined for natural selection were met:

1. Variation: The Australian rabbit population consisted of individuals that varied genetically in their ability to withstand the virus causing myxomatosis.
2. Environmental condition: Myxomatosis virus was introduced into the environment, making some of the variations naturally present in the population of rabbits more valuable than others.
3. Differential reproduction: Rabbits that happened to have variations allowing them to survive this viral disease reproduced more than others, leaving more copies of their genes in future generations. Eventually the population of Australian rabbits consisted of individuals that were more likely to have the beneficial variation. When myxomatosis virus again was introduced into the environment, fewer rabbits were killed.

Natural selection involves adaptation: having characteristics that allow an organism to survive and reproduce in its environment. Which characteristics increase or decrease in the population through time depends on the value of the characteristic, and that depends on the particular environment—adaptation is not one size fits all. Because environments can change, it is difficult to precisely predict which characteristics will increase or decrease, though general predictions can be made. (No evolutionary biologist would predict that natural selection would produce naked mole rats in the Arctic, for example.) As a result, natural selection is sometimes defined as adaptive differential reproduction. It is differential reproduction because some individuals reproduce more or less than others. It is adaptive because the reason for the differential in reproduction has to do with a value that a trait or set of traits has in a particular environment.

Natural Selection and Chance. The myxomatosis example illustrates two important aspects of natural selection: it is dependent on the genetic variation present in the population and on the value of some of the genes in the population. Some individual rabbits just happened to have the genetically based resistance to myxomatosis virus even before the virus was introduced; the ability to tolerate the virus wasn't generated by the need to survive under tough circumstances. It is a matter of chance which particular rabbits were lucky enough to have the set of genes conferring resistance. So, is it correct to say that natural selection is a chance process?

Quite the contrary. Natural selection is the opposite of chance. It is adaptive differential reproduction: the individuals that survive to pass on their genes do so because they have genes that are helpful (or at least not negative) in a particular environment. Indeed, there are chance aspects to the production of genetic variability in a population: Mendel's laws of genetic recombination are, after all, based on probability. However, the chance elements are restricted to affecting the genetic variation on which natural selection works, not natural selection itself. If indeed evolution is driven primarily by natural selection, then evolution is not the result of chance.

Now, during the course of a species' evolution, unusual things may happen that are outside anything genetics or adaptation can affect, such as a mass extinction caused by an asteroid that strikes Earth, but such events—though they may be dramatic—are exceedingly rare. Such contingencies do not make evolution a chance phenomenon any more than your life is governed by chance because there is a 1 in 2.8 million chance that you will be struck by lightning.

Natural Selection and Perfection of Adaptation. The first batch of Australian rabbits to be exposed to myxomatosis virus died in droves, though some survived to reproduce. Why weren't the offspring of these surviving rabbits completely resistant to the disease? A lot of them died, too, though a smaller proportion than that of the parent's generation. This is because natural selection usually does not result in perfectly adapted structures or individuals. There are several reasons for this, and one has to do with the genetic basis of heredity.

Genes are the elements that control the traits of an organism. They are located on chromosomes, in the cells of organisms. Because chromosomes are paired, genes also come in pairs, and for some traits, the two genes are identical. For mammals, genes that contribute to building a four-chambered heart do not vary—or at least if there are any variants, the organisms that have them don't survive. But many genetic features do vary from individual to individual. Variation can be produced when the two genes of a pair differ, as they do for many traits. Some traits (perhaps most) are influenced by more than one gene, and similarly, one gene may have more than one effect. The nature of the genetic material and how it behaves is a major source of variation in each generation.

The rabbits that survived the first application of myxomatosis bred with one another, and because of genetic recombination, some offspring were produced that had myxomatosis resistance, and others were produced that lacked the adaptation. The latter were the ones that died in the second round when exposed to the virus. Back in Darwin's day, a contemporary of his invented a sound bite for natural selection: he called it "survival of the fittest," with *fit* meaning best adapted—not necessarily the biggest and strongest. Correctly understood, though, natural selection is survival of the fit enough. It is not, in fact, only the individuals who are most perfectly suited to the environment that survive; reproduction, after all, is a matter of degree, with some rabbits (or humans or spiders or oak trees) reproducing at higher than the average rate and some at lower than the average rate. As long as an individual reproduces at all, though, it is fit, even if some are fitter than others.

Furthermore, just as there is selection within the rabbit population for resistance to the virus, so there is selection among the viruses that cause myxomatosis. The only way that viruses can reproduce is in the body of a live rabbit. If the infected rabbit dies too quickly, the virus doesn't have a chance to spread. Viruses that are too virulent tend to be selected against, just as the rabbits that are too susceptible will also be selected against. The result is an evolutionary contest between host and pathogen, which reduces the probability that the rabbit species will ever be fully free of the virus but also reduces the likelihood that the virus will wipe out the rabbit species.

Another reason natural selection doesn't result in perfection of adaptation is that once there has been any evolution at all (and there has been considerable animal evolution since the appearance of the first metazoan), there are constraints on the direction in which evolution can go. As discussed elsewhere, if a vertebrate's forelimb is shaped for running, it would not be expected to become a wing at a later time; that is one kind of constraint. Another constraint is that natural selection has to work with structures and variations that are available, regardless of what sort of architecture could best do the job. If you need a guitar but all you have is a toilet seat, you could make a sort of guitar by running strings across the opening, but it wouldn't be a perfect design. The process of natural selection works more like a tinkerer than an engineer (Monod 1971), and these two specialists work quite differently.

Evolution and Tinkering. Some builders are engineers and some are tinkerers, and the way they go about constructing something differs quite a lot. An engineering approach to building a swing for little Charlie is to measure the distance from the tree branch to a few feet off the ground; to go to the hardware store to buy some chain, hardware, and a piece of wood for the bench; and to assemble the parts, using the appropriate tools: measuring devices, a drill, a screwdriver, screws, a saw, sandpaper, and paint. Charlie ends up with a really nice, sturdy swing that avoids the "down will come baby, cradle, and all" problem and that won't give him slivers in his little backside when he sits on it. A tinkerer, on the other hand, building a swing for little Mary, might look around the garage for a piece of rope, throw it over the branch to see if it is long enough, and tie it around an old tire. Little Mary has a swing, but it isn't quite the same as Charlie's. It gets the job done, but it certainly isn't an optimal design: the rope may suspend little Mary too far off the ground for her to be able to use the swing without someone to help her get into it; the rope may be frayed and break; the swing may be suspended too close to the trunk so Mary careens into it—you get the idea. The tinkering situation, in which a structural problem is solved by taking something extant that can be bent, cut, hammered, twisted, or manipulated into something that more or less works, however crudely, mirrors the process of evolution much more than do the precise procedures of an engineer. Nature is full of structures that work quite well—but it also is full of structures that just barely work, or that, if one were to imagine designing from scratch, one would certainly not have chosen the particular modification that natural selection did.

Several articles by Stephen Jay Gould have discussed the seemingly peculiar ways some organisms get some particular job done. An anglerfish has a clever "lure" resembling a wormlike creature that it waves at smaller fish to attract them close enough to eat. The lure, actually a modified dorsal fin spine, springs from its forehead (Gould 1980a). During embryological development, the panda's wrist bone is converted into a sixth digit, which forms a grasping hand out of the normal five fingers of a bear paw plus a "thumb" that is jury-rigged out of a modifiable bone (Gould 1980b). Like a tinkerer's project, it gets the job done, even if it isn't a great design. After all, natural selection is really about survival of the fit enough.

Natural selection is usually viewed as a mechanism that works on a population or sometimes on a species to produce adaptations. Natural selection can also bring about adaptation on a very large scale through adaptive radiation.

Adaptive Radiation

To be fruitful and multiply, all living things have to acquire energy (through photosynthesis or by consuming other living things), avoid predation and illness, and
reproduce. As is clear from the study of natural history, there are many different
ways that organisms manage to perform these tasks, which reflects both the variety of
environments on Earth and the variety of living things. Any environment—marine,
terrestrial, arboreal, aerial, subterranean—contains many ecological niches that provide means that living things use to make a living. The principle of adaptive radiation
helps to explain how niches get filled.

The geological record reveals many examples of the opening of a new environment and its subsequent occupation by living things. Island environments such as
the Hawaiian Islands, the Galápagos Islands, Madagascar, and Australia show this
especially well. The Hawaiian archipelago was formed as lava erupted from undersea
volcanoes, and what we see as islands actually are the tips of volcanic mountains.
Erosion produced soils and land plants—their seeds or spores blown or washed in—
subsequently colonized the islands. Eventually land animals reached the islands as
well. Birds, insects, and a species of bat were blown to Hawaii or rafted there from
other Pacific islands on chunks of land torn off by huge storms.

The Hawaiian honeycreepers are a group of approximately twenty-three species
of brightly colored birds that range from four to eight inches long. Ornithologists
have studied them extensively and have shown them to be very closely related.
Even though they are closely related, honeycreeper species vary quite a bit from one
another and occupy many different ecological niches. Some are insectivorous, some
suck nectar from flowers, others are adapted to eating different kinds of seeds—one
variety has even evolved to exploit a woodpecker-like niche. The best explanation for
the similarity of honeycreepers in Hawaii is that they are all descended from a common
ancestor. The best explanation for the diversity of these birds is that the descendants
of this common ancestor diverged into many subgroups over time as they became
adapted to new, open ecological niches. Honeycreepers are, in fact, a good example
of the principle of adaptive radiation, by which one or a few individual animals arrive
in a new environment that has empty ecological niches, and their descendants are
selected to quickly evolve the characteristics needed to exploit these niches. Lemurs on
Madagascar, finches on the Galápagos Islands, and the variety of marsupial mammals
in Australia and prehistoric South America are other examples of adaptive radiation.

A major adaptive radiation occurred in the Ordovician period (about 430 million
years ago), when plants developed protections against drying out and against ultraviolet
radiation, vascular tissue to support erect stems, and other adaptations allowing for
life out of water (Richardson 1992). It was then that plants could colonize the dry
land. The number of free niches enabled plants to radiate into a huge number of ways
of life. The movement of plants from aquatic environments onto land was truly an
Earth-changing event. Another major adaptive radiation occurred about 400 million
years ago in the Devonian, when vertebrates evolved adaptations (lungs and legs) that
permitted their movement onto land. One branch of these early tetrapods radiated
into the various amphibians and another branch into reptiles and mammals. A major
difference between the reptile and mammal branch and amphibians was the amniotic

egg: an adaptation that allowed reproduction to take place independent of a watery environment.

During the late Cretaceous and early Cenozoic, about 65 million years ago, mammals began adaptively radiating after the demise of the dinosaurs opened up new ecological niches for them. Mammals moved into gnawing niches (rodents), a variety of grazing and browsing niches (hoofed quadrupeds, the artiodactyls and perissodactyls), insect-eating niches (insectivores and primates), and meat-eating niches (carnivores). Over time, subniches were occupied: some carnivores stalk their prey (lions, saber-toothed cats), and others run it down (cheetahs, wolves); some (lions, wolves, hyenas) hunt large-bodied prey, and some (foxes, bobcats) hunt small prey.

If a particular adaptive shift requires extensive changes, such as greatly increasing or reducing the size or number of parts of the body, the tendency is for that change to occur early in the evolution of the lineage rather than later. Although not a hard-and-fast rule, it follows logically from natural selection that the greatest potential for evolutionary change will occur before specializations of size or shape take place. Early in evolutionary history, the morphology of a major group tends to be more generalized, but as adaptive radiation takes place, structures are selected to enable the organisms to adapt better to their environments. In most cases, these adaptations constrain, or limit, future evolution in some ways. The forelimbs of perch are committed to propelling them through the water and are specialized for this purpose; they will not become grasping hands.

We and all other land vertebrates have four limbs. Why? We tend to think of four limbs as being "normal," yet there are other ways to move bodies around on land. Insects have six legs and spiders have eight, and these groups of animals have been very successful in diversifying into many varieties and are represented in great numbers all over the world. So, there is nothing especially superior about having four limbs, although apparently, because no organism has evolved wheels for locomotion, two or more limbs apparently work better. But all land vertebrates have four limbs rather than six or eight because reptiles, birds, and mammals are descended from early four-legged creatures. These first land vertebrates had four legs because the swimming vertebrates that gave rise to them had two fins in front and two in back. The number of legs in land vertebrates was constrained because of the number of legs of their aquatic ancestors. Imagine what life on Earth would have looked like if the first aquatic vertebrates had had six fins! Might there have been more ecological niches for land creatures to move into? It certainly would have made sports more interesting if human beings had four feet to kick balls with—or four hands to swing bats or rackets.

We see many examples of constraints on evolution; mammalian evolution provides another example. After the demise of the dinosaurs, mammals began to radiate into niches that had previously been occupied by the varieties of dinosaurs. As suggested by the shape of their teeth, mammals of the late Cretaceous and early Paleocene were small, mostly undifferentiated creatures that occupied a variety of insectivorous, gnawing, and seed-eating niches that dinosaurs were not exploiting. As new niches became available, these stem mammals quickly diverged into basic mammalian body plans: the two kinds of hoofed mammals, the carnivores, bats, insectivores, primates, rodents, sloths, and so on. Once a lineage developed (for example, carnivores), it radiated within the basic pattern to produce a variety of different forms (for example,

Figure 2.5
Vertebrate forelimbs all contain the same bones, although these bones have evolved over time for different locomotor purposes, such as running, swimming, flying, and grasping. Courtesy of Janet Dreyer.

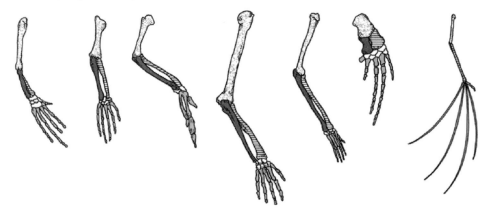

cats, dogs, bears, raccoons) in many sizes and shapes, all of which inherited basic dental and skeletal traits from the early carnivore ancestor. Once a lineage is "committed" to a basic way of life, it is rare indeed for a major adaptive shift of the same degree to take place. Although both horses and bats are descended from generalized quadrupedal early mammalian ancestors, the bones in a horse's forelimbs have been modified for swift running: some bones have been greatly elongated, others have been lost completely, and others have been reshaped. A bat has the same basic bones in its forelimb, but they have been greatly modified in other ways: some bones have been elongated, others have been lost, and yet others have been reshaped for flight (Figure 2.5).

Humans belong in the primate group of mammals, and primates are characterized by relatively fewer skeletal changes than have occurred in other mammal lineages. A primate doesn't have the extensive remodeling of the forelimb and hand that resulted in a bat's wing or a horse's hoof. We primates have a relatively basic "four on the floor" quadruped limb pattern of one bone close to the body (the femur in the leg and the humerus in the arm), two bones next to that one (the tibia and fibula in the leg and the radius and ulna in the arm), a group of small bones after this (tarsal or ankle bones in the leg and carpal or wrist bones in the arm), and a fanlike spray of small bones at the end of the limb (metatarsals and toe bones in the leg and metacarpals and finger bones in the arm) (Figure 2.5). Most primates locomote using four limbs; we human primates have taken this quadrupedal pattern and tipped it back so that our hind limbs bear all our weight (and not too successfully, as witnessed by hernias and the knee and lower-back problems that plague our species). Being bipedal, though, meant that we did not have to use our hands for locomotion, and they were thus freed for other purposes, like carrying things and making tools. Fortunately for human beings, dependent on tools and brains to survive, our early primate ancestors did not evolve to have specialized appendages like those of horses or bats.

Which is better, to be generalized or to be specialized? It's impossible to say without knowing more about the environment or niche in which a species lives. Specialized organisms may do very well by being better able to exploit a resource than are their possible competitors, yet generalized organisms may have an advantage in being able to adjust to a new environmental challenge.

Speciation

A species includes all the individuals that are capable of exchanging genes with one another. Some species are composed of very few individuals located in a restricted area, and others have millions of members spread out over large areas of the world. Some plant species are restricted to small areas of rain-forest habitat, while rats and humans live on literally every continent. It is more likely that an individual will mate with another individual that lives close by than farther away, and as a result, most species can be divided into smaller populations. Sometimes geographical factors, such as rivers or mountains or temperature gradients in different depths of water, naturally carve species into populations.

Because of geographical differences among populations, natural selection tends to result in populations varying from one another. A typical widespread species may be divided into many different populations. As long as they exchange genes at least at intervals, populations are likely to remain part of the same species. But how do new species form? New species form when members of a population or subdivision of a species no longer are able to exchange genes with the rest of the species. This is more likely to happen at the edges of the species range than in the center. We can say that speciation has occurred when a population becomes reproductively isolated from the rest of the species.

If a population at the end of the geographic range of a species is cut off from the rest of the species, through time it may become different from other populations. Perhaps natural selection is operating differently in its environment than it is in the rest of the species range, or perhaps the population has a somewhat different set of genes than other populations of the species. Just by the rules of probability, a small population at one end of the range of a species is not likely to have all the variants of genes that are present in the whole species, which might result in its future evolution taking a different turn.

No longer exchanging genes with other populations of the species, and diverging genetically through time from them, members of a peripheral, isolated population might reach the stage at which, were they to have the opportunity to mate with a member of the parent species, they would not be able to produce offspring. Isolating mechanisms, most of which are genetic but some of which are behavioral, can arise to prevent reproduction between organisms from different populations. Some isolating mechanisms prevent two individuals from mating; in some insects, for example, the sexual parts of males and females of related species are so different in shape or size that copulation cannot take place. Other isolating mechanisms come into effect when sperm and egg cannot fuse for biochemical or structural reasons. An isolating mechanism could take the form of the prevention of implantation of the egg or of disruption of the growth of the embryo after a few divisions. Or the isolating mechanism could kick in later: mules, which result from crossing horses and donkeys, are healthy but sterile. Donkey genes thus are inhibited from entering into the horse species, and vice versa. When members of two groups are not able to share genes because of isolating mechanisms, we can say that speciation between them has occurred. (Outside of the laboratory, it may be difficult to determine whether two species that no longer live in the same environment are reproductively isolated.)

The new species would of course be very similar to the old one—in fact, it might not be possible to tell them apart. Over time, though, if the new species manages successfully to adapt to its environment, it might also expand and bud off new species, which would be yet more different from the parent—now grandparent—species. This branching and splitting has, through time, given us the variety of species that we see today.

We can see this process of speciation operating today. Speciation in the wild usually takes place too slowly to be observed during the lifetime of any single individual, but there have been demonstrations of speciation under laboratory conditions. The geneticist Dobzhansky and his colleagues isolated a strain of Venezuelan fruit fly and bred it for several years. This strain of flies eventually reached a point of differentiation where it was no longer able to reproduce with other Venezuelan strains with which it had formerly been fertile. Speciation had occurred (Dobzhansky and Pavlovsky 1971).

Although not observed directly, good inferential evidence for speciation can be obtained from environments that we know were colonized only recently. The Hawaiian and Galápagos islands have been formed within the last few million years from undersea volcanoes and acquired their plants and animals from elsewhere. The Galápagos flora and fauna derive from South America, whereas the native Hawaiian flora and fauna are more similar to those of the Pacific islands, which in turn derive mostly from Asia. But Hawaiian species are reproductively isolated from their mainland counterparts.

One of the most dramatic examples of speciation took place among cichlid fish in the East African great lakes: Lake Victoria, Lake Malawi, and Lake Tanganyika. Geological evidence indicates that about twenty-five thousand years ago, Lake Tanganyika underwent a drying spell that divided the lake into three separate basins. Perhaps as a result of this and similar episodes, the cichlid fish that had entered the lake from adjacent rivers and streams underwent explosive adaptive radiation. There are at least 175 species of cichlid fish found in Lake Tanganyika and nowhere else. Similar speciation events took place in Lake Victoria and Lake Malawi—only over shorter periods of time (Goldschmidt 1996). Large lakes like these can be watery versions of an island: interesting biological things can go on.

Occasionally speciation can take place very quickly. The London subway, known as "the Tube," was built during the 1880s. At that time, some mosquitoes found their way into the miles of tunnels, and they successfully bred in the warm air and intermittent puddles—probably several times per year. Because they were isolated from surface mosquitoes, differences that cropped up among them would not have been shared with their relatives above, and vice versa. In the late 1990s, it was discovered that the Tube mosquitoes were a different species from the surface species. One major, if unfortunate, difference is that the surface mosquitoes, *Culex pipiens*, bite birds, whereas the related Tube species, *Culex molestus*, has shifted its predation to people. What is surprising about this discovery is that it shows that at least among rapidly breeding insects like mosquitoes, speciation does not require thousands of years but can occur within a century (Bryne and Nichols 1999). Natural selection, adaptation, adaptive radiation, and speciation—these are the major principles that help us explain the pattern and understand the process of evolution. These principles have resulted in an immense proliferation of living things over time that occupy a mind-boggling array of ecological niches.

A famous anecdote: asked by a member of the clergy what his study of nature had revealed to him about the mind of God, the biologist J. B. S. Haldane is supposed to have answered, "An inordinate fondness for beetles." And in fact one-fifth of the known animal species are species of beetles. Because there are so many different kinds of organisms, and not just beetles, human beings have always sought to make some sense of them by grouping them in various ways. All human cultures attempt to group plants and animals according to various schemes, which often have to do with how they can be used. In the Bible, the dietary laws of the Jews divided animals into clean and unclean, the latter being unsuitable for eating. Plants might be grouped according to whether they are for human consumption, for animal consumption, used for making dyes, or for some other purpose. Students of nature, naturalists, of the 1700s and 1800s sought to group animals and plants according to similarities and differences independent of their utility. The science of systematics, the study of the relationship among organisms, dates to a Swedish scientist known by his Latinized name, Carolus Linnaeus.

ORGANIZING THOSE BEETLES

Linnaeus classified a huge number of plants and animals during his lifetime. His rationale was overall similarity: the more similar organisms were, the more closely together they were placed in the ranked (hierarchical) system familiar to anyone who has taken middle school or high school biology. The highest Linnaean ranking is kingdom, followed by phylum, class, order, family, genus, and species. (There are a variety of mnemonics to remember their order, such as Kings Play Chess On Fine Golden Sets, or Kids Playing Chicken On Freeways Get Smashed.) Any plant or animal can be assigned a series of labels reflecting its membership in a group from each of these categories. Species was the smallest category, consisting of organisms that have the greatest similarity. But all members of the genus—a group of species—have certain characteristics in common as well, and the same can be said for family, and for every other category all the way up to kingdom. Here are the Linnaean classifications for house cats, chimpanzees, and human beings:

	House cats	Chimpanzees	Human beings
Kingdom:	Animalia	Animalia	Animalia
Phylum:	Chordata	Chordata	Chordata
Class:	Mammalia	Mammalia	Mammalia
Order:	Carnivora	Primates	Primates
Family:	Felidae	Pongidae	Hominidae
Genus:	Felis	Pan	Homo
Species:	cattus	troglodytes	sapiens

House cats, chimpanzees, and humans all belong to the same kingdom, phylum, and class; they have very many characteristics in common, and their Linnaean classification reflects this. Among other characteristics, they lack chlorophyll, so they are animals; they have a notochord, so they are chordates; and they have a single bone in the lower jaw, so they are all mammals. But chimpanzees and humans have more characteristics in common than either one has with cats, and Linnaeus grouped chimps and humans into the same order and cats into a different order. Humans and chimps were separated at the level of family, indicating that they were quite similar to one another.

Linnaeus's classification is useful, but classifying organisms on the basis of their similarities alone does not truly get at the underlying reality of nature. Why is it that all mammals have a single bone in the lower jaw? Why is it that humans and chimps are able to swing their arms over their heads but horses cannot? Organisms often have the same traits because they share genes. You and your brother or sister are more similar to each other than you are to your cousins because you and your siblings share more of your gene sequences with one another than you share with your cousins. Genes have a lot to do with important traits that an organism exhibits: they are why insects have six legs and spiders have eight, and why you walk on two legs and a monkey on four.

You and your siblings and cousins are similar in some traits (perhaps hair color, or stature, or blood type) because you share genes, and you share genes because you have a genealogical relationship to one another. You have descended with modification from common ancestors: parents in the case of your siblings and grandparents in the case of your first cousins. Similarly, all species are kin to one another in varying degrees because of common descent. The history of life is a branching and splitting genealogy of species changing through time. The Linnaean system, based on similarity and differences, provides an overall shape of this huge family tree of life, but it is not based on the underlying genealogical relationship of species—and thus does not always reflect the true relationship of organisms.

Ideally, a classification scheme would reflect genealogical relationships of organisms rather than just similarity, because similarity can be relatively superficial. Consider dolphins and tuna: both have an overall streamlined shape because that shape is very useful for getting around at high speeds in an open, watery environment. Yet there are many interior differences between dolphins and tuna: the skeletal systems, the circulatory systems, nervous systems, digestive systems, and so on. So, just because creatures are similar in overall shape does not mean that they are very closely related.

A late-twentieth-century classification method that has largely replaced the Linnaean system among biologists today is cladistics. *Clade* is a Greek word for "branch," and cladistics focuses on the branching of lineages through time. Both cladistics and the classical Linnaean system look at similarities among organisms to establish their relationships, but cladistics seeks in addition to reflect the actual results of evolution. In cladistics, the only groups of organisms that are considered natural are monophyletic, that is, groups comprising a single common ancestral species and all of its descendants. In terms of the tree of life, monophyletic groups correspond to whole branches that can be separated from the tree with a single cut. In contrast, the class of reptiles in the Linnaean system is not monophyletic because it excludes birds, which are descended from reptiles (as I discuss later). Similarly, a group consisting of warm-blooded animals (e.g., birds and mammals) also would not be monophyletic because all warm-blooded animals do not share a recent common ancestor.

Table 2.1
Ancestral traits, derived traits

	Trait a	Trait b	Trait c	Trait d	Trait e	Trait f	Trait g
	Warm blood	Hair	Diversified dentition	Fingernails	Grasping hands	Flat chest	Shoulder mobility
Chimps	x	x	x	x	x	x	x
Humans	x	x	x	x	x	x	x
Monkeys	x	x	x	x			
Cats	x	x	x				

Letters indicate characteristics. Traits a–c are found in all mammals, traits d–e additionally are found in all primates, and traits f–g are found in chimps and humans. From the standpoint of humans and chimps, traits a–e are ancestral traits, inherited from earlier mammal and primate ancestors. From the standpoint of chimps and humans, traits f and g are shared derived traits, inherited from a more recent ancestor. Looking at traits as ancestral or derived can help us reconstruct the evolutionary relationships of groups.

Unlike the Linnaean system, cladistic taxonomy encourages naming and using only monophyletic groups. For that reason, cladistics focuses on a particular kind of trait (i.e., derived traits) as indicators of evolutionary (phylogenetic) relationships. A cladistic analysis divides traits into two kinds, ancestral and derived, and then constructs evolutionary trees based on the distribution of derived traits. Let me give an example of how that works.

Consider that humans, monkeys, cats, and chimps have many characteristics in common: they all have warm blood; hair; and incisor, canine, premolar, and molar teeth that come in different shapes (compared to, for example, a crocodile, whose teeth all have pretty much the same conical shape). These three traits (Table 2.1, traits a–c) cannot differentiate among these species because they are common to all, and being common to all, they must have been present in the common ancestor of all of these mammals; we call such traits ancestral traits. But note that monkeys, humans, and chimps have traits that cats lack: fingernails rather than claws and hands that can grasp rather than paws (Table 2.1, traits d–e). These traits are associated with the common descent of primates after they separated from other mammalian groups such as the cats and are therefore not shared with cats or other nonprimate mammals. Similarly, a broad, flat chest and the ability to move the arm in a circle at the shoulder (Table 2.1, traits f–g) are traits that chimps and humans share but monkeys lack, which provides evidence that chimps and humans form a separate branch from monkeys. These are derived traits.

Traits are ancestral or derived not in an absolute sense but relative to one group or another. Having fingernails is a derived trait of primates relative to mammals, but having fingernails—common to all primates—can be considered an ancestral trait and thus not useful when one is trying to determine the relationships among different primates, such as between monkeys and apes.

Being able to differentiate ancestral and derived traits makes it possible to recon-struct the evolutionary relationships among organisms. To do so, one must look at the

Figure 2.6
A Cladogram of Primates: A cladogram shows the evolutionary relationship of organisms on the basis of their possession
of shared and ancestral traits. Warm blood, hair, and diversified dentition are found in all of the organisms in the diagram;
they are ancestral traits. Flat chests and shoulder mobility are
found only in the two groups above the mark: chimps and
humans. These would be shared derived traits of chimps and
humans. Courtesy of Alan Gishlick.

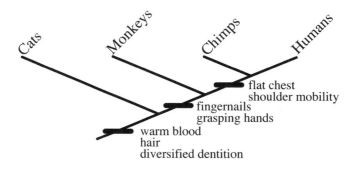

presence or absence of traits across a group of organisms, much as we did above with
some traits of cats, monkeys, chimps, and humans. When enough traits are examined,
certain traits emerge that indicate when a new lineage (a branch of the tree of life)
appears—and these obviously are the most informative for reconstructing the tree of
life. To determine the traits indicating a separate lineage, it is necessary to find an
outgroup: a species or other group that is related to the group you are studying and
that shows ancestral traits. To figure out the evolutionary relationships of monkeys,
humans, and chimps, we can use cats as an outgroup: cats and primates both are mammals, and cats exhibit the mammalian ancestral traits. This allows us to set aside a very
large number of traits that primates and cats share (like warm blood and diversified
dentition) and focus on those derived traits that distinguish monkeys, humans, and
apes from one another.

We can illustrate the relationships among these animals using a diagram called
a cladogram, which indicates the characteristics that distinguish clades (Figure 2.6).
Traits apply to all species to the right of where their labels appear, so mammalian
characteristics such as warm blood and hair will occur in the animals named at the
bottom of the diagonal line, because they are found in all of the organisms on the
diagram: cats, monkeys, chimps, and humans. Cats lack fingernails and grasping hands,
though, and because those characteristics set off primates from other mammals, they
are shared derived traits of primates. Only humans and chimps have the flat chest and
mobile shoulders that allow the arm-over-arm locomotion called brachiation, so these
traits are shared derived traits for humans and chimps.

Mammals form a clade because they share a common ancestor, all of whose descendants are mammals; primates are a clade within the mammal clade because they
share a common ancestor, all of whose descendants are primates, and humans and
chimps form a clade within primates because they share a common ancestor, all of

whose descendants are hominids (the technical name for the animals descended from the last common ancestors of chimpanzees and humans). We often will lack the fossil evidence for an actual ancestor of a lineage, but by using cladistic reconstruction, we can reconstruct many of the traits this ancestor would have had. For example, the first member of the lineage leading to humans, separate from chimpanzees, would be a biped, because that is a derived trait of our lineage, as is the presence of relatively small canine teeth. Such reconstructive expectations also help us interpret fossil remains.

This is a very brief and necessarily incomplete introduction to cladistic taxonomy. I used anatomical characteristics, but one can also use biochemical similarities, genetic or chromosomal similarities, and even developmental (embryological) characteristics to form cladograms of evolutionary relationships. Most evolutionary biologists use the cladistic approach to classify organisms because it avoids grouping organisms together on the basis of characteristics that do not reflect evolutionary relationships. When I was in high school and college, the Linnaean system was used. Birds were considered a separate branch of vertebrate life at the same level (i.e., class) as mammals or reptiles, partly because they had warm blood. Yet cladistic analysis, which separates ancestral from derived traits, shows that birds have a large number of traits that they share with a group of dinosaurs, and evolutionarily are closer to them than to mammals. Indeed, because cladistic taxonomy produces nested monophyletic groups, birds *are* dinosaurs—think about that during your next Thanksgiving dinner! Warm blood turns out to be a trait that has evolved more than once in the lineage of tetrapods (the descendants of the fish that adopted a terrestrial lifestyle about 365 million years ago). So, warm blood is a derived trait of both the mammal lineage and the reptile lineage that gave rise to dinosaurs and birds. Warm blood is a trait birds and mammals share— but not a trait that indicates close relationship. The division of traits into ancestral and derived clears up the confusion. To classify birds as a separate class, parallel to mammals and amphibians, would not reflect what really happened in evolutionary history. If we want all of our clades to reflect monophyly, we need to include birds as a subgroup of reptiles.

So cladistics is preferred to traditional Linnaean taxonomy because, by forcing us to classify according to monophyletic relationships, it better reflects the true genealogical relationship of living things. It also focuses on clades, or branches of the tree of life, and especially on the traits that distinguish clades, rather than on difficult-to-obtain ancestors. Cladistics is also considered superior to the Linnaean system because it does not depend on hunches about relationships among species, but rather allows—and requires—rigorous testing of hypotheses of evolutionary relationships. If you are interested in cladistic analysis, a good place to begin is the Web site of the University of California Museum of Paleontology (http://www.ucmp.berkeley.edu/IB181/VPL/Phylo/PhyloTitle.html).

DID MAN EVOLVE FROM MONKEYS?

So, to end with the question we began with, Did man evolve from monkeys? No. The concept of biological evolution, that living things share common ancestry, implies that human beings did not descend from monkeys, but shared a common ancestor with them, and shared a common ancestor farther back in time with other mammals, and farther back in time with tetrapods, and farther back in time with fish, and farther back in time with worms, and farther back in time with petunias. We are not descended from petunias, worms, fish, or monkeys, but we shared common ancestors with all of these creatures, and with some more recently than others. The inference of common ancestry helps us make sense of biological variation. We humans are more similar to monkeys than we are to dogs because we shared a common ancestor with monkeys more recently than we shared her a common ancestor with dogs. Humans, dogs, and monkeys are more similar to one another (they are all mammals) than they are to salamanders, because the species that provided the common ancestor of all mammals lived more recently than the species providing the common ancestors of salamanders and mammals. This historical branching relationship of species through time allows us to group species into categories such as primates, mammals, and vertebrates, which allows us to hypothesize about other relationships. Indeed, the theory of evolution, as one famous geneticist put it, is what "makes sense" of biology: "Seen in the light of evolution, biology is, perhaps, the most satisfying science. Without that light it becomes a pile of sundry facts, some of them more or less interesting, but making no comprehensible whole" (Dobzhansky 1973: 129). Evolution tells us why biology is like it is: living things had common ancestors, which makes a comprehensible whole of all those facts and details.

REFERENCES

Bryne, Katharine, and Richard A. Nichols. 1999. *Culex pipiens* in London underground tunnels: Differentiation between surface and subterranean populations. *Heredity* 82: 7–15.

Deamer, David, Jason P. Dworkin, Scott A. Sandford, Max P. Bernstein, Louis J. Allamandola. 2002. The first cell membranes. *Astrobiology* 2(4): 371–381.

Dennett, Daniel C. 1995. *Darwin's dangerous idea.* New York: Simon and Schuster.

Dobzhansky, Theodosius. 1973. Nothing in biology makes sense except in the light of evolution. *American Biology Teacher* 25: 125–129.

Dobzhansky, Theodosius, and O. Pavlovsky. 1971. Experimentally caused incipient species of *Drosophila. Nature* 230: 289–292.

Felton, Michael J. 2000. Survival of the fittest in drug design. *Drug* 3(9): 49–50, 53–54.

Goldschmidt, Tijs. 1996. *Darwin's dreampond: Drama in Lake Victoria,* trans. S. Marx-MacDonald. Cambridge, MA: MIT Press.

Gould, Stephen Jay. 1980a. Double trouble. In *The panda's thumb: More reflections on natural history.* New York: Norton.

Gould, Stephen Jay. 1980b. The panda's thumb. In *The panda's thumb: More reflections on natural history.* New York: Norton.

Gould, Stephen Jay. 1987. *Time's arrow, time's cycle: Myth and metaphor in the discovery of geological time.* Cambridge, MA: Harvard University Press.

Hazen, Robert M., Timothy R. Filley, and Glenn A. Goodfriend. 2001. Selective adsorption of L- and D-amino acids on calcite: Implications for biochemical homochirality. *Proceedings of the National Academy of Sciences* 98(10): 5487–5490.

Huxley, Thomas Henry. 1888. On the reception of the "Origin of Species." In *The life and letters of Charles Darwin,* ed. F. Darwin. London: Thomas Murray.

Joyce, Gerald F. 1991. The rise and fall of the RNA world. *The New Biologist* 3: 399–407.

Lewis, Ricki. 1997. Scientists debate RNA's role at beginning of life on Earth. *The Scientist* 11(7): 11.

Margulis, Lynn. 1993. *Symbiosis in cell evolution*, 2nd ed. New York: Freeman.

Miller, Stanley L. 1992. The prebiotic synthesis of organic compounds as a step towards the origin of life. In *Major events in the history of life*, ed. W. J. Schopf, 1–28. Boston: Jones and Bartlett.

Monod, Jacques. 1971. *Chance and necessity*. New York: Knopf.

Nelson, Kevin E., Matthew Levy, and Stanley L. Miller. 2000. Peptide nucleic acids rather than RNA may have been the first genetic molecule. *Proceedings of the National Academy of Sciences* 97(8): 3868–3871.

Richardson, John B. 1992. Origin and evolution of the earliest land plants. In *Major events in the history of life*, ed. J. W. Schopf, 95–118. Boston: Jones and Bartlett.

Rode, Bernd Michael. 1999. Peptides and the origin of life. *Peptides* 20: 773–786.

Schopf, J. William, ed. 1992. *Major events in the history of life*. Boston: Jones and Bartlett.

Shapiro, Robert. 2007. A simpler origin for life. *Scientific American* 296: 46–53.

CHAPTER 3

• • • • • • • • • • • • •

Beliefs:
Religion, Creationism, and Naturalism

Because the methodology of science works so well, you will find people from every nation, religion, and culture using it. Science is recognized internationally as the best way to find out about the natural world. But the natural world is not the only thing that human beings ask questions about, are concerned about, or think about. In fact, in every known human society, from the most complex urban civilization to the simplest community of hunter-gatherers, most people believe that there is a universe or world or something beyond or other than this material one. Gods, spirits, ancestors, or other nonmaterial beings populate this something. Science doesn't tell us anything about this world; this transcendent world—if it exists—is the province of religion.

RELIGION

Americans are most familiar with the Middle Eastern monotheistic traditions of the Jews, Christians, and Muslims. These are known as Abrahamic religions because all three revere the patriarch Abraham, and their practitioners worship a single God who reveals himself through sacred writings (the Torah, the Bible, and the Koran). All human societies have religious beliefs, however, and it is important not to let our understanding of a human universal such as religion be limited only to that which is familiar to us. To understand religion, one must look beyond, as well as at, the great Abrahamic religions.

All human societies have some belief system that can be called religion. Some of these are believed in by hundreds of millions of people, such as Christianity, Islam, Confucianism, and Hinduism, whereas others are believed in by tribal groups whose numbers are reckoned in thousands or even fewer. With such a disparity of beliefs, can we find any commonalities?

One thing all religions appear to have in common is a belief in something beyond the material world, an ultimate or absolute or transcendent reality beyond the earthly. A sense of sacredness, awe, or mystery about this beyond is common to religious beliefs and practices, and almost universal is the notion of spiritual (rather than mortal) beings that inhabit this realm and have special powers. These include gods, witches, powerful spirits, and the like. Most religions, though not all, include the concept of life after death, and most include a component of worship—ritual behavior associated with spiritual beliefs.

Intermediaries (such as priests and shamans) between people and the spiritual world are often very powerful and authoritative. Commonly there are special places for worship (such as temples, churches, holy sites) that are set apart from other sites (Stevens 1996). In virtually all religions, knowledge (about the supernatural; about where people, animals, and other natural objects came from; and about moral and ritual conduct) is obtained partly by revelation from supernatural sources. The gods of the Greeks revealed information through oracles, and the god of the Hebrews gave the Ten Commandments to Moses. Sometimes this revealed truth is recorded in texts that believers consider holy, such as the Koran of the Muslims, the Hindu Vedas, the Book of Mormon, or the New Testament of the Christians. Believers may dispute among themselves as to the proper interpretation of these holy texts.

How believers in a particular religion conceive of the ultimate varies enormously, from views similar to the Christian personal God to the considerably more diffuse Hindu conception of Brahma, a generalized "spirit behind, beneath, and beyond the world of matter and energy" (Raman 1998–1999: 6). Even within Christianity, the concept of God varies widely from an anthropomorphic creator God, such as that portrayed by Michelangelo on the ceiling of the Sistine Chapel, to a generalized force undergirding the universe that, although a source of awe, some Christians neither regard as a person nor pray to.

Human societies could not function without ethical systems—rules for behavior toward other people—and usually, though not universally, religion determines or at least strongly influences these systems. In many human societies, it is believed that rules for behavior are divinely revealed, such as the Ten Commandments, which Christians, Jews, and Muslims believe that God gave to Moses. Others may ascribe the rules for proper behavior to directives from ancestors, and still others have no supernatural source for their rules but attribute the origin of such rules to custom and tradition.

RELIGION AND EXPLANATION

Although the primary function of religion is to mediate between people and the gods or forces beyond everyday existence, it may additionally provide explanations of the natural world. In many human societies, natural phenomena are frequently explained by reference to supernatural causation. The sun shines or rain falls, but some sort of personal causation is involved in producing this effect. For example, the Brazilian Kuikuru people "know it was the wind that blew the roof off a house, but they carry the search for explanation one step further and ask, 'Who sent the wind?'" A human or spirit personality "had to direct the natural force of the wind to produce its effect" (Carneiro 1983). Sickness; death; meteorological phenomena such as rain or tornadoes; the existence and location of mountains and other landforms; earthquakes; volcanoes; the passage of seasons; and the positions of the sun, stars, and planets also frequently have religiously based explanations. In fact, for most people living in tribal, nonindustrial settings, the natural world and the spiritual world are not divided but are blended, in contrast to the modern Western cultural view.

In earlier times in Western society, it was common for biblical statements about the natural world to be accepted as authoritative and for God to be viewed as the direct cause of natural events. If plague struck a community or if a comet blazed across the sky, the event was attributed to the direct action of God, specially intervening in God's created world. Gradually, though, some of these statements in the Bible were discarded as they were found to be inaccurate—for example, that Earth is a circle (reflecting early civilization's belief that the world was disk shaped rather than spherical). Livestock breeders found that coat color in cattle was not affected by watering them at troughs in which peeled sticks had been placed (as claimed in Genesis 30:35–39), and thus the Bible came to be taken less as a source of information about the natural world and more as a guide to understanding the relationship of man to God. St. Augustine, among other early church leaders, argued in the fourth and fifth centuries that it was bad theology to accept biblical statements about the natural world uncritically if such statements contradicted experience. He felt that too-strict adherence to biblical literalism regarding statements about the natural world would diminish the credibility of proselytizers:

> Usually, even a non-Christian knows something about the earth, the heavens, and the other elements of this world, about the motion and orbit of the stars and even their size and relative positions, about the predictable eclipses of the sun and moon, the cycles of the years and the seasons, about the kinds of animals, shrubs, stones and so forth, and this knowledge he holds to as being certain from reason and experience.
>
> Now, it is a disgraceful and dangerous thing for an infidel to hear a Christian, while presumably giving the meaning of Holy Scripture, talking nonsense on these topics. . . . If they find a Christian mistaken in a field which they themselves know well, and hear him maintaining his foolish opinions about the Scriptures, how then are they going to believe those Scriptures in matters concerning the resurrection of the dead, the hope of eternal life, and the kingdom of heaven? How indeed, when they think that their pages are full of falsehoods on facts which they themselves have learnt from experience and the light of reason? (Augustine 1982: 42–43)

During the seventeenth and eighteenth centuries, science was developing as a methodology of knowing about the natural world. Natural philosophy, the study of nature, was regarded equally as a means to understand the mind of God and a means to understand the natural world. A considerable increase in knowledge about the natural world was obtained through the systematic methodology of science, in which natural phenomena were explained as instances of natural laws or theories. God was by no means ignored, but the focus was on discovering the laws that God had created. Isaac Newton, for example, was a highly religious man who sought to discover the natural laws by which God governed the universe. He felt that a God who worked through his created natural laws was a God more worthy of awe and worship than one who constantly intervened to maintain the universe. To Newton, God was more awesome if God caused planets to orbit about the sun using gravity than if God directly suspended them. Of course, as an omnipotent being, God could intervene at any time in the operation of the universe—miracles were possible—but it was not considered blasphemous to conclude that God acted through secondary causes (interpreted to be God's laws).

By the mid-nineteenth century, the success of science as a way of understanding the natural world was clear. It was possible to explain geological strata, for example, by reference to observable forces of deposition, erosion, volcanism, and other processes rather than by reliance on the direct hand of God to have formed the layers. By the late nineteenth century, science was well on its way to avoiding even the occasional reliance on God as immediate cause and to invoking only natural causes in explaining natural phenomena. This change in emphasis occurred not because of any animosity toward religion; rather, limiting science only to natural causes came about because it worked: a great deal was learned about the natural world by applying materialist (matter, energy, and their interactions) explanations.

Twentieth- and twenty-first-century scientists limit themselves to explaining natural phenomena using only natural causes for another practical reason: if a scientist is "allowed" to refer to God as a direct causal force, then there is no reason to continue looking for a natural explanation. Scientific explanation screeches to a halt. If there were a natural explanation, perhaps unknown or not yet able to be studied given technological limits or inadequate theory, then it would never be discovered if scientists, giving up in despair, invoked the supernatural. Scientists are quite used to saying, "I don't know yet."

But perhaps the most important reason scientists restrict themselves to natural explanations is that the methods of science are inadequate to test explanations involving supernatural forces. Recall that one of the hallmarks of science is the ability to hold constant some variables to be able to test the role of others. If indeed there is an omnipotent force that intervenes in the material world, by definition it is not possible to control for—to hold constant—its actions. As one wag put it, "You can't put God in a test tube"; and, one must add, you can't keep God out of one, either. Such is the nature of omnipotence—by definition. So, because God is unconstrained, any test of an explanation that involves God would be impossible to set up: all results or outcomes of the test are compatible with God's acts.

As a result, scientists do not consider supernatural explanations scientific. We will encounter a contrary opinion when we discuss intelligent design. Of course, limiting scientific explanation to natural causes has been extraordinarily fruitful. In the spirit of the adage "if it ain't broke, don't fix it," scientists continue to seek explanations in natural processes when doing science, whether they are believers or nonbelievers in an omnipotent power.

A topic to which we will return at the end of the chapter concerns a difference between a rule of science and a philosophical view—between methodological naturalism and philosophical naturalism. We have been discussing a rule of science that requires that scientific explanations use only material (matter, energy, and their interaction) cause; this is known as methodological naturalism. To go beyond methodological naturalism to claim that the universe consists only of matter and energy—that is, that there is no God or, more generally, no supernatural entities—is philosophical naturalism. The two views are logically decoupled because one can be a methodological naturalist but not accept naturalism as a philosophy. Scientists who are theists are examples: in their scientific work they explain natural phenomena in terms of natural causes, even if in their personal lives they believe in God, and even that God may intervene in nature.

Christianity and many other religions rely at least in part on truth revealed from God. When a revelation-based claim about the natural world is made, it may come into conflict with knowledge gained from experience—as St. Augustine described in the quote earlier in this chapter. A classic example of revealed truth conflicting with scientific interpretation is the seventeenth-century debate regarding the relationship of Earth to the other planets and the sun. Traditionally, the Bible was interpreted as reflecting a geocentric, or Earth centered, model of the universe. The sun and the other planets revolved around Earth. Early astronomers such as Copernicus and Galileo challenged the geocentric view, based on their empirical observations, inferences, and mathematical calculations, holding instead the heliocentric view that Earth and other planets revolved around the sun. The Catholic Church rejected these conclusions partly on scientific grounds, but primarily because heliocentrism contradicted the accepted interpretation of the Bible that Earth had to be the center of the universe. God had created humankind to worship him, and, in turn, had made the whole universe for us. Because Earth was the place where human beings lived, logically it would be the center of the universe. Bible passages such as Joshua 10:12–13 reinforced this view. In this passage, Joshua requests God to lengthen the day so his soldiers might win on the battlefield; God lengthens the day by stopping the sun, reflecting the geocentric model of the universe extant when the book of Joshua was set down. Although at one time, heliocentrism was considered blasphemous, today only a tiny fraction of Christians interpret the Bible as a geocentric document; for the vast majority of Christians, it is no longer necessary to interpret the Bible as presenting a geocentric cosmology.

CREATIONISM

Just as with *evolution*, the word *creationism* has a broad and a narrow definition. Broadly, *creationism* refers to the idea of creation by a supernatural force. To Christians, Jews, and Muslims, this supernatural force is God; to people of other religions, it is other deities. The creative power may be unlimited, like that of the Christian God, or it may be restricted to the ability to affect certain parts of nature, such as heavenly bodies or certain kinds of living things.

The term *creationism* to many people connotes the theological doctrine of special creationism: that God created the universe essentially as we see it today, and that this universe has not changed appreciably since that creation event. Special creationism includes the idea that God created living things in their present forms, and it reflects a literalist view of the Bible. It is most closely associated with the endeavor of "creation science," which includes the view that the universe is only 10,000 years old. But the most important aspect of special creation is the idea that things are created in their present forms. In intelligent design creationism, for example, God is required to specially create complex structures such as the bacterial flagellum or the body plans of animals of the Cambrian period, even though many if not most intelligent design proponents accept an ancient Earth.

It is important to define terms and use them consistently. In this book, the usual connotation of *creationism* will be the Christian view that God created directly. Special creationism is the most familiar form of direct creationism, but some Christians view God as creating sequentially rather than all at once. Later in this chapter, readers will be introduced to a range of religious views about creationism and evolution that will help clarify these relationships.

ORIGIN MYTHS

All people try to make sense of the world around them, and that includes speculating about the course of events that brought the world and its inhabitants to their present state. Stories of how things came to be are known as origin myths. They are tied to the broad definition of creationism.

Now, just as the word *theory* is used differently in science than in casual conversation (see chapter 1), so the word *myth* is a term of art in the anthropological study of cultures. The common connotation of *myth* is something that is untrue, primitive, or superstitious—something that should be discounted. Yet when anthropologists talk of myths, it is to describe stories within a culture that symbolize what members of the culture hold to be most important. A culture's myths are unquestionably important, and *myth* is not a term of denigration.

Rather than being dismissible untruths, myths express some of the most powerful and important ideas in a society. In societies dependent on oral tradition rather than writing, myths reinforce values and ideals and help to transmit them from generation to generation. Myths in this sense are true even if they are fantastic and deal with impossible events or have actors who could not have existed—like talking steam engines. Because myths encapsulate important cultural truths, anthropologists recognize that they are vitally important to a society and deserve respect. In the anthropological study of cultures, the term *myth* is not pejorative. Myths are of great importance.

Although myths tend to be more common in nonliterate societies, they occur even in developed countries like our own. The children's story of *The Little Engine That Could*, for example, is a classic myth that expresses an important value in American culture: persevering in the face of adversity. The Horatio Alger myth of the poor but plucky youth who achieves success through hard work, pulling himself up by his bootstraps, is classically American. Both of these secular myths also express the American value of individualism—something quite characteristic of our culture. Mythical elements arise around historical and popular heroes as well: there are many myths associated with Abraham Lincoln and George Washington, for example.

Some myths are secular and others are religious, but all involve a symbolic representation of some societal or human truth. In the mythology of the ancient Greeks, the goddess Persephone joins her husband Hades below the surface of Earth for part of the year. When she is gone, her mother, Demeter, the goddess of growing things, laments her absence, and winter comes. In the spring, when Persephone rejoins her mother, the world becomes green and fertile again. The story of Persephone and Hades not only symbolizes the passage of seasons but also is a metaphor of the human realities of death and birth. Chinese culture reflects a strong sense of the importance of balancing opposites: yin and yang, light and dark, hot and cold, good and evil, wet and dry, earth and sky, female and male—there are many examples of this duality. A Chinese origin myth reflects this important cultural concern of balance: the creator god Pan Gu separates chaos into these opposites and establishes a series of dualities, including the separation of earth from sky, and other elements of the physical universe.

Some cultures have myths about creator figures or heroes who establish legitimacy for tribes or kin groups within a tribe by giving certain people particular lands, objects, or rituals that only they can use (Leeming and Leeming 1994). The telling of these myths may be incorporated in rituals that remind people of the relationships among people in society, as well as relationships between groups. They can also be art forms: myths are often a form of literature as well as a means to promote the continuity of a culture. And in truth, stories are more meaningful and much easier to remember than lectures—a principle doubtlessly recognizable to anyone who has been a student!

Just as do tools and language, myths spread from people to people in a process anthropologists call diffusion. Humans necessarily must live near water, and after agriculture was invented, human settlements tended to congregate in river valleys, where control of water for agriculture often was the basis for political and religious power. Floods are not uncommon in such environments, and overflowing rivers may be a source of the fertility that attracts people to such settings. So, it is not surprising to find that the early agricultural societies of the Middle East all possessed versions of a flood myth and a hero who survived it on a raft or boat: the Babylonians (Utnapishtim), Sumerians (Ziusudra), Indians (Manu), Greeks (Deucalion and Pyrrha), and Hebrews (Noah). Similarities in the flood myths of all of these groups suggest considerable diffusion—but there are differences as well, which presumably reflect individual cultural elements. After all, myths are symbolic of what is important to a people—and what was important to the Babylonians differed from what was important to the Hebrews, to take just one pair.

Sometimes as cultures come in contact with one another, new ideas and practices replace old ones, but more frequently cultural elements are borrowed and recombined. When the African Efe people encountered Christian versions of creation from Genesis, what eventually emerged was a combination origin myth incorporating a traditional female moon figure who helps the high god create human beings. He commands the people not to eat the fruit of the tahu tree, but one of the women disobeys. The moon sees her and reports her to the high god, who punishes human beings with death. If you are familiar with the biblical Adam and Eve story, you can see how the Efe adapted components of this creation myth.

Types of Origin Myths

Although origin myths are quite varied, they can be grouped into types. The origin myth of the Cubeo people of Colombia presents the world as always having existed, without a specific origin event, but most myths include a beginning time or event. Several cultures believe that in the beginning was a "cosmic egg," which either breaks like a familiar bird's egg to let forth a creator god (the Chinese Pan Gu, the Polynesian Ta'aroa, or the Hindu Prajapati) or is itself laid by a deity and hatches into elements of the universe. The myth of the Pelasgians of ancient Greece, for example, featured a cosmic egg laid by the goddess Eurynome, which hatched into the sun, moon, and stars as well as plants and animals (Leeming and Leeming 1994).

The beginning period might be a time of chaos, usually watery and dark, with supernatural beings emerging from a void. Perhaps reflecting a normal human preference for order and predictability over disorder and chaos, many origin myths attempt to explain how an orderly, understandable world emerged from frightening, formless disorder. Many traditions, such as that of the Native American Hopi people, speak of a time when human beings lived underground and emerged to the upper world when led there by a spirit figure or god. Many origin myths describe the creation of Earth as resulting from the dismembering of a god or previous spirit: the Norse god Odin creates the mountains, seas, and other geographical features from the body of the slain giant Ymir; the Babylonian god Marduk creates the world from the body of the slain mother figure Tiamat.

The origin myths of North American Indian groups frequently include the earth-diver motif, in which a god or messenger is commanded to dive into the formless waters and bring up mud or silt, which is made into dry land. Earth-diver myths are common, ranging from Eastern Europe throughout Asia and into North America. The motif is even found in some Melanesian tribes of the Pacific.

Genesis Symbolism

The story of Creation in the biblical book of Genesis symbolizes many things to people of Abrahamic faiths. Because they were migratory, and because they were located at a geographical crossroads, ancient Hebrews encountered many other Middle Eastern groups; as is typical in culture contact, they borrowed from neighbors and shared their own heritage. Origin myths of most of the Middle Eastern cultures, for example, included the motifs of the creation of humans from clay, as well as a primordial, chaotic state composed of water. The Genesis creation story derives in part from earlier Middle Eastern traditions from Babylonia and Persia, but with important differences.

According to the theologian Conrad Hyers, the ancient Hebrews found themselves surrounded by other tribes that worshipped multiple gods, a practice called polytheism. Of central importance to the Hebrews, and their major distinction among their neighbors, was their belief in one god (monotheism), and maintaining this belief (especially in the face of conquest) was difficult. The Hebrews were variously conquered by Egyptians, Assyrians, Babylonians, and Persians, which meant that remaining true to their traditions and avoiding absorption was a constant challenge. There was much pressure on the Hebrews to adopt the gods and idols of their neighbors. According to Hyers, the religious meaning of Genesis is largely to make a statement to both Hebrews and surrounding tribes that the one god of Abraham was superior to the false gods of their neighbors: sky gods (the sun, the moon, and stars), earth gods, nature gods, light and darkness, rivers, and animals (Hyers 1983). As Hyers (1983: 101) puts it, "Each day of creation takes on two principal categories of divinity in the pantheons of the day, and declares that these are not gods at all, but . . . creations of the one true God who is the only one. . . . Each day dismisses an additional cluster of deities, arranged in a cosmological and symmetrical order."

So on day 1 ("Let there be light"), God vanquishes the pagan gods of light and darkness. Similarly, gods of the sky and seas are displaced on day 2, while Earth gods and gods of vegetation are done away with on the third day. On the fourth day God creates the sun, moon, and stars, thereby establishing his superiority to them, and the fifth day removes divinity from the animal kingdom. Finally, on the sixth day God specially creates human beings, which takes away from the divinity of kings and pharaohs—but because God creates humans as his own special part of creation (in God's image), all human beings are in some degree divine.

Genesis also described the nature of the Hebrew God. Unlike the gods of other Middle Eastern groups, the Hebrew God was ever present. Unlike the high god Marduk of the Mesopotamians, the Hebrew God did not originate from the actions of some other god or preexisting force. Genesis also suggests that God is omnipotent; unlike the Mesopotamian or Sumerian gods, the Hebrew God does not require preexisting materials from which to assemble creation but speaks (wills) the universe into being. God is also moral, being concerned with good and evil, which contrasts strongly with the gods of the Hebrews' neighbors, who seem to govern in a universe that has little meaning or purpose. The Bible's God also is not part of nature, as some of the gods of others, but stands outside of nature as its creator (Sarna 1983).

Genesis also tells of the nature of humankind, "a God-like creature, uniquely endowed with dignity, honor, and infinite worth, into whose hands God has entrusted mastery over His creation" (Sarna 1983: 137). God forms the universe, making Earth the most important component and humans its most important creature, having been given dominion over all other creatures and Earth itself. Humanity's responsibility is to husband the Earth but also to worship and obey God. Much of Genesis, especially the stories of Adam and Eve and of Noah and the Flood, reflect these themes; Adam and Eve are cast out of Paradise for disobeying God, and Noah is rewarded for his obedience and faith by being chosen to survive the Flood.

Thus, Genesis reflects the character of a classic origin myth: it presents in symbolic form the values ancient Hebrews felt were most important: the nature of God, the nature of human beings, and the relationship of God to humankind. Hebrews distinguished their God from those of their neighbors and presented God's deeds in their oral traditions and, eventually, in written form. Some of these writings were selected over time to become the Old Testament of the Bible.

Modern Jews, Christians, and Muslims all revere the Bible as a sacred book, but each of the Abrahamic faiths has different interpretations of many of the events depicted—and differences of interpretation occur within the three faiths as well. For example, in contrast to the early Hebrew view, some modern Christians and Jews do not necessarily see God as separate from God's creation. There are also differences in beliefs among sects as to the amount that God intervenes in the world, and the nature and even the existence of miracles. Yet as did the ancient Hebrews, the Abrahamic faiths generally agree that God is omnipotent and good and that human beings are responsible to God. As will be discussed later, there are vast differences among believers as to specifics of faith, such as how literally the Bible should be read. Christians, Jews, and Muslims all have constituent sects that demand that the holy texts (Bible, Koran, or Torah) be read literally, and all have sects that feel many or most passages should be read symbolically.

AMERICAN RELIGIONS

Americans practice a large number of religions, but the religion with the most adherents by far is Christianity. According to several polls, upward of 85 percent of Americans describe themselves as Christian. Scholars at the City University of New York (CUNY) conducted the largest survey of American religious views in 1990. In the National Survey of Religious Identification (NSRI), researchers conducted a telephone survey of 113,723 adults, randomly chosen, with results statistically weighted to reflect American demographic characteristics (Kosmin and Lachman 1993). The percentage of error in a survey of this size is less than 0.5 percent.

Respondents were asked a simple question—"What is your religion?"—and answers, as well as information on geographic location, age, sex, income, and so on, were tabulated. The results of the survey are presented in Table 3.1.

The religious profile of Americans in the 1990 NSRI study is echoed in other surveys conducted during that decade. In a 1996 poll conducted by the humanist publication *Free Inquiry*, 90.7 percent of Americans stated that they have a religion, with 83.8 percent identifying as either Catholic or Protestant (Free Inquiry, 1996). A Gallup poll conducted in December 1999 similarly found that 94 percent of Americans identified themselves as believing in God on a higher power, and only 5 percent stated that they did not (New Port 1999).

However, a 2001 follow-up survey by the NSRI investigators showed some changes in this religious profile. Using a smaller but still very large sample of 50,281 individuals, investigators found that the percentage of Americans professing belief in God had declined from 89.5 percent to 80.2 percent, as had the percentage of Christians (from 86.2 percent to 76.5 percent) (Kosmin, Mayer, and Keysar 2002). The largest increase was in the percentage of nonbelievers, which increased from 8.2 percent in 1990 to 14.1 percent in 2001. The American population might be becoming more secular, although another possible explanation for the different results might be a change in how the question about religious adherence was asked. In 1990, the question asked was "What is your religion?" In 2001, the question was, "What is your religion, if any?" Perhaps being reminded of the option of not being religious might have increased the number of people who thus classified themselves (see Table 3.1 for these more recent data).

Similar results were found in a survey conducted in 2007 by the Pew Research Foundation (Pew Forum on Religion and Public Life 2008); they are presented in Table 3.1. The Pew U.S. Religious Landscape Survey was another large telephone survey involving about thirty-six thousand adults. Interviewers asked respondents, "What is your present religion, if any?" and then prompted the respondent with a list of denominations. All three surveys found high percentages of Americans professing religion, and high percentages identifying themselves specifically as Christian. The two most recent surveys suggest that secularism may be increasing; the percentage claiming no religion, although relatively small, is greater than it was in 1990. With samples as large as these, the margin of error is less than 1 percent, which makes the results quite reliable.

But whether the percentage of Christians is near 80 percent or 70 percent, it is nonetheless true that Christians are the largest religious group in the United States. It

Table 3.1
American Religious Profiles

	1990 (%)	2001 (%)	2007 (%)
Religious	89.5	80.2	82.1
Christian	86.2	76.5	78.4
Non-Christian	3.3	3.7	4.7
Jewish	1.8	1.3	1.7
Muslim	0.5	0.5	0.6
Other non-Christian	1.0	1.9	2.4
No religion	8.2	14.1	16.1
Refused to state	2.3	5.4	0.8

Source: 1990: Kosmin and Lachman, 1993; 2001: Kosmin et al., 2001; 2007: Pew Forum on Religion and Public Life, 2008.

is also true that in international comparisons, Americans rank highly in the percentage of adults who believe in God.

Christians can be further broken down into conservative or born-again Christians on the one hand and mainstream Christians on the other. Conservative Christians are those who believe that they have a personal relationship with Jesus and who tie salvation to this belief. A greater percentage of conservative Christians than mainstream Christians regard the Bible as being literally true, according to a poll conducted by the Barna organization (Barna 2007). Most conservative Christians are Protestants, but some Catholics hold the same beliefs, especially those who embrace charismatic Catholicism.

Antievolutionism in North America is rooted in religiously conservative Christianity; there are few if any activist Jews or Muslims who oppose evolution in North America, and only small antievolution movements in Islamic countries such as Turkey and in the Jewish state of Israel. Although minority religions are growing in the United States, it is clear that Christianity is now, and for the near and intermediate future will be, the predominant American religious tradition. Because of their numbers and their prominence in the antievolution movement, the rest of this chapter will concentrate on Christians.

Many people are under the impression that there is a dichotomy between evolution and Christianity, a line in the sand between two incompatible belief systems. These people believe that a person must choose one side of the line or the other. In reality, Christians hold many views about evolution, and Christian views actually range along a continuum rather than being separated into a dichotomy.

THE CREATION/EVOLUTION CONTINUUM

Figure 3.1 presents a continuum of religious views with creationism at one end and evolution at the other. The most extreme views are, of course, at the ends of the continuum. The creation/evolution continuum reflects the degree to which the Bible is interpreted as literally true, with the greatest degree of literalism at the top.

Figure 3.1
The relationship between evolution and creationism in Christianity
is a continuum, not a dichotomy between two choices. Courtesy of
Alan Gishlick.

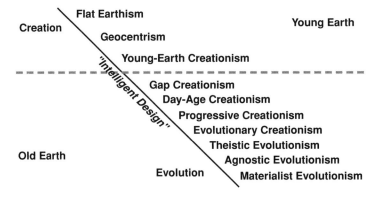

Although it is a continuum of religious and philosophical beliefs, the cre-
ation/evolution continuum inversely reflects how much of modern science holders
of these different views accept. I will begin with the strictest biblical literalists, the flat
earthers. (For readers not familiar with the Bible, references take the form of book,
chapter: verse; thus, Genesis 1:4 refers to the book of Genesis, chapter 1, verse 4.)

Flat Earthism

Until his death in March 2001, Charles K. Johnson of Lancaster, California, was
the head of the International Flat Earth Research Society, an organization with a
claimed membership of 3,500 (Martin 2001) that may not long outlive its leader's
demise. Johnson—and we assume the members of his society—were very serious about
their contention that the shape of Earth is flat rather than spherical, because they are
the most strict of biblical literalists. Few other biblical literalists hold to such stringent
interpretations of the Bible. To flat earthers, many passages in the Bible imply that
God created an Earth that is shaped like a coin, not a ball: flat and round at the edges.
Earth's disklike (not spherical) shape reflects biblical passages referring to the "circle"
of the Earth (Isaiah 40:22) and permits one to sail around the planet and return to
one's starting point: one merely has to sail to the edge of Earth and make the circuit.

Because their theology requires the Bible to be read as literally true, flat earthers
believe Earth must be flat (Schadewald 1991). The Englishman responsible for the
nineteenth-century revival of flat earthism, Samuel Birley Rowbotham, "cited 76
scriptures in the last chapter of his monumental second edition of *Earth Not a Globe*"
(Schadewald 1987: 27). Many of these refer to "ends of the Earth" (Deuteronomy
28:64, 33:17; Psalms 98:3, 135:7; Jeremiah 25:31) or "quadrants" (Revelation 20:8).
For flat earthers—and other literalists—the Bible takes primacy over the information
provided by science; thus, because modern geology, physics, biology, and astronomy
contradict a strict biblical interpretation, these sciences are held to be in error.

Geocentrism

Geocentrists accept that Earth is a sphere but deny that the sun is the center of the solar system. Like flat earthers, they reject virtually all of modern physics and astronomy as well as biology. Geocentrism is a somewhat larger, though still insignificant, component of modern antievolutionism. At the Bible-Science Association creationism conference in 1985, the plenary session debate was between two geocentrists and two heliocentrists (Bible-Science Association 1985). Similarly, as recently as 1985, the secretary of the still-influential Creation Research Society was a published geocentrist (Kaufmann 1985).

Both flat earthers and geocentrists reflect to a greater or lesser degree the perception of Earth held by the ancient Hebrews, which was that it was a disk-shaped structure (Figure 3.2). They believed that the heavens were held up by a dome (*raqiya* or firmament) that arched over the land and that water surrounded the land. The firmament was perceived as a solid, metal-like structure that could be hammered and shaped (as in Job 37:18: "Can you, like him, spread out the skies, hard as a molten mirror?" [All biblical quotes are from the Revised Standard Bible, Zondervan, 1981]). The surface of the firmament is solid enough that God can walk on it (as in Job 22:14: "Thick clouds enwrap him, so that he does not see, he walks on the vault of heaven"). The sun, moon, and stars were attached to the firmament, which means that these heavenly bodies circled Earth beneath the firmament and, hence, were part of a geocentric universe. Further support for the idea of a solid sky and a geocentric solar system is found in Revelation 6:13–16: "and the stars of the sky fell to the earth as the fig tree sheds its winter fruit when shaken by a gale; the sky vanished like a scroll that is rolled up." Stars were regarded as small, bright objects rather than massive suns hugely larger than Earth. They could fall on Earth because they were below the firmament, a solid object that, if rolled aside, would reveal the throne of God (Schadewald 1987, 1981–1982).

The Bible also speaks of the waters above the firmament; ancient Hebrews conceived of the firmament supporting a body of water that came to Earth as rain through the "windows of heaven" and was the source of the forty days and nights of rain that began Noah's Flood.

Ancient and modern geocentricity reflects the idea that because the Earth and its creatures—especially humans—are central to God. To symbolize this importance, God would have made Earth the center of the universe. Taking Earth out of this central position reduces its importance, which reduces (according to their interpretation) man's place as the most important element in creation. Although not actively supporting geocentrism, young-Earth creationist astronomer D. Russell Humphreys has promoted the idea of the centrality of Earth and humans by claiming that Earth is at the center of the universe (Humphreys 2002). His conception of cosmology has the central Earth surrounded by galaxies and ultimately a sphere of water that is light-years in diameter (the "waters which were above the firmament" of Genesis 1:7) (Humphreys 2007; see Figure 3.3).

The next group of creationists on the continuum are less biblically literalist than the previous two, but all three endorse the theological doctrine of special creationism, which stresses the view that God created the universe, Earth, plants, and animals,

Figure 3.2
An early twentieth-century conceptualization of ancient cosmology. Early Hebrews conceived of the universe as consisting of a disk-shaped Earth that was the center of the cosmos,
in which a domelike sky was supported by pillars of heaven.
From Robinson (1913), frontispiece.

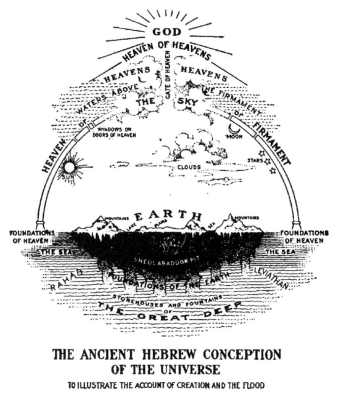

**THE ANCIENT HEBREW CONCEPTION
OF THE UNIVERSE**
TO ILLUSTRATE THE ACCOUNT OF CREATION AND THE FLOOD

and humans in essentially their present form. The most common form of special
creationism holds that the creation event took place relatively recently, and is thus
called young-Earth creationism.

Young-Earth Creationism

Few proponents of young-Earth creationism interpret the flat Earth and geocentric
passages of the Bible literally. They accept heliocentrism but reject the conclusions of
modern physics, astronomy, chemistry, and geology concerning the age of Earth, and
they deny biological descent with modification. Earth, in their view, is between 6,000
and 10,000 years old. They reject the Big Bang theory and postulate catastrophic
mechanisms as the cause of most of the world's geological features. The Flood of
Noah, for example, is allegedly responsible for carving the Grand Canyon and other
geological features.

Figure 3.3
Humphreys's model of the universe. The young-Earth creationist Russell Humphreys envisions the cosmos as spherical, with galaxies and all other phenomena surrounded by a layer of water. This view is derived from the biblical reference to "waters above the firmament." Courtesy Sarina Bronson.

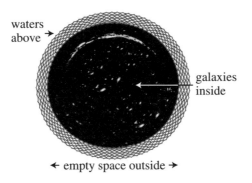

waters above →

galaxies inside

← empty space outside →

Young-Earth creationists (YECs) reject the inference that earlier forms of life are ancestral to later ones. Instead, they embrace the special creation of separate "kinds" of plants and animals, as stated in Genesis. The definition of kinds is inconsistent among YECs but usually refers to a higher taxonomic level than species. Most YECs accept that God created creatures possessing at least as much genetic variation as occurs within a biological family (for example, the cat family Felidae, the cattle family Bovidae) and then considerable evolution within a kind occurred. The created cat kind thus would have possessed sufficient genetic variability to differentiate into lions, tigers, leopards, pumas, bobcats, and house cats, through the normal microevolutionary processes of mutation and recombination, natural selection, genetic drift, and speciation. Most YECs view the basic body plans of major phyla that appear in the Cambrian explosion as evidence of special creation.

The term *young-Earth creationist* is often associated with the followers of Henry Morris, founder of the Institute for Creation Research (ICR) and arguably the most influential creationist of the second half of the twentieth century. He and John C. Whitcomb Jr. published *The Genesis Flood*, a seminal work that claimed to provide a scientific rationale for young-Earth creationism (Whitcomb and Morris 1961). As the title suggests, the authors read Genesis literally, including not just the special, separate creation of humans and all other kinds of plants and animals but also the historicity of Noah's Flood. Whitcomb and Morris proposed that there is scientific evidence to demonstrate the truth of special creationism: Earth is young, the universe appeared in essentially its present form about 10,000 years ago, and plants and animals appeared in their present forms as created kinds rather than having evolved over millions of years through common ancestors. Although efforts were made during the eighteenth and nineteenth centuries to claim that a literal interpretation of the Bible is compatible with science, *The Genesis Flood* was the first twentieth-century effort to attract a large following. Religious antievolutionists were greatly encouraged by the thought that there might be evidence that evolution was not only religiously objectionable but also scientifically flawed. Creation science has been augmented by hundreds of books and pamphlets written by Morris and those inspired by him (McIver 1988). More on Morris and young-Earth creationism can be found in Chapter 5.

Old-Earth Creationism

As mentioned, the idea that Earth is ancient was well established in science by the mid-1800s and was not considered a radical idea in either the Church of England or the Catholic Church (Eiseley 1961). From the mid-1700s on, the theology of special creationism has been partly harmonized with scientific data and theory showing that Earth is ancient. To many Christians, the most critical element of special creation is God's personal involvement in Creation; precise details of how God created are considered secondary. The present may indeed be different from the past, but old-Earth creationists (OECs) see God as a direct causal agent of the observed changes.

The creation/evolution continuum, like most continua, has few sharp boundaries. Although there is a sharp division between YECs and OECs, the separation among the various OEC persuasions is less clear cut. Even though OECs accept most of modern physics, chemistry, and geology, they are not very dissimilar to YECs in their rejection of biological evolution. There are several religious views that can be classed as OEC.

Gap Creationism. One of the better-known nineteenth-century accommodations allowing Christianity to accept the science of its time was gap or restitution creationism, which claimed that there was a large temporal gap between verses 1 and 2 of chapter 1 of Genesis (Young 1982). Articulated from approximately the late eighteenth century on, gap creationism assumes a pre-Adamic creation that was destroyed before Genesis 1:2, when God re-created the world in six days and created Adam and Eve. A time gap between two separate creations allows for an accommodation of special creationism with the evidence for an ancient age of Earth. In gap creationism, the six days of Genesis 1:2 and following are considered twenty-four-hour days.

Day-Age Creationism. Another attempt to accommodate science to a literal, or mostly literal, reading of the Bible is the day-age theory, which was more popular than gap creationism in the nineteenth century and the earlier part of the twentieth (Young 1982). Here religion is accommodated to science by having each of the six days of creation be not twenty-four hours but long periods of time—even thousands or millions of years. This allows for recognition of an ancient age of Earth but still retains a quite literal interpretation of Genesis. Many literalists have found comfort in what they interpret as a rough parallel between organic evolution and Genesis, in which plants appear before animals, and human beings appear afterward. Anomalies such as flowering plants being created before animals and birds occurring before land animals—incidents unsupported by the fossil record—are usually ignored.

Progressive Creationism. Although some modern activist antievolutionists may still hold to day-age and gap views, the view held by the majority of today's OECs is some form of progressive creationism (PC). The PC view accepts more of modern science than do day-age and gap creationism: progressive creationists do not dispute scientific data concerning the Big Bang, the age of Earth, or the long period of time it has taken for Earth to come to its current form. Indeed, some cite the Big Bang as confirmation of Genesis, in that the Big Bang is viewed as the origin of matter, energy, and time, which in the PC view is equivalent to creation ex nihilo, the doctrine of creation out of nothing. As in other forms of old-Earth creationism, although theories of modern physical science are accepted, PC incorporates only parts of modern biological science.

For example, the fossil record shows a consistent distribution of plants and animals through time: mammals are never found in the Cambrian, for example, and flowering plants are never found in the Devonian. However, YECs believe that flowering plants, dinosaurs, humans, and trilobites were all created at the same time and therefore all lived at the same time. They regard the orderly distribution of fossils in strata around the world to be an artifact of Noah's Flood, which is thought to have differentially sorted organisms into groups, even if they all died at the same time. In contrast, PCs generally accept the fossil distribution of organisms as "real" because they believe that God created kinds of animals sequentially. To PCs, the geological column reflects history: God first created simple, single-celled organisms, then more complex single-celled life, then simple multicellular organisms, then more complex ones, and so on up until the present time. With PC, there is no difficulty that seed-bearing plants appear after ferns and cycads: God created the more "advanced" plants at a later time. However, progressive creationists do not accept that the kinds evolved from one another, though they are no more specific than YECs about what constitutes a kind. As in young-Earth creationism, though, a kind is viewed as genetically limited: as a result, one kind cannot change into another.

Evolutionary Creationism. Despite its name, evolutionary creationism (EC) is actually a type of evolution. Here, God the Creator uses evolution to bring about the universe according to God's plan. From a scientific point of view, evolutionary creationism is hardly distinguishable from theistic evolution, which follows it on the continuum. The differences between EC and theistic evolution lie not in science but in theology, with EC being held by more conservative (Evangelical) Christians, who view God as being more actively involved in evolution than do most theistic evolutionists (Lamoureux 2008).

Intelligent design creationism has been positioned on the continuum as overlapping YEC and OEC because some of its proponents can be found in each camp; old-Earthers among the intelligent design creationists have not categorically denied the scientific validity of YEC.

Intelligent Design Creationism

Intelligent design creationism (IDC) is the newest manifestation of American creationism, and yet it resembles a much earlier idea. In most ways, IDC is a descendant of William Paley's argument from design (Paley 1802), which argued that God's existence could be proved by examining God's works. Paley used a metaphor: if one found a watch, it was obvious that such a complex object could not have come together by chance; the existence of a watch implied a watchmaker who had designed the watch with a purpose in mind. By analogy, the finding of order, purpose, and design in the world was proof of an omniscient designer.

The vertebrate eye was Paley's classic example of design in nature, well known to educated people in the nineteenth century. Because of its familiarity, Darwin deliberately used the vertebrate eye in *On the Origin of Species* to demonstrate how complexity and intricate design could indeed come about through a natural process; complexity in nature did not require divine intervention.

Structures and organs that accomplish a purpose for the organism—allowing capture of prey, escape from predators, or attracting a mate—could be designed directly by an omniscient designer, or they could be "designed" by a natural process that produces the same effect. As will be discussed in more detail elsewhere in this book, Darwin's argument that a natural process such as natural selection could explain apparent design was theologically offensive to those who believed that God created directly.

In IDC one is less likely to find references to the vertebrate eye and more likely to find molecular phenomena such as DNA structure or complex cellular mechanisms held up as too complex to have evolved "by chance." The IDC high school biology supplemental textbooks *Of Pandas and People* (Davis and Kenyon 1993) and *Explore Evolution* (Meyer, Minnich, Moneymaker, Nelson, and Seelke 2007) both attempt to prove that DNA is too complex to explain through natural causes by weaving allusions to information theory into an exposition of the "linguistics" of the DNA code.

Following creationist tradition, IDC proponents accept natural selection but deny that mutation and natural selection are adequate to explain the evolution of one kind to another, such as chordates from echinoderms or humans and chimps from a common ancestor. The emergence of major anatomical body types and the origin of life, to choose just two examples popular among IDC followers, are phenomena supposedly too complex to be explained naturally; thus, IDC demands that a role be left for the intelligent designer—God. Chapter 7 discusses IDC in more detail.

Theistic Evolution

Theistic evolution is a theological view in which God creates through the laws of nature. Theistic evolutionists (TEs) accept all the results of modern science, in anthropology and biology as well as in astronomy, physics, and geology. In particular, it is acceptable to TEs that one species give rise to another; they accept descent with modification. However, TEs vary in whether and how much God is allowed to intervene—some believe that God created the laws of nature and allows events to occur with no further intervention. Other TEs believe that God intervenes at critical intervals during the history of life (especially in the origin of humans). A 2003 book presents an entire continuum of TEs; clearly, there is much variation among Christians regarding this theological view (Peters and Hewlett 2003). In one form or another, TE is the view of creation taught at the majority of mainline Protestant seminaries, and it is the position of the Catholic Church. In 1996, Pope John Paul II (1996) reiterated the Catholic version of theistic evolution, in which God created, evolution happened, humans may indeed be descended from more primitive forms, but the hand of God was required for the production of the human soul. The current pope, Benedict XVI, has reiterated the evolution-friendly Catholic view, stressing the importance of rejecting philosophical naturalism (Lawton 2007).

Agnostic Evolutionism

Although poll data indicate that most Americans have a belief in God or some higher power, a (perhaps growing) minority do not (Pew Forum on Religion and Public Life 2008). The term *agnostic* was coined by "Darwin's bulldog," the nineteenth-century scientist Thomas Henry Huxley, to refer to someone who suspended judgment about the existence of God. Huxley felt that human beings, part of the material universe, would be unable to grasp ultimate reality; therefore, neither belief in nor rejection of the existence of God is warranted. To Huxley, the thoughtful person should suspend judgment. Huxley was a strong supporter of science and believed that knowledge and beliefs should be based on empirical knowledge—and that science would eventually supplant supernaturalism. But he felt it was more honest not to categorically reject an ultimate force or power beyond the material world:

> I have no doubt that scientific criticism will prove destructive to the forms of supernat-uralism which enter into the constitution of existing religions. On trial of any so-called miracle the verdict of science is "Not proven." But true Agnosticism will not forget that existence, motion, and law-abiding operation in nature are more stupendous miracles than any recounted by the mythologies, and that there may be things, not only in the heavens and earth, but beyond the intelligible universe, which "are not dreamt of in our philosophy." The theological "gnosis" would have us believe that the world is a conjuror's house; the anti-theological "gnosis" talks as if it were a "dirt-pie" made by the two blind children, Law and Force. Agnosticism simply says that we know nothing of what may be beyond phenomena. (Huxley 1884)

Agnostics believe that, in this life, it is impossible to know truly whether there is a God, and although they believe that it is not probable that God exists, they tend not to be dogmatic about this conclusion. One can find individuals who accept the scientific evidence that evolution occurred but do not consider important the question of whether God is or was or will be involved. We can call this belief agnostic evolutionism. Holders of this view differ from the next position on the continuum by not categorically ruling out the involvement of God, although they tend to side with those who doubt the existence of God and whether God acts in the world.

Materialist Evolutionism

Before discussing materialist evolutionism, I need to distinguish between two uses of the term *materialism* (or *naturalism*). As I mentioned in chapter 1, modern science operates under a rule of methodological naturalism that limits it to attempting to ex-plain natural phenomena using natural causes. Philosophical materialists (sometimes

Figure 3.4
The Relationship between Methodological and Philosophical Naturalism. All philosophical naturalists are methodological naturalists, but it is not accurate to say that all methodological naturalists are philosophical naturalists. One can thus be a scientist practicing methodological naturalism but still be a theist.

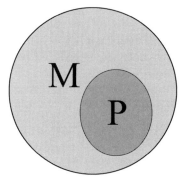

referred to as philosophical naturalists) go beyond the methodological naturalism of science to propose not only that material (matter and energy) causes are sufficient to explain natural phenomena but also that the supernatural does not exist. To a philosophical naturalist, there is no God. The philosophy of humanism is a materialistic philosophy, as is atheism. As discussed earlier in this chapter, philosophical naturalism is distinct from the practical rules of how to do science.

This is an important distinction to the subject of this book because some antievolutionists accuse evolution and science in general of being not only methodologically naturalistic but also philosophically naturalistic. This is a logical error, as Figure 3.4 shows. It is very likely the case that all philosophical naturalists are simultaneously methodological naturalists (all Ps are Ms). It does not follow that all methodological naturalists are philosophical naturalists (not all Ms are Ps). It might be the case—if both circles were the same size and right on top of one another—but this would have to be determined empirically, not logically. In fact, such a claim is empirically falsified, for there are many scientists who accept methodological naturalism in their work but who are theists and therefore not philosophical naturalists. Gregor Mendel—the monk whose research became the foundation of genetics—is a classic case of a scientist who was a methodological naturalist but not a philosophical one, and there are many scientists today who, like him, are methodological but not philosophical naturalists.

As mentioned, there are varieties of belief within the various theistic positions on the continuum, and this is true for materialists as well. For example, although materialists share a high opinion of science and accept evolution, they do not all share the same attitudes toward religion. Agnostics are materialists who do not consider that the question of whether God created can be answered. Humanists have a philosophy of life and an ethical code that holds, "Humanism is a progressive lifestance that, without supernaturalism, affirms our ability and responsibility to lead meaningful, ethical lives capable of adding to the greater good of humanity" (American Humanist Association 2002). The two major humanist organizations are the American Humanist Association, with approximately 5,000 members at the time of this writing, and the Council for Secular Humanism, with approximately 4,000 members.

Atheists, the third major group within materialists, reject the existence of God but tend to be more actively antireligious than the other two. There are about 2,200 members of the best-known atheist group, the American Atheists. Clearly, any single theist organization has far more members than all the materialist organizations combined. If nonbelievers make up between 10 percent and 14 percent of the population, as some polls suggest, the vast majority of them do not join groups of like-minded individuals. Someone holding to materialist evolutionism, then, believes that evolution occurred but that there was absolutely no supernatural entities or forces affecting it, because such forces do not exist. As we will see later in this book, creationists consider materialist evolutionism the true enemy of religion; actually, although all material evolutionists reject the involvement of God in evolution, not all material evolutionists are antireligious.

This presentation of Christian and materialist views regarding creation and evolution is simplified—as was the earlier presentation of the nature of science in chapter 1 and the presentation of the science of evolution in chapter 2. It is possible to go into far more detail on any of these beliefs, but a shorthand version will have to suffice to introduce the topic.

RELIGION, SCIENCE, AND PHILOSOPHICAL NATURALISM

What are the relationships among religion, science, and philosophical naturalism? Everyone recognizes that there are differences, but there are similarities as well. All three of these terms refer to ways of knowing: a field of study that philosophers call epistemology. The epistemology we call science is primarily a methodology that attempts to explain the natural world using natural causes. Although individual scientists may be concerned with moral and ethical issues or rules of conduct, science as a way of knowing is not concerned with these things. The methodology of testing natural explanations against the natural world will not tell us whether it is immoral for coyotes to kill rabbits or whether members of one sex or another should keep their heads covered in public, or whether marrying your father's brother's child is immoral but marrying your father's sister's child is not. Science is a limited way of knowing, with limited goals and a limited set of tools to use to accomplish those goals.

Philosophical naturalism relies on science and is inspired by science, but it differs from science in being concerned with rules of conduct, ethics, and morals. When a scientist makes a statement like, "Man is the result of a purposeless and natural process that did not have him in mind" (Simpson 1967: 344), it is clear that he or she is speaking from the perspective of philosophical naturalism rather than from the methodology of science itself. As anthropologist Matt Cartmill (1998: 83) has observed, "Many scientists are atheists or agnostics who want to believe that the natural world they study is all there is, and being only human, they try to persuade themselves that science gives them grounds for that belief. It's an honorable belief, but it isn't a research finding." Only a minority of Americans embrace philosophical naturalism—perhaps as few as 10 to 16 percent or so—but it has had a long history in Western culture, going back to some of the pre-Socratic philosophers of ancient Greece.

Religion concerns the relationship of people with the divine, but it also may include explanations of the natural world and the origin of natural phenomena. Religious views almost universally derive from revelation, but this does not rule out the use of empirical and logical approaches to theology. In fact, many Christian denominations pride themselves on their reliance on logic and reason as a means both to understand the natural world and to evaluate theological positions. But an ultimate reliance on revelation can place religion into conflict with science, as discussed earlier in this chapter. When revealed truth conflicts with empirical knowledge, how does one choose?

Different religious traditions provide different interpretations of revealed truth—all held with equal fervor—and within the same religious tradition the documents that are considered authoritative can be, and usually are, interpreted differently by different adherents. Reform and Hasidic Jews interpret the Torah differently, Muslims of the Shiite and Sunni traditions have some different interpretations of the Koran, and Catholics and Protestants use Bibles with different books. Which tradition is more faithful to the sacred documents is ascertained differently by different factions, and unless agreement can be reached on criteria of judgment, different factions will be unable to determine whose interpretation is correct.

For example, some Christians interpret the Bible as indicating that the Flood of Noah was an actual historical event that covered the entire Earth, and they believe that the receding floodwaters cut the Grand Canyon. Other Christians interpret the Bible differently and argue that the Flood was not a universal historical event and could not have carved the Grand Canyon. Proponents of different biblical interpretations tend not to persuade one another because their religious assumptions are different; to some it is not a matter of logic or empirical evidence (as will be illustrated in the readings in part 3).

In science, on the other hand, there is no revealed truth. Although some explanations are believed to be very solidly grounded, it is understood that even well-supported theories can be modified and, in rare circumstances, even replaced by other explanations. For the limited purpose of explaining the natural world, science has a major advantage over religion in that individuals of different philosophical, religious, cultural, and/or ideological orientations, using the methodology of science, can debate their differences on the basis of repeatable—and repeated—empirical investigations. Different scientists, using different techniques, technologies, and observational approaches, provide validation not possible through revelation.

Scientists looking at geological and biological data can piece together a natural history of the Grand Canyon and test one another's explanations against the lay of the land itself. The ability to go back to nature—again and again—to test explanations, rework them, and retest them is one of the strengths of science and a major contributor to the amount of empirical knowledge exponentially amassed over the past three hundred years. To some, though, the open-endedness of science is a weakness: they seek definite answers that will never change. For them, Ashley Montagu's (1984: 9) definition of science as "truth without certainty" is insufficient; for others, it is science's greatest strength.

Just as attempts to explain the natural world through revelation cause friction with scientists, so materialist scientists cause friction with religious people when they make statements about the ultimate nature of reality—when they speak as if they speak for science itself. On reflection it should be recognizable that if science has the limited goal of explaining the natural world using natural causes, it lacks the tools to make justifiable statements about whether there is or is not a reality beyond the familiar one of matter and energy. As will be clear in some of the readings to follow, both supporters and deniers of evolution argue erroneously that because science uses methodological naturalism (and quite successfully), science therefore also incorporates philosophical naturalism. Unfortunately, such confusion makes communication about science and religion, or creationism and evolution, more difficult.

REFERENCES

American Humanist Association. 2002. *Definitions of humanism*. Accessed October 13, 2008. Available from http://www.americanhumanist.org/humanism/definitions.htm.

Augustine. 1982. *The literal meaning of Genesis (De Genesi ad litteram)*, trans. J. H. Taylor. Vol. 1. New York: Newman Press.

Barna, George. 2007. Most Americans take well-known Bible stories at face value. Barna Research Group, October 21, 2007. Accessed October 28, 2008. Available from http://www.barna.org/flexPage.aspx?page=Barnaupdate&BarnaupdateID=282.

Bible-Science Association. 1985. *Bible-Science Association conference schedule*. Cleveland, OH: Bible-Science Association.

Carneiro, Robert. 1983. *Origin myths* (pamphlet). Berkeley, CA: National Center for Science Education Inc.

Cartmill, Matt. 1998. Oppressed by evolution. *Discover* (March): 78–83.

Davis, Percival W., and Dean H. Kenyon. 1993. *Of pandas and people*, 2nd ed. Dallas, TX: Haughton.

Eiseley, Loren. 1961. *Darwin's century*. Garden City, NY: Doubleday.

Free Inquiry. 1996. Religious belief in America: A new poll. *Free Inquiry* (Summer): 34–40.

Humphreys, D. Russell. 2002. The battle for the cosmic center. *Impact* (August): a–c.

Humphreys, D. R. 2007. Creationist cosmologies solve spacecraft mystery, Institute for Creation Research. Accessed April 8, 2008. Available from http://www.icr.org/article/3472/.

Huxley, Thomas H. 1884. Agnosticism: A symposium. *The agnostic annual* 1. Available from http://aleph0.clarku.edu/huxley/UnColl/Rdetc/AgnAnn.html.

Hyers, Conrad. 1983. Biblical literalism: Constricting the cosmic dance. In *Is God a creationist? The religious case against creation-science*, ed. R. M. Frye. New York: Scribner's.

John Paul II. 1996. Magisterium. *L'osservatore romano*, October 30, 3, 7.

Kaufmann, D. A. 1985. Geocentricity: A creationist consideration. *The Christian News* 23(21): 7.

Kosmin, Barry A., and Seymour P. Lachman. 1993. *One nation under God*. New York: Crown.

Kosmin, Barry A., Egon Mayer, and Arieta Keysar. 2002. *American Religious Identification Survey 2001*. Graduate Center of the City University of New York, 2001. Accessed October 3, 2008. Available from http://www.gc.cuny.edu/faculty/research_briefs/aris.pdf.

Lamoureux, Denis O. 2008. *Evolutionary creation: A Christian approach to evolution*. Eugene, OR: Wipf and Stock.

Leeming, David A., and Margaret Leeming. 1994. *A dictionary of creation myths*. New York: Oxford University Press.

Lawton, M. 2007. In new book, Pope quoted as seeing no conflict between faith, science. *Catholic News*. April 16.

Martin, Douglas. 2001. Charles Johnson, 76, proponent of flat Earth. *New York Times*, March 25, 44.

McIver, Thomas. 1988. *Anti-Evolution: An annotated bibliography*. Jefferson, NC: McFarland.

Montagu, M. F. Ashley, ed. 1984. *Science and creationism*. New York: Oxford University Press.

Meyer, Stephen C., Scott Minnich, Jonathan Moneymaker, Paul A. Nelson, and Ralph Seelke. 2007. *Explore Evolution*. Melbourne and London, Hill House Publishers.

Newport, Frank. 1999. *Americans remain very religious, but not necessarily in conventional ways*. Gallup News Service. Accessed October 28, 2008. Available from http://www.gallup.com/poll/3385/Americans-remain-very-religious-necessarily-conventional-ways.aspx.

Paley, William. 1802. *Natural theology; or, Evidences of the existence and attributes of the deity, collected from the appearances of nature*, 5th ed. London: R. Faulder.

Peters, Ted, and Martinez Hewlett. 2003. *Evolution from Creation to New Creation*. Nashville, TN: Abingdon Press.

Pew Forum on Religion and Public Life. 2008. *U.S. Religious Landscape Survey*. Washington, DC: Pew Research Center.

Raman, V. V. 1998–1999. A Hindu view on science and spirituality. *Science and Spirit* 9(5): 6–7.

Robinson, G. L. 1913. *Leaders of Israel*. New York: Association Press.

Sarna, Nahum M. 1983. Understanding creation in Genesis. pp. 155–173. In *Is God a creationist? The religious case against creation-science*, ed. R. M. Frye. New York: Scribner's.

Schadewald, Robert J. 1981–1982. Scientific creationism, geocentricity, and the flat Earth. *Skeptical Inquirer* 6(2): 41–48.

Schadewald, Robert J. 1987. The flat-Earth bible. *Bulletin of the Tychonian Society* 44 (July): 27–39.

Schadewald, Robert J. 1991. Introduction. pp. 1–2. In *A reparation*, C. S. De Ford. Oakesdale, WA: Ye Galleon.

Simpson, George Gaylord. 1967. *The meaning of evolution*, rev. ed. New Haven, CT: Yale University Press.

Stevens, P., Jr. 1996. Religion. In *Encyclopedia of cultural anthropology*. New York: Holt.

Whitcomb, John C., Jr. and Henry R. Morris. 1961. *The Genesis flood: The biblical record and its scientific implications*. Phillipsburg, NJ: Presbyterian and Reformed Publishing Company.

Young, Davis A. 1982. *Christianity and the age of the Earth*. Grand Rapids, MI: Zondervan.

Zondervan. 1981.*The New layman's parallel Bible: the King James version; the New International Version; the Living Bible; the Revised Standard version*. Grand Rapids, MI, Zondervan Publishers.

PART II

• • • • • • • • • •

A History of the Creationism/Evolution Controversy

The history of—and potential conflict between—creationism and evolution extends back hundreds of years. The current controversy has its roots in conflicting ideas of stasis and change that reach back beyond the Middle Ages. Darwin is unquestionably a central figure in the development of ideas of biological evolution, but he of course built on ideas of an ancient and changing physical universe and world that astronomers and geologists had proposed during the previous 150 years.

Part 2 provides an introduction to this history, beginning in chapter 4 with pre-Darwinian ideas about evolution and continuing with Darwin's ideas and their reception. Chapter 5 picks up at the beginning of the twentieth century with the antievolution movement that culminated in the Scopes trial, which was followed by a long period during which evolution was largely ignored in the public schools. With evolution's return to textbooks and the classroom in the late 1960s, we encounter the period of creation science. Chapters 6 and 7 present the origin and current status of the neocreationist movement, which employs a set of antievolutionist strategies designed to avoid the legal decisions that hamstrung the earlier creation science movement.

CHAPTER 4

Before Darwin to the Twentieth Century

BEFORE DARWIN: STASIS VERSUS CHANGE

Some think of the creationism/evolution controversy as "God did it" versus "natural processes did it," but that is a false dichotomy. As discussed in chapter 3, many religious people conceive of creation as the result of God working through natural processes. Historically, a more accurate distinction between creationism and evolution focuses on what happened rather than who done it. Special creationists view nature as largely static after the events of the Creation, whereas those who accept evolution view the universe as having a history: the present universe, the planet Earth, and the living things on it are different from the universe, Earth, and life of the past: change through time has taken place. However, the two models of stasis and change are not relegated to the current creationism/evolution controversy: they have deep historical roots.

Nature through Time

The ideas of the philosophers Plato and Aristotle, as well as those of many Christian thinkers, have shaped Western concepts of nature and time—even though sometimes these ideas were not in complete harmony with one another. From Plato came the idea of idealism, the concept that the world and its objects as perceived by our senses were not "real" but only an imperfect copy of what existed in a "transcendent world of pure and immutable forms" (Durant 1998: 269). In *The Republic*, Plato uses the metaphor of our images of the world being similar to flickering shadows cast on a wall, unreal in themselves, with the true reality being the light that produces the shadows. Christian scholars reformulated idealism in terms of the Creation: God created the world according to a plan; there was an ideal form that lodged in the mind of God, and what we see in the real world are merely variants—imperfect copies in some cases—of that ideal. Dwelling as an idea in the mind of God is the ideal rabbit, human being, or barnacle: the variation in size, shape, color, and so on, that we see in nature is less important than the true essence of rabbits, humans, and barnacles that resides in the mind of God. (A linguistic fossil of this view—though not carrying the same meaning—is the biological term *type specimen*.)

From Aristotle came a view of nature that focused not only on form but also on function. Aristotle wrote of the purpose of nature: why something existed, not just what form it took. The rain falls to make the grass grow. Deer have long legs to run away from predators. These ideas also influenced Christian theology: humans exist because they had been created by God to worship God. Explaining something by its purpose is known as teleology. It is understandable that form is related to purpose: form follows (and contributes to) function, after all. To allow it to escape from predators, the deer has long legs: the legs of a deer were designed to enable it to survive, as its teeth were designed to allow it to eat woody shoots—as the teeth of the wolf were designed to eat meat. Thus, purpose (teleology) and design were linked concepts.

Reflecting his view of immutable forms, Aristotle classified plants and animals in terms of kinds of organisms that could be ranked in a linear "great chain of being," or "scale of nature" (*scala naturae*). This essentialist view fit very comfortably with the Christian doctrine of special creationism. God created all creatures great and small, and simple to complex, and the span of created beings could be ranked hierarchically. Humans were almost at the top of the great chain of being, right beneath angels, which in turn were second to God, who was at the very top of the hierarchy.

The doctrine of special creationism incorporated these Greek ideas—the hierarchical ordering of nature and of design and purpose—and included the Christian idea of an omnipotent, omniscient creator who stood outside of nature. In the theology of special creationism, God created the universe at one time (taking six days in the most common view, although, as discussed in chapter 3, gap creationism considered two special creations) in essentially its present form. God created living things as we see them today for a particular environment and way of life. God also created stars and galaxies as we see them, and the planet Earth as we see it today, as the home of human beings and the creatures over which we have been given dominion and stewardship.

For most of European history, educated people blended the Christian and Greek views and concluded that the world was stable and largely unchanging. In such a conception, the age of Earth was unimportant: it was not until theologians untangling the genealogies of the Bible calculated that Earth was approximately 6,000 years old that anyone considered the question of the age of Earth even worth asking: the specially created, essentialist universe of stars, planets, Earth, and its inhabitants had come into being in its present form, and was assumed to be virtually the same as it had been at the Creation. The notion that Earth—much less living things—could have had a history was not entertained throughout the Middle Ages. Stasis extended even to medieval and feudal social life: everyone's place in society was determined by birth. Serfs were to serve, the nobility were to rule them, and kings had a divine right— God given—to rule. The sociopolitical stasis of society mirrored the conception of an unchanging natural world, all created the way it was by God, for God's purposes, whatever they may be.

But there was growing evidence that things might not be static after all—both socially and in the natural world. By the Renaissance in the 1500s, a middle class began growing and society was rather less static (though Shakespeare's *Henry V* still reflects enthusiasm for the old view of the divine right of monarchs to rule). The expansion of societal boundaries continued through the Enlightenment of the 1700s, as cities grew, the merchant class expanded, and democratic ideas began to replace those of the divine right of the church and hereditary monarchs to rule.

The conception of nature as stable—and known—was troubled by the European discovery and exploration of North and South America and Oceania from about 1500 to 1800. The age of exploration exposed Europeans to huge unknown natural areas. Even if Columbus died thinking he had discovered a route to the Orient, it soon became clear to others that the animals, plants, people, and geological features he had encountered were truly from the New World. During the 1700s and through the 1800s, the study of nature—natural history—was a popular pastime of not only educated individuals but also ordinary citizens. The Swedish natural historian Carl von Linné (whose name is Latinized as *Linnaeus*) developed a useful classification system for plants and animals that grouped them into gradually broader categories: species were grouped into genera, genera into families, families into orders, orders into classes, and so on. He received specimens to classify from all over the world, sent not only by captains of exploring ships but even by traders and common seamen. Another "new world" became apparent with the invention of the microscope in the early 1600s and the discovery of microorganisms. Europeans of the Enlightenment experienced an expansion of knowledge of the natural world that disrupted old ways of thinking, much as new economic and political systems disrupted the social systems of the day.

The discoveries of natural history had implications for Christian religious beliefs. Europe, Africa, and Asia were mentioned in the Bible, but the New World was not; thus, the Bible did not contain all knowledge. Puzzles appeared: there were animal and plant species in North America and other new lands that were not found in the Old World, such as opossums, llamas, tobacco, tomatoes, potatoes, and corn. Had the newly discovered species been created at the same time as known ones? Had they merely died out in some places? In the early 1800s, the French comparative anatomist Georges Cuvier had determined that fossil bones found in Europe were indeed sufficiently similar to living forms to be classified as mammals or reptiles, and even more narrowly as elephants and other known entities. Yet these bones were sufficiently different that it was clear that they came from species that no longer existed. The disappearance of huge reptiles (dinosaurs) and certain mammals, such as mammoths and saber-toothed cats, was unexplained. The notion that some kinds had become extinct was theologically troubling because of the implication that the Creation might not have been perfect, which in turn generated problems for the concept of the original sin of Adam and Eve. Perhaps the species represented by the European fossils were actually still living in the New World—that would solve some theological problems. One of the instructions Thomas Jefferson gave to the explorers Meriwether Lewis and William Clark, in fact, was to keep watch for mammoths and other animals known only from the fossil record as they explored the western reaches of the North American continent. Cuvier himself argued that extinctions of some species had occurred and were the result of a series of environmental catastrophes. To some of the scientists of the day, the most recent of these catastrophes was Noah's Flood.

Even more difficult to explain—and creating theological problems in their own right—were the human inhabitants of the new lands. The Bible did not mention

Native Americans, Polynesians, and other peoples new to Europeans. Wild tales were told of one-eyed races, of people who barked like dogs or who were part animal, and other monstrous creatures. But real, undeniable human beings were encountered as well. How could they be explained? Were they also the children of Adam? Or were they creatures of Satan? Were they human? Did they have souls? Could they become Christian? In 1537, Pope Paul III declared that the Indians of the New World were indeed human and not animals—and therefore should not be enslaved (Gossett 1965: 13). They thus had souls and were fit subjects for Christianization. But how did they come to be living where they were found? If Noah's ark had landed at Ararat, how did Native Americans get to the New World?

In 1665, Isaac La Peyrère produced the first version of gap creationism (see chapter 3), proposing an explanation for these newly discovered peoples that was compatible with the Bible. He proposed that Genesis records two creations, the first being described in Genesis 1, and the second—the Adam and Eve creation—in Genesis 2. Native Americans, Polynesians, Australian Aborigines, and anyone else not specifically mentioned in the Bible were descendants of the first, or preadamite, creation. The preadamites were also the source of Cain's wife—solving another theological problem. In the second, Adamic creation, Genesis 2 and following, God created anew, and Adam and Eve were the progenitors of the more familiar human beings in Europe, Asia, and Africa. Unfortunately, this theological view generated problems of its own, raising the issue of whether preadamites were innocent of original sin. Presumably so—as they were unrelated to Adam—but then, were they in need of salvation by Jesus? The discovery of the New World required the rethinking of many Christian doctrines, as new facts had to be fit into old frameworks.

More new facts were forthcoming from the study of Earth in the late 1700s. In Great Britain, William Smith was given the task of surveying the countryside preparatory to the excavation of a canal system across England (Winchester 2001). It was clear that Great Britain consisted of a variety of types of geological formations, some of which held water better than others, and it behooved the young surveyor to be able to identify and classify the various layers to ensure that canals functioned properly. He did a superb job, tracing strata for sometimes hundreds of miles across the countryside and making detailed maps. He made a discovery (which Cuvier and French geologists confirmed): different strata consistently contained different fossils, and he could classify a stratum if he knew what kinds of fossils it contained, regardless of where it was found. He also found that the deeper the layer, the more different the fossils were from living plants and animals. Many fossils were no longer represented by living animals—especially the deeper ones. It seemed logical that, by and large, bottom layers were older than top layers; thus, there were older animals that differed from more recent ones, and extinct animals that had lived long ago. Estimates could be made of the length of time it took for a valley to erode or for a chain of mountains to lift up. Through careful description and logic, Smith demonstrated the principle that rocks reflect time and change (Winchester 2001).

An appreciation also grew for the nature of geological processes such as sedimentation and erosion; the understanding that nature was dynamic rather than static began to grow as knowledge of the natural world—from geology as well as biology—increased through the 1700s and 1800s. Arguably, the view of nature as dynamic required the amassing of a critical amount of accurate information about the natural world, which hadn't accumulated until the early 1800s. A relationship among geology, biology, and time began to be appreciated: by the mid-nineteenth century, Darwin's time, the once-radical idea that Earth was really quite old, and had changed through time, was becoming well accepted in the scientific community and by educated people in general—including the clergy. If Earth had changed, couldn't other aspects of nature also have changed? Darwin's contribution to the growing appreciation that nature was dynamic rather than static was to add living things to the list of natural phenomena that changed through time.

WHAT DARWIN WROUGHT

Charles Darwin was a respected scholar and scientist well before the 1859 publication of his best-known book, *On the Origin of Species*. He made his original reputation as a geologist, by providing a plausible (and correct) hypothesis about the formation of coral reefs. He then wrote about other geological topics such as volcanoes before turning his hand to biology. Darwin was a meticulous observer of nature (as seen in his four-volume study of the anatomy and physiology of barnacles, and in his research on orchids) but also an experimentalist: at his country estate he had not only a small laboratory but also sufficient land to conduct experiments that required growing plants. He maintained voluminous correspondence with scientists of his day, and because he was so meticulous in his record keeping, much of it remains for scholars to study (Burkhardt and Smith 2002).

On the Origin of Species was Darwin's ninth book of an eventual total of nineteen books and monographs. The first printing of 1,250 copies of *Origin* sold out rapidly, bought not only by scientists but also by educated laity and clergy. It sold steadily over the years, which allowed Darwin to make corrections and small modifications in subsequent editions. There were six editions in all.

The Scientific Response to *On the Origin of Species*

Darwin made two major points in *Origin*: that living things had descended with modification from common ancestors and that the main mechanism resulting in evolution was the mechanism he had discovered, which he called natural selection (see chapter 2). As described by the historian Ronald Numbers (1998), in the late nineteenth and early twentieth centuries, scientists in the United States largely responded positively to Darwin's ideas. The idea of evolution itself was less controversial than Darwin's mechanism of natural selection to explain it.

The scientific knowledge of the time was insufficient to provide support for a full-fledged theory of natural selection, primarily because of a lack of understanding of heredity. Although the Austrian monk Gregor Mendel had discovered the basic principles of heredity, he labored in obscurity, his insights unknown to other scientists of his time. How organisms passed information from generation to generation was a puzzle. Many theories of the day involved the idea that some activity of the individual animal caused organic change that was subsequently passed to offspring—by mechanisms only guessed at. Darwin himself favored a blending type of inheritance in which particles (which he called gemmules) from all parts of the parents' bodies would flow to the reproductive organs, where they would be blended and passed on to offspring.

But natural selection could not be combined with blending inheritance or various models on which acquired characteristics are inherited because such mechanisms would reduce genetic variation each generation. Natural selection is based on the fact that individuals in a population vary in hereditary characteristics, and that organisms that have characteristics most suitable to a particular environment are the ones that tend to survive and reproduce. Natural selection thus requires that variation be continually renewed each generation; both blending inheritance (if true) and natural selection itself would reduce variation. Adaptation would be unlikely to occur. In Darwin's day, many (though not all) scientists concluded that there were critical problems with natural selection as a mechanism of evolution because there was no consensus among scientists on how new variation could be produced every generation.

It was not until the early twentieth century that it became clear that variability does not reduce each generation and that a mechanism to explain it was postulated. Gregor Mendel's rediscovered (and confirmed) research on pea plants showed that whatever it was that was passed on from generation to generation (later to be called genes, and even later to be recognized as DNA-encoded instructions), it did not blend in the offspring but remained separate, even if it was hidden for one or more generations. Heredity material acts like particles and does not blend each generation. Furthermore, genetic information is shuffled each time a sperm fertilizes an egg. Given the particulate nature of inheritance, the mixing up of genes among sexually reproducing organisms, and the existence of phenomena such as dominance and recessiveness, it was clear that natural selection would have sufficient variation on which to operate.

In the late nineteenth and early twentieth centuries, natural selection nonetheless competed with alternate explanations of evolution (Bowler 1988: 7), including a brief revival in popularity of Jean-Baptiste Lamarck's views of the inheritance of acquired characteristics. Lamarckism pointed to observable change: the activities in which an individual engaged during its life could affect its size, shape, and even other characteristics. If these characteristics could be passed on to its offspring, a mechanism would exist to bring about adaptive change. A rabbit living in a cold climate grew a thicker coat; did it pass on its thicker coat to its offspring? There seemed to be evidence of such things: the blacksmith developed large muscles, and the blacksmith's son also tended to be well muscled—but was this a result of the blacksmith's passing down the big muscles acquired from swinging a hammer at the forge? Or was there another explanation, such as the son's going into the family business (and having inherited the potential to develop large muscles under conditions of strenuous exercise)? Without a better knowledge of how heredity operated, evolution by natural selection seemed no more plausible than Lamarckism and other teleological explanations.

In the 1890s, the German biologist August Weismann performed an experiment that was instrumental in convincing most scientists that Lamarckian evolution was untenable. First, he cut the tails off of a number of rats and then bred them with one another. When the rat pups were born, all of them had normal tails; so he cut them off and again bred the offspring with one another. The next generation of rats was also born with normal-length tails. Weismann continued his experiment for twenty generations of rats, and in each and every new generation, there was no inheritance of the acquired trait of cropped tails. The combination of reduced confidence in Lamarckism together with experimental demonstration of Mendelian principles of heredity moved Mendelian genetics to the forefront of heredity studies during the 1930s.

In the 1940s, Darwinian natural selection and Mendelian genetics came together as scientists recognized the powerful support that Mendelian genetics provided to the basic Darwinian model of evolution by natural selection. Called the neo-Darwinian synthesis or neo-Darwinism, it remains a basic approach to understanding the mechanisms of evolution. Neo-Darwinism further has been expanded by the second genetic revolution of the twentieth century, the discovery of the molecular basis of heredity. Since the 1953 discovery by James Watson and Francis Crick of the structure of DNA, the hereditary material of cells, investigation of the molecular basis of life has expanded almost exponentially to become perhaps the most active—and certainly the best funded—area of biological research. Such knowledge has also informed our understanding of the relationships among living things. The big idea of descent with modification—that the more recently two forms have shared a common ancestor, the more similar they will be—is reflected not only in anatomy and behavior but also in proteins.

It is safe to say that by the mid-twentieth century, mainstream science in both Europe and the United States was unanimous in accepting not only the common ancestry of living things but also natural selection as the main—though not the only—force bringing about evolution. The late-twentieth-century advances in biochemistry and molecular biology have further substantiated these conclusions.

Darwin's Science

In addition to the idea of evolution by natural selection, *On the Origin of Species* illustrated a somewhat different way of looking at biology and a different philosophy of science from that familiar to Darwin's contemporaries (Mayr 1964: xviii).

A New Conception of Biology. For Darwin, transmutation of species was a natural phenomenon: it neither required a guiding hand nor resulted in a predetermined goal. Species changed as a result of the need to adapt to immediate environmental circumstances. Because the geology of the planet, and thus environmental circumstances, changed over time, there could not be an ultimate goal toward which creation was heading. It was not possible to predict future changes in living organisms. Darwin's view of science restricted scientific explanations to natural causes. In this he was preceded and influenced by changes that had taken place during the previous one hundred years or so in the field of geology (Gillespie 1979: 11).

In the late 1700s, the Scottish geologist James Hutton proposed a view that became known as uniformitarianism: that Earth was ancient, and its surface could be explained by processes we see taking place today—sedimentation, erosion, faulting, flooding, and the like. There was no need to invoke the direct hand of God to explain the building up of mountains, the presence of seas modern or ancient, or the accumulation of layers of strata. Geology could be understood through natural processes. Darwin's mentor and friend Charles Lyell promoted uniformitarianism in the 1830s and beyond, and the view came to predominate—though not without opposition.

Uniformitarian geologists eventually won the day, but biologists lagged behind; a seminal uniformitarian text by the Scottish scientist John Playfair, *Illustrations of the Huttonian Theory*, was published in 1802—the same year that William Paley published his argument for design, *Natural Theology*. But the seeds for a naturalistic foundation for biology had been planted: geology, after all, has consequences for biology, as fossils partly define the geological column. Different strata are regularly marked by the disappearance of some life-forms and the appearance of new ones (even if they are similar to previous ones). How can these be explained? Creationist geologists required that God re-create life-forms after every catastrophic geological change. Darwin viewed the appearance of new species in a stratum as the result of evolutionary change, of descent with modification from earlier ancestors. His mechanism of natural selection likewise reinforced the conclusion that the fossil record and current diversity of life could be explained without recourse to divine intervention. Darwin's bold naturalism applied to biology proved difficult for many critics to take. Many scientists and theologians objected to Darwin's removal of the need for divine intervention in the biological sciences—much as critics of uniformitarian geology had protested a century before.

Similarly, because there was so much evidence that species had indeed changed through time, and because Darwin's and other scientists' studies of both wild and domesticated animals and plants had demonstrated great variation of form within species, equally untenable were typological species concepts in which species were conceived of as reflections of a Platonic eidos (or idea). Darwin practiced what the modern biologist Ernst Mayr (1964) calls population thinking, in which the object of study of biology is actual individual-to-individual variations rather than an abstract concept of an ideal form.

Perhaps because Darwin was fundamentally a naturalist with broad knowledge of living plants and animals, he was able to conceive of species as having almost unlimited variation, which allowed him to speculate about variation as a source of gradual adaptation and eventual transmutation.

A New Conception of Science. The expectation of scientists in the mid-nineteenth century was that the goal of science was the accumulation of certain knowledge. A successful scientific explanation resulted in positive finality. Anything less than certitude was deficient (Moore 1979: 194).

According to this inductivist approach, the scientist who properly performs his or her craft is one who patiently collects facts, assembles them in a logical and orderly fashion, and lets explanations arise out of this network of ideas. "The outcome of repeated inductions would be a series of propositions, decreasing in number, increasing in generality, and culminating in 'those laws and determinations of absolute actuality' which can be known to be certainly true" (Moore 1979: 194; internal quote from Losee 1972: 164–167). A scientific explanation was considered to have been proved when it accounted for all the facts and thus was a complete and certain law of nature.

Of course, such an ideal is hardly ever obtainable. It is the nature of science that new discoveries cause us to rethink our conclusions and rework our explanations. Today no one thinks that there is ever final certainty to a scientific explanation, but in the late eighteenth century, such a view was common—though not universal—among scientists and other educated individuals. This was not Darwin's approach, however.

Darwin recognized that the world is not static and, with his abundant knowledge of natural history, knew that variability characterizing natural phenomena would make the certainty sought by the strict inductivist approach highly improbable. How could one account for all the facts if new facts were continually being generated? "The lesson was plain: induction, no matter how rigorous, could never rule out the possibility of alternative explanations" (Moore 1979: 196).

Darwin's approach to science indeed was to collect facts (and there is an abundance of them in *Origin*—Darwin was a skilled natural historian and experimenter), but to collect them with a hypothesis or tentative explanation in mind. He used those hypotheses that were not factually disproved to generate additional hypotheses, which he then tested against the facts. He thereby established a network of inferences. Darwin was careful to state how his hypotheses and generalizations could be tested by listing what sort of observations would have to be made to disprove his views—but he also firmly asserted that, until that time, his explanations were the best available.

Rather than the more familiar approach of presenting his views when they were proved or certain, Darwin's approach was to present a coherent set of supported inferences, arguing that the lack of counterevidence gave them the highest probability of being an accurate or true explanation. The probabilistic approach to science, reflecting a dynamic universe, was a sharp contrast to the older approach of many of Darwin's contemporaries, many of whom viewed the universe as specially designed and largely static. According to Moore (1979), Darwin's approach to science itself was one of the major reasons that the concept of evolution by natural selection presented in *On the Origin of Species* was rejected. Darwin's great work was denounced as speculative, probabilistic, unsupported, and far from proven. Yet Darwin's way of doing science—probabilities and all—is much more familiar to us in the twenty-first century than is that of his contemporaries.

The Religious Response to *On the Origin of Species*

Christians who reject evolution tend to reject it for one or both of two reasons. Common descent conflicts with biblical special creation. The Bible in one literal reading tells of the universe's creation in six days, yet data from physics, astronomy, geology, and biology support a picture of the universe unfolding over billions of years. First there was the Big Bang, then gas clouds, then stars, and only about 4.5 billion years ago did planet Earth form. Life did not appear for another billion years or so, and then not all at once (see chapter 2). The Bible read literally also suggests that this creation event occurred a relatively short time ago, geologically speaking—a span measured over thousands rather than billions of years. Yet data from physics and geology firmly support the inference that Earth is ancient. A literal reading of Genesis has animal kinds appearing in their present form, and varying only within the kind, whereas biology, genetics, and geology strongly support the inference that species change through time. The perspective of special creationism holds to a sudden, recent, unchanging universe, whereas the perspective of evolution is that of a gradually appearing, ancient, changing universe. It is not surprising that two such different perspectives clash.

Modern mainstream Christians generally are not biblical literalists and thus do not regard the incompatibility of evolution with biblical literalism as a reason to reject the former. Not believing in created kinds, they have no theological objection to living things descending with modification from common ancestors. But there is a second reason that Christians reject evolution, shared by literalists and nonliteralists alike, and this is the issue of design, purpose, and meaning.

The Problem of Design and Purpose. In Aristotelian philosophy, the purpose or end result of something is thought to be a cause. Explaining something by its purpose, as I mentioned earlier in this chapter, is known as teleology. Up until the nineteenth century, the cause of the marvelous wonders of nature, including the intricacies of anatomical structure, was widely considered to be God's purposive design. Thus, the fit of an organism to its environment was the result of the special creation of its features.

While he was a college student, Darwin read William Paley's 1802 *Natural Theology; or, Evidences of the Existence and Attributes of the Deity, Collected from the Appearances of Nature*, which he thought a splendid book. Paley's view was that God specifically designed complex structures to meet the needs of organisms. *Natural Theology* was also an apologetic, or religious proof of the existence of God; Paley's version of the argument from design is considered a classic. God's existence could be proved, said Paley, by the existence of structural complexity in nature. In a famous analogy, he compared finding a stone on a heath to finding a watch. The former could have been there forever; it was a natural object and did not require any special explanation. But the watch was obviously an artifact—its springs, wires, and other components had been assembled to mark the passage of time. Structural complexity that achieved a purpose was evidence for design and therefore of a designer. When we see a natural structure such as the vertebrate eye, which accomplishes the purpose of allowing sight, we can similarly infer design and hence a designer. The existence of structures such as the vertebrate eye is evidence for the existence of God, according to this analogy.

Paley contrasted design with chance, and it was clearly as absurd to believe that something like the vertebrate eye could assemble by chance as it was to believe that the parts of a watch might come together and function as a result of random movements of springs and wires. Modern creationists take the same view, equating evolution with chance (in the sense of being unguided and purposeless, and therefore random and chaotic) and contrasting it with guided design. A favorite creationist argument is quite similar to that of Paley: many cite astronomer Fred Hoyle's estimate of the possibility of life forming "by chance" as equivalent to a Boeing 707 airplane's being assembled by a whirlwind passing through a junkyard (Hoyle 1983). A current YEC book, in fact, is entitled *Tornado in a Junkyard* (Perloff 1999).

But even before Darwin's *Origin*, the argument from design was proving to be not useful in understanding the natural world. This was partly because of increased knowledge of the natural world during the 1700s and early 1800s. As naturalists examined the world and its creatures more carefully, it became clear that William Paley's ideas of the perfection of structural complexity didn't match reality. Although there were many wonderful structures that admirably suited organisms to their environments—the waterproof feathers of ducks, or the hollow bones of birds that provide strength with lightness—there were also curious constructions that didn't seem to make survival more probable, like reduced wing size in kiwis and similar flightless birds. Other structural oddities seemed unnecessarily complex, such as the migration of the eyes of young flounders from a normal position on either side of the head to both eyes on one side of the head. If flounders are to be adapted to living flat on the ocean floor, why are they not born with both eyes on the same side of the head? Examples can be multiplied (for examples from a modern author, see Gould 1980), but the point was recognized even before Darwin that there were many examples of odd structures that didn't appear to have been the direct creation of an omniscient, benevolent God. The weight of natural historical observations weakened the argument from design.

Natural selection, of course, provided a natural means to explain complex structures that adapted their owners to their environments. As discussed in chapter 2, those organisms with structures that better suited them to a particular environment were more likely to leave descendants than were those that lacked the useful structures. Populations would thus change over time, as their members became better adapted to their environments. Paley was correct to choose design over chance, but he did not know that there was a natural as well as a transcendent source of design.

But how to choose between transcendent design and natural selection? Obviously an omniscient creator could specially create structures such as the vertebrate eye—but so could natural selection. Either the direct hand of God or natural selection could explain well-designed structures. In fact, in *Origin*, Darwin used Paley's example of the vertebrate eye to illustrate how a complex structure might plausibly result through natural selection. More difficult for the supporters of the argument from design was explaining those structures that just barely worked or were obviously cobbled together from disparate parts having other functions in related species. Natural selection can operate only on available variations, so if the "right" variation is not available, either the population dies out or some other structure will have to be modified into an adaptation. So nature is full of oddities like antennae modified into fishing lures, or jawbones turned into hearing structures—things that don't look so much engineered as tinkered with (Jacob 1977).

Along the same lines, some structures are not "fearfully and wonderfully made" (Psalm 139:14) but seem to barely work; that's tough to explain through an omniscient, benevolent designer. But because natural selection is the survival of the "fit enough," it is not expected that "perfect," optimal structures will always be the end result. Thus, natural selection can account for both well-designed (in the sense of good or efficient operation) and poorly designed structures. On the other hand, for God to have deliberately created jerry-rigged, odd, or poorly designed structures is of course possible, but it is theologically unsatisfying and empirically untestable. Natural selection, in fact, offered a theological way out to those concerned with this issue: God could work through natural selection and thus not be stuck with accusations of deliberately creating bad design.

The power of natural selection to explain the oddities of nature drew people away from design as a scientific explanation. It became possible to explain structural complexity and adaptation through natural causes. Still, there remained a theological problem: if Darwin was right, and natural selection explained design, the implication was clear: God did not need to create humans directly. But if God did not create humans directly, did this mean that humans were less special to God? Traditional teleological views held that humans existed because God created them with a specific purpose. If humans were the result of a natural process that didn't *require* the direct involvement of God, did that negate an ultimate meaning or purpose to life? (Chapter 12 presents theological responses to this question.)

Both biblical literalism and problems with design and purpose played roles in the reception of Darwin's ideas in the nineteenth century. In the early nineteenth century, both arguments were raised against evolution, at a time when links between science and religion were still strong.

Science and Religion. In the United States, early nineteenth-century religious intellectuals (including clergy and theologians, as well as religious scientists and laypeople) embraced science as providing proof of design, the existence of God, and other Christian theological positions. Many nineteenth-century scientists worked within a theological framework and frequently referred to religious views in discussing scientific positions. "The existence of God, the reality of His providential concern for his creation, the veracity of miracles, the importance of humanity as the focus of divine plan—all these doctrines appeared to be legitimate inferences from the clearest disclosures of scientific investigation" (Roberts 1988: 13).

As geologists explored the fossil record in the early half of the century, the sequence of changing forms through time was seen to reflect separate creations—and progressively improved ones, as well. This, too, harmonized with the Christian view that there was a divine providential plan unfolding through time. That these now-extinct creatures were also adapted to their environments reinforced the argument from design. It was, however, a time of rapid growth of scientific knowledge, and these same geological observations encouraged an alternative explanation: the transmutation of species (the changing of one species into another).

Science itself was evolving into a more naturalistic methodology, as natural explanations provided more testable and reliable inferences than supernatural causes. One of the early presentations of the idea of transmutation of species appeared in Robert Chambers's *Vestiges of the Natural History of Creation*, published anonymously in 1844—just about the time Darwin was beginning to work in earnest on the principle of natural selection. In Chambers's view, living things adapted to their environments in response to God-created law rather than having been specially created for that purpose. His hands-off, non-miracle-generating Creator was widely rejected by many clerics because it did not reflect the personal God with which they were familiar (Roberts 1988). Furthermore, scientists were unimpressed by the somewhat wispy scientific mechanisms that Chambers proposed.

By midcentury, then, transmutation and changes in ideas of how science should be done were in the air. Darwin's science pushed the boundaries much farther than did Chambers, and *On the Origin of Species* subsequently experienced an even stronger reaction from the religious community. But the seeds of change had been sown: the concept of a dynamic rather than a static world, already accepted in astronomy and growing in geology, would eventually wash over biology as well. But if the universe were in a state of change, of evolution, did this negate the Christian view that there was an overarching purpose for the universe? For humankind? There were many theological issues affected by Darwin's views, some of which are still being grappled with today.

Religious Responses in Context. The religious response to Darwin's ideas is not easily summarized. According to Roberts (1988), there was no pressure on clerics to modify traditional views until the scientific community agreed on transmutation of species; the scientific community had rejected previous theories of species change such as Chambers's, thereby obviating the need for the theological community to grapple with contradictions between the new science and traditional Christianity. However, by the mid-1870s, the scientific community became convinced of transmutation, and religious leaders were forced to consider evolution seriously (Roberts 1988). The initial reaction of the British clergy to Darwin's ideas was mixed, but eventually the strength of the science and the need for coherence between science and theology brought about sufficient accommodation.

Still, the process took decades, and in some denominations, resolution was not achieved until the twentieth century. The acceptance of Darwinian ideas was neither uniform nor simple, reflecting not only the growing tension between religion and science but also local issues that at some times and places were more important than the scientific or theological ones. Even within a religious tradition, there could be significant differences in the degree to which evolution was accepted. In Presbyterian Edinburgh, Scotland, for example, during the 1870s, evolution was generally accepted by the leading clerics, who were less concerned about evolution than with the growth of German modernism and biblical criticism, and their consequences for Presbyterian theology. In the late 1870s and 1880s, Presbyterians in the United States, led by the Princeton Theological Seminary theologian, Charles Hodge, generally accepted evolution but rejected Darwin's mechanism of natural selection. On the other hand, in Belfast, Ireland, Presbyterians spoke out strongly against both evolution and the natural selection mechanism (Livingstone 1999).

The Catholic response to evolution was similarly complex, reflecting social movements and consequent church politics over the struggle of American Catholics to define a Catholic identity that reflected their national and cultural needs. The late nineteenth and early twentieth centuries were periods of increased immigration to the United States by European Catholics, primarily from Ireland, Germany, Italy, and Poland. They faced a great deal of prejudice both because of their nationalities and their religion in what was a largely Anglo-Protestant America: the No Irish Need Apply signs had their equivalent for other foreigners as well. Progressive American Catholic leaders were eager to integrate the largely working-class newcomers into American society, to educate and Americanize them. Americanizing the newcomers would also Americanize Catholicism, and vice versa. Progressive Catholics sought to define Catholicism in terms of American tradition and history. However, the Vatican considered Americanism, which included the separation of church and state, support of labor unions (important to immigrants), individual liberty, and material as well as spiritual progress, a threat (Appleby 1999). Progressive Catholics promoting

Americanism were also more accepting of biblical criticism, science, and evolution, thus tainting evolution in the eyes of the Vatican. The fact that liberal Protestants largely accepted evolution made evolution even less palatable to conservative Catholics.

Turn-of-the-century Catholic immigration generated a nativist backlash from the Anglo-Protestant majority, and this, too, became entangled with attitudes toward evolution. Evolution and natural selection were incorporated into the anti-immigration arguments and into the subsequent eugenics movement of the early twentieth century. Thus, evolution became associated in some minds with anti-Catholicism, sterilization of the "feeble-minded" (read: Catholic immigrants), birth control, and racism (Appleby 1999). To be sure, there was doctrinal opposition to evolution, but because the acceptance or rejection of evolution was so embedded in other issues, as these other issues evolved, it made it easier for the Catholic Church to eventually accept evolution:

> First, the debate over Darwinism itself, for all its virulence, was actually an occasion for American Catholics to work out a number of identity-defining issues facing the immigrant community. Second, the debate did not leave a strong antievolutionist legacy to future Catholic educators in quite the way Protestant fundamentalism did. The advent of evolutionary theory, in other words, served as a catalyst for the resolution of internal Catholic issues rather than as a sustained evaluation of Darwinism and evolutionary theory in itself. Despite the seeming victory of the conservatives, moreover, the scientific theory of evolution was never formally condemned, and Roman Catholicism modified its general anti-evolutionist stance several times in the twentieth century; this was a process that culminated in the conditional approval of the theory by Pope Pius XII in 1950. (Appleby 1999: 179)

In addition to Pius XII, subsequent popes have offered an accommodation of Catholic theology to science along the lines sketched in the nineteenth century by the Catholic scientist St. George Jackson Mivart: God directly infuses the human soul, but the body has evolved from animal predecessors. Catholic high schools thus routinely teach evolution, as it has no formal doctrinal conflict with Catholic theology.

In both the United States and Great Britain, religious objection to evolution was spurred on by the anticlericalism of some of Darwin's early defenders, especially Thomas Henry Huxley and Herbert Spencer. It was easy for religious intellectuals to reject evolution by natural selection when some of its supporters presented it as compelling atheist belief. The active support of evolution in the 1860s and 1870s of a number of American scientists who were also active churchmen, such as Asa Gray, greatly helped to defuse the idea that evolution was an inherently atheistic idea (Numbers 1998).

By the mid-twentieth century in Great Britain, Europe, and North America, the scientific community no longer questioned whether evolution occurred. The neo-Darwinian revolution of the 1930s and 1940s had been successful (see chapter 3). In Great Britain and Europe, but not in North America, evolution was included matter-of-factly in textbooks and curricula of education systems. In the United States, however, evolution was a topic consistently taught only at the college level—largely absent from the K–12 curriculum. Understanding this difference requires a closer look at American history.

BACKGROUND TO CONFLICT

Why was evolution absent from American schools in the early twentieth century? To understand, we need to reflect on both American religious history and the educational structure of the United States.

America's Decentralized Educational System

Consider the settlement history of the United States: beginning with northern European (English, Dutch) contact in the northeastern part of the continent and southern European (Spanish, Portuguese) exploration of the south and west, the movement of people began at the continental coasts and worked inward. After the initial trappers and explorers mapped out the territory, settlers filled in the river valleys, using the vast interior waterways as arteries for trade and communication. People preceded government: territorial or state governmental services we today take for granted, such as police, courts, and the rule of law, and maintenance of public facilities such as roads and bridges, usually lagged well behind the expansion of people into new territories. The contributions of state or territorial governing bodies were rarely felt in newly settled areas; hardly ever were federal agencies functional in these early settlements. This, in fact, paralleled the experiences of the earliest European settlers, deposited with no support from their governments on the shores of a new land—which they more often than not must have viewed with very mixed feelings of both opportunity and foreboding as the ships that had brought them sailed back to civilization.

Because of this lack of connection with government agencies, and the independent structure of states relative to the national government, frontier communities were generally responsible for setting up their own school systems largely independent of state and federal agencies. Local communities determined whether there would be a school, constructed the building—if there was one—and determined who should teach, what he or she would be paid, and even the content of what the teacher would teach. Local control of education began as a necessity, and through custom it became enshrined as a right.

To this day, American education remains remarkably decentralized. The federal government has a role to play in education, but that role is dwarfed by the responsibility and activity of states and local school districts. In some states, a large percentage—even a preponderance—of the budget is devoted to education, and states rigidly insist on their right to determine the structure and content of the educational system, with a minimum of interference from the federal government. There is a similar tension between most state governments and local school districts. These local districts—which may be cities, regions incorporating more than one city, or smaller units corresponding to neighborhoods or other subdivisions of cities—are governed by locally elected school boards consisting of interested citizens who may or may not know much about the field of education but who, by virtue of being from the community, maintain a localized focus on education. Many states have state-level education standards that are used to guide curricular development in local school districts, but in most states, the districts have the final say as to how much of the state standards in a given field will be used in their schools.

For decades, as more money for education has come out of Washington, the U.S. government has argued that it has the right to oversee how its money is spent, though the federal government has been quicker to stress financial accountability than academic content. In 1989, the first Bush administration's Department of Education proposed the establishment of national standards for history, mathematics, and science—but, reflecting the emphasis in American education on local control, such standards were to be only advisory, not mandatory. The National Science Education Standards were published seven years later (National Committee on Science Education Standards and Assessment, National Research Council 1996).

The decentralization of American education is a source of wonder to Europeans and the Japanese, for example, who have curricula that are uniform across all communities in their nations. In France, for example, the curriculum in any particular grade is virtually the same from week to week in any classroom in any city. In the United States, even schools within the same district may not teach the same subjects in the same order, or even in the same year.

America's Decentralized Religious History

American religious history reflects an equally decentralized, "frontier" orientation. The nation initially was settled largely by religious dissidents, who came here at least partly for their own religious freedom—though once here, they generally discriminated against people who practiced other faiths! The first East Coast settlers were mostly Protestant and generally came from Congregational traditions in which most decisions were made at the level of the individual church rather than imposed hierarchically from church bureaucracies. The nature of the frontier reinforced this tendency: pioneers establishing new settlements had to establish not only police and educational systems but also churches, if they wanted them: certainly the government was not going to do so. As a result, churches took on a regional flavor, often diverging theologically from other churches that were nominally the same.

The United States also has been the nursery for a wide variety of spontaneously generated, independent sects, often inspired by charismatic leaders. It was in the United States that the Seventh-Day Adventists, the Church of Jesus Christ of Latter-Day Saints (Mormons), Jehovah's Witnesses, Christian Scientists, and now-extinct sects such as Shakers and Millerites were founded, reflecting our decentralized, nonhierarchical religious past. But perhaps the most important reason that modern antievolutionism developed here rather than in, say, Europe, was the founding in 1910–1915 of fundamentalism, a Protestant view that stresses the inerrancy of the Bible. Fundamentalism was not successfully exported to Europe or Great Britain, but it formed the basis in the United States for the antievolutionism of the 1920s Scopes era as well as the present day.

AMERICAN ANTIEVOLUTIONISM

Antievolutionism in the United States can be divided into three periods. During the first, antievolutionists worked to pass legislation that would eliminate evolution from the classroom and textbooks. When laws restricting the teaching of evolution were eventually struck down, creation science developed, bringing about the second major period of American antievolutionism. These two periods are the subject of the next chapter. When laws promoting equal time for creation science were eventually struck down, antievolution forces regrouped under a diverse set of schemes, including various repackagings of creation science as well as some new offerings. Chapter 6 will describe these changes.

REFERENCES

Appleby, R. Scott. 1999. Exposing Darwin's "hidden agenda": Roman Catholic responses to evolution, 1875–1915. In *Disseminating Darwinism: The role of place, race, religion, and gender*, ed. R. L. Numbers and J. Stenhouse, 173–208. Cambridge: Cambridge University Press.

Bowler, Peter J. 1988. *The non-Darwinian revolution: Reinterpreting a historical myth*. Baltimore: Johns Hopkins University Press.

Burkhardt, Frederick, and Sydney Smith, eds. 2002. *1865*. Vol. 13 of *The correspondence of Charles Darwin*. Cambridge: Cambridge University Press.

Chambers, Robert. 1844. *Vestiges of the natural history of creation*. London, J. Churchill.

Durant, John R. 1998. A critical-historical perspective on the argument about evolution and creation. In *An evolving dialogue: Scientific, historical, philosophical and theological perspectives on evolution*, ed. J. B. Miller, 265–280. Washington, DC: American Association for the Advancement of Science.

Gillespie, Neil C. 1979. *Charles Darwin and the problem of creation*. Chicago: University of Chicago Press.

Gossett, Thomas F. 1965. *Race: The history of an idea in America*. New York: Schocken Books.

Gould, Stephen Jay. 1980. The panda's thumb. In *The panda's thumb: More reflections on natural history*, 19–26. New York: Norton.

Hoyle, F. (1983). *The intelligent universe*. New York: Holt, Rinehart and Winston.

Jacob, François. 1977. Evolution and tinkering. *Science* 196 (4295): 1161–1166.

Livingstone, David N. 1999. Science, region, and religion: The reception of Darwinism in Princeton, Belfast, and Edinburgh. In *Disseminating Darwinism: The role of place, race, religion, and gender*, ed. R. L. Numbers and J. S. Stenhouse, 7–38. Cambridge: Cambridge University Press.

Losee, John. 1972. *A historical introduction to the philosophy of science*. London: Oxford University Press.

Mayr, Ernst. 1964. Introduction to *On the Origin of Species by Charles Darwin: A Facsimile of the First Edition*, vii–xxviii. Cambridge, MA: Harvard University Press.

Moore, James R. 1979. *The post-Darwinian controversies: A study of the Protestant struggle to come to terms with Darwin in Great Britain and America, 1870–1900*. Cambridge: Cambridge University Press.

National Committee on Science Education Standards and Assessment, National Research Council. 1996. *National Science Education Standards*. Washington, DC: National Academy Press.

Numbers, Ronald L. 1998. *Darwinism comes to America.* Cambridge, MA: Harvard University Press.

Paley, William. 1802. *Natural theology; or, Evidences of the existence and attributes of the deity, collected from the appearances of nature.* London: R. Faulder.

Perloff, James. 1999. *Tornado in a junkyard: The relentless myth of Darwinism.* Arlington, MA: Refuge Books.

Roberts, Jon H. 1988. *Darwinism and the divine in America: Protestant intellectuals and organic evolution, 1859–1900.* Madison: University of Wisconsin Press.

Winchester, Simon. 2001. *The map that changed the world: William Smith and the birth of modern geology.* New York: HarperCollins.

CHAPTER 5

• • • • • • • • • • • • • •

Eliminating Evolution, Inventing Creation Science

A GROWING CRISIS

As discussed in chapter 4, evolution had become well accepted by the scientific community by the turn of the twentieth century. It thereafter began to be included in college and secondary school textbooks. The late nineteenth century was not a period of extensive religious hostility to evolution, partly because of the efforts of American scientists who accepted evolution and who also were active church members. It was not until the twentieth century that the antievolution movement became organized, active, and effective. Three trends converged to produce the first major manifestation of antievolutionism in the twentieth century: the growth of secondary education, the appearance of Protestant fundamentalism, and the association of evolution with social and political ideas of social Darwinism that became unpopular after World War I.

Although textbooks at the turn of the century included evolution, few students were exposed to the evolution contained in these books: in the late nineteenth century, high school education was largely limited to urban dwellers and the elite. In 1890, for example, only 3.8 percent of children aged fourteen to seventeen attended school—about 202,960 students (Larson 2003: 26). But high school enrollment approximately doubled during each subsequent decade, so that by 1920, there were almost 2 million students attending high school. The practical effect of this was that more students were being exposed to evolution—and parents who felt uneasy about evolution for religious or political reasons rallied around the politician William Jennings Bryan to protest the teaching of evolution to their children.

Fundamentalism

The fundamentalist movement in American Protestantism is named for a theological perspective developed during the first few decades of the twentieth century. It was encapsulated in a series of small booklets collectively called *The Fundamentals*, published between 1910 and 1915 (Armstrong 2000: 171). Its roots, however, go back to earlier conservative Protestant movements. Fundamentalism is partly a reaction to the theological movement called modernism that began in Germany in the 1880s. Modernism reflected a technique of biblical interpretation called higher criticism, which proposed looking at the Bible in its cultural, historical, and even literary contexts. Creation and Flood stories, for example, were shown by comparison of ancient texts to have been influenced by similar stories from earlier non-Hebrew religions. With such interpretations, the Bible could be viewed as a product of human agency—with all that suggests of the possibilities of error, misunderstanding, and contradiction—as well as a product of divine inspiration.

Christians who were more conservative preferred a more traditional interpretation on which the Bible was considered inerrant (wholly true and free from error—though some individuals qualified inerrancy as applying only to the version of the Bible God gave to the original authors). Passages were to be taken at face value when at all possible rather than be "interpreted." Fundamentalists stressed "(1) the inerrancy of Scripture, (2) the Virgin Birth of Christ, (3) Christ's atonement for our sins on the cross, (4) his bodily resurrection and (5) the objective reality of his miracles" (Armstrong 2000: 171).

Financed by millionaires who had founded a conservative evangelical college in Los Angeles (the Bible Institute of Los Angeles, now Biola), millions of copies of *The Fundamentals* were printed and distributed "free of charge, to every pastor, professor, and theology student in America" (Armstrong 2000: 171). Different essays treated evolution in different ways: some of the authors rejected evolution, but some accepted various forms of theistic evolution. In some, natural selection was rejected but not common ancestry itself (Larson 1997). Some writers allowed for animal evolution, but not human, and some even allowed for human evolution, though not through natural selection. Natural selection was opposed because it replaced God's direct action with natural causes, and thus indicated to some a less personal, hands-on, involved God, unacceptable to fundamentalist theology. Most of the authors of *The Fundamentals* were day-age creationists, who allowed for an old Earth but insisted on a recent appearance of humans. Although not all *The Fundamentals* were antievolutionary, the fundamentalist position toward evolution hardened fairly quickly. Fundamentalists became the ground troops for the campaign to rid schools of evolution. They were motivated by religious sentiments and by a concern that evolution was the source of many negative and even corrosive social trends.

Evolution as Social Evil

The second decade of the twentieth century was a time of considerable social unrest and psychological unease. The appalling death, brutality, destruction, and devastation of World War I led many citizens, including many conservative Christians, to conclude that civilization itself had failed. Conservative Christians sought a solution in a return to biblical authority and in the literal interpretation of Scripture. Their views were further reinforced by Germany's having been the main source of both higher criticism, viewed as an attack on religion, and World War I militarism, viewed as an attack on civilization (Armstrong 2000; Marsden 1980).

Conservative American Christians felt that German militarism, theories of racial superiority, and eugenics were directly related to the acceptance of evolution by Germany at the end of the nineteenth century. In reality, German views of evolution were quite different from those of Darwin, largely rejecting natural selection as a mechanism of change, biological or societal. Evolution by natural selection did not fit German militaristic views of the inevitability of Teutonic triumph; natural selection relies on selection of the "most fit" in terms of a particular environment. It does not support the idea that Germans or anyone else inevitably would be superior to all others, regardless of environmental circumstance.

In the early twentieth century, evolution was also credited with providing the foundation for laissez-faire capitalism, as robber barons of the late nineteenth and early twentieth centuries sometimes claimed that natural selection justified their exploitative labor policies and cutthroat business practices:

> The price which society pays for the law of competition, like the price it pays for cheap comforts and luxuries is also great; but the advantages of this law are also greater still for it is to this law that we owe our wonderful material development, which brings improved conditions in its train. But whether the law be benign or not, we cannot evade it; no substitutes for it have been found, and while the law may be sometimes hard for the individual, it is best for the race, because it insures the survival of the fittest in every department. (Carnegie 1889: 653)

Thus, fundamentalists, led by the famous progressive politician and champion of the worker William Jennings Bryan, had many reasons to oppose the teaching of evolution to their children, whether or not these reasons were justified. Beginning in the early 1920s, several state legislatures took up Bryan's call to outlaw evolution, and finally, on March 23, 1925, Tennessee passed the Butler Act. This set in motion events that would culminate in the so-called trial of the century.

SHOWDOWN IN DAYTON

"It shall be unlawful for any teacher to teach any theory that denies the Story of Divine Creation of man as taught in the Bible, and to teach instead that man has descended from a lower order of animal" declared Tennessee's Butler Act (Larson 2003: 54). The Tennessee House of Representatives passed the act with virtually no debate, but the Senate heard considerable testimony both for and against it. Scientists almost uniformly opposed it, and the religious community was split, with fundamentalists strongly supporting the bill as a means of preserving children's faith and liberals opposing it on the grounds that the state should not favor one religious position over another. Public sentiment in Tennessee was so strong, though, that the state's senators felt great pressure to pass the bill, and they did.

Almost immediately, the young American Civil Liberties Union in New York took up the challenge of testing the new law. Because of restrictions on civil liberties imposed by the government during and after World War I, the ACLU was particularly concerned with free speech. The early 1920s and the preceding decade were a time of social unrest, economic insecurity, and agitation for workers' rights. Strikes in mills and mines over long hours, poor working conditions, and low pay resulted in considerable unrest. Many Americans feared that European anarchism and socialism were taking root in American soil, and establishing the rights of workers was a struggle. The ACLU focused on the free speech and other rights of workers—and teaching school qualified as labor. The ACLU leaders believed that the Butler Act infringed on the free speech rights of teachers by restricting what they could teach.

The ACLU took out advertisements in Tennessee newspapers, offering to defend any teacher willing to volunteer to be the defendant in a legal challenge to the Butler Act. Businessmen in the small town of Dayton concocted a plan to bring publicity and business to their community as the site of a high-visibility trial challenging a controversial law. They persuaded John T. Scopes, a young science teacher, to be the ACLU's test case. He would be accused of teaching evolution, the trial would be held, the law would be struck down, and Dayton would receive publicity and a welcome economic shot in the arm. The scenario played out almost as planned.

Scopes was a young man of twenty-four who taught science at the high school. As the tale is told, town leaders called him in from a tennis game to pitch the idea of challenging the Butler Act in Dayton.

> Scopes was the ideal defendant for the test case. Single, easy-going, and without any fixed intention of staying in Dayton, he had little to lose from a summertime caper—unlike the regular biology teacher, who had a family and administrative responsibilities. Scopes also looked the part of an earnest young teacher, complete with horn-rimmed glasses and a boyish face that made him appear academic but not threatening. Naturally shy, cooperative, and well-liked, he would not alienate parents or taxpayers with soapbox speeches on evolution or give the appearance of a radical or ungrateful public employee. Yet his friends knew that Scopes disapproved of the new law and accepted an evolutionary view of human origins. (Larson 1997: 90–91)

The amiable Scopes agreed, a warrant was sworn out, and Scopes was duly charged with the crime of violating the Butler Act, after which he returned to his tennis game. Plans were made to hold the trial in Dayton, the seat of Rhea County.

The plan to bring publicity to Dayton succeeded beyond the businessmen's wildest expectations, and certainly beyond what the young schoolteacher had anticipated. The 1925 trial was truly the trial of the century, being the first trial to be covered not only by the print media but also through live radio broadcasts. The trial would have received a lot of attention on its own merits: the Butler Act had received national publicity, and already battle lines had been drawn over the merit of passing antievolution laws. The unexpected appearance of two political giants of the day, William Jennings Bryan for the prosecution and Clarence Darrow for the defense, only heightened public interest. All these factors transformed the trial into a three-ring circus.

Bryan was one of the nation's most famous and popular public figures. He had been three times the Democratic Party's candidate for president and had served as secretary of state under Woodrow Wilson. He had made his political reputation as a well-known promoter of progressive causes such as women's suffrage, pacifism, and better working conditions for workers. Always a devout man, in his later years he became known as much for his fundamentalist Christian views as for his progressivism. Today—largely because of the Scopes trial—much of his political progressivism has been forgotten. But in the late 1910s and 1920s, laissez-faire capitalism—the source of poor working conditions, child labor, and worker exploitation—was believed to be supported by evolution, which in addition was believed to be antireligious. Of course, robber barons like Andrew Carnegie were neither the first nor the last to latch onto science to promote an ideological view; the fact that natural selection theory has been used to support both Marxism and laissez-faire capitalism suggests that the link is more in the eyes of the proposers than in reality. But Bryan's combination of political progressivism and fundamentalist Christian antievolutionism well fit the social and political views of the times

On Scope's side was Clarence Darrow, the most famous defense attorney in the country. Like Bryan, he was a political progressive; he had supported Bryan in the latter's early attempts to become president. Darrow was also a pacifist and a supporter of free speech—which were not uniformly popular positions in the second decade of the twentieth century. He also was a well-known atheist, thus contrasting sharply with Bryan. With two such giants squaring off against one another, the public found the trial irresistible.

The Scopes trial originally had been conceived of as a test of the truth of evolution. Both sides—especially the antievolution side—were eager to testify regarding evolution's validity and to whether evolution was inherently anti-Christian or led to immoral or unethical behavior. But the prosecution began to lose enthusiasm for this approach when it became clear that the defense was quickly lining up scientists and theologians who would affirm that evolution was scientific and assert that it was not necessarily anti-Christian. The prosecution had a difficult time finding scientists who rejected evolution (Larson 1997: 130). It then switched its strategy to argue the case narrowly: did or did not Scopes break the law? Fortunately for the prosecution, once the trial began, the judge quickly took its point of view, ruling that, indeed, the trial would focus only on whether Scopes had broken the law (i.e., had taught evolution). The defense's carefully chosen scientific witnesses (from biology, anthropology, and geology) and its three theologians were not permitted to testify.

One of the most memorable moments in the trial involved a daring legal move: Darrow requested that Bryan take the witness stand as an expert on religion. Bryan accepted, against the advice of his cocounsels. He planned to use the opportunity to witness his Christian faith to both supporters and those not yet converted, and to defend Christianity against the atheist Darrow. Unfortunately for him, however, it became clear that he was an expert neither on the Bible nor on comparative religion, and he was certainly no expert on science or evolution.

Throughout Bryan's examination, Darrow sought to show that certain passages of the Bible cannot rationally be accepted as literally true. Bryan fell for this scheme by admitting that despite his reputation as a promoter of fundamentalism, he had no explanation for how Joshua lengthened the day by making the sun stand still. Similarly, he could not answer Darrow's questions about whether the Noachian Flood that allegedly destroyed all life outside the ark also killed fish, where Cain got his wife, and how the snake that tempted Eve moved before God made it crawl on its belly as punishment. Bryan acknowledged his acceptance of a long Earth history and a day-age interpretation of the Genesis account, which of course allowed enough time for evolution to take place. Further undermining his stance against evolution, Bryan confessed that he knew little about comparative religion or science (Hileary and Metzger 1990). The cross examination was widely viewed as a public relations disaster for Bryan, although it had little effect on the outcome of the trial.

And in fact, given the narrow grounds on which testimony was allowed, it was a foregone conclusion that Scopes would lose. Both sides anticipated the verdict; to a large degree, both sides viewed the Dayton trial as a preliminary step toward the appeals process: eventually the Supreme Court would test the legality of antievolution laws. Scopes was convicted of having taught evolution (though in reality he may never have actually taught that chapter of the textbook). The defense also lost its appeal to the Tennessee Supreme Court: the ACLU's concern that individual freedom should take priority over the government's authority over public employees was rejected in favor of the state's right to set conditions for employment.

In a surprise move, however, the Tennessee Supreme Court then reversed the Scopes conviction on a technicality. The trial judge (as was not uncommon in such minor cases) had assigned the $100 fine, but the law required the jury to set the penalty. On those grounds, the Supreme Court threw out Scopes's conviction, which made further appeal moot. The ACLU's plan to appeal the case to the U.S. Supreme Court was thwarted.

AFTEREFFECTS OF THE SCOPES TRIAL

Although antievolution forces prevailed during the Scopes trial, the aftereffects of the trial of the century were not as clear cut. Antievolution laws continued to be submitted, but few passed. Mississippi and Arkansas passed antievolution bills in 1926, but in 1927 bills were defeated in Arkansas, Oklahoma, Missouri, West Virginia, Delaware, Georgia, Alabama, North Carolina, Florida, Minnesota, and California (Holmes 1927). The Supreme Court eventually would have the opportunity to rule on the constitutionality of antievolution laws—but not until the 1960s.

After the Scopes trial, antievolutionism became associated in the popular imagination with conservative religious views—and with the most negative stereotypes of such views. Antievolutionists and fundamentalists in general were portrayed as foolish, unthinking, religious zealots. Particularly effective in contributing to this stereotype were the Dayton dispatches of the acerbic reporter for the *Baltimore Sun*, H. L. Mencken, but accounts written in the 1930s and afterward also reinforced the view that antievolutionism was a campaign of backward (or at best premodern), uneducated religious fanatics. Although many leaders of the pre-Scopes antievolution movement were from Northern states, after the Scopes trial, antievolutionism became more regionalized, retaining momentum in the South and rural areas of the country, where fundamentalism remained strong. Where fundamentalists held political power, school boards imposed regulations to restrict the teaching of evolution. But the demographics of fundamentalism were changing, as it moved from the cities of its origin to the rural South—where it largely disappeared from the view of the mainstream (East Coast, urban) press (Marsden 1980: 184).

Jerome Lawrence and Robert E. Lee were inspired by the issues raised in the Scopes trial when writing their 1955 Broadway play *Inherit the Wind*. This play and the movies based on it have strongly shaped public images of the Scopes trial and contributed to the negative public image of fundamentalists. Although the authors explicitly distanced themselves from the Scopes trial in the introduction of the play ("It is not 1925. The stage directions set the time as 'Not too long ago.' It might have been yesterday. It could be tomorrow") and argued that their motivation for writing the play was to consider issues of free speech, the closeness of the story line in *Inherit the Wind* to the events of the Scopes trial was obvious. The play featured a young teacher tried and imprisoned for teaching evolution and thereby violating an antievolution law. Two prominent political figures—one a fundamentalist and one a freethinker—lined up on the prosecution and defense sides, respectively. Issues of fundamentalism and modernism (science) were constantly present. The play even included an H. L. Mencken–like cynical reporter with an acid tongue. The circus atmosphere of the trial in the play certainly paralleled that of the actual trial. The play's strong characters and memorable writing have made it a classic, often read and performed in high schools as a vehicle for discussing issues of free speech and the role in society of minority and majority views.

Of course there were major differences between the Scopes trial and *Inherit the Wind*; the goal of the playwrights was to present a dramatic narrative rather than a historical account. Modern antievolutionists particularly object to the treatment of the character based on Bryan, who is bombastic and a caricature of a religious bigot. Viewed as history, *Inherit the Wind* is clearly inaccurate: although the Scopes-like character in the play goes to jail, Scopes himself was never imprisoned, and a fundamentalist minister who rails against evolution and science—and who is the father of the young teacher's girlfriend—did not have a Scopes trial counterpart but was added for plot reasons. *Inherit the Wind* was intended, according to its authors, as a metaphor for 1950s McCarthyist politics, which threatened free speech and freedom of conscience.

Perhaps the biggest difference between the picture painted by the play and movies and the actual trial was the false image presented in the fictionalized account that "the light of reason had banished religious obscurantism" (Larson 1997: 246). Neither fundamentalism nor the antievolutionist campaign disappeared after 1925, though the latter abated somewhat. This was primarily because antievolutionism became largely unnecessary: evolution remained effectively absent from science instruction until the 1960s.

Scopes lost; the antievolution laws remained on the books, and even increased in number. In the South, states and local school districts restricted the teaching of evolution, and teachers and parents who chose textbooks preferred ones that slighted evolution. The economic pressures were effective: textbook publishers knew they had to remove, downplay, or qualify evolution if they wanted sales, and they did. Books tailored for the Southern markets were of course sold elsewhere, and evolution disappeared from textbooks all over the nation (Grabiner and Miller 1974). Because of the influence textbooks have on curricula, with evolution absent from the textbooks, it quickly disappeared from the classroom. By 1930, only five years after the Scopes trial, an estimated 70 percent of American classrooms omitted evolution (Larson 2003: 85), and the amount diminished even further thereafter. Its return sparked the next chapter in American antievolutionism, as creationists lashed back at the reintroduction of evolution in American schools.

CREATION SCIENCE EVOLVES

Even though most consider the Scopes trial a victory for evolution, fewer high school teachers taught evolution after the Scopes trial than before. The amount of evolution in textbooks decreased rapidly after 1925 (Grabiner and Miller 1974; Skoog 1979). This remained the case until the late 1950s, when a federally funded campaign to improve precollege science education brought evolution back into textbooks. As evolution eased back into the science curriculm, antievolutionists reacted and creation science appeared on the scene. The appearance of a small metal sphere in the heavens helped to kick-start the process.

The Sputnik Scare

In October 1957, the Soviet Union launched Sputnik, the world's first artificial satellite. The United States was shocked: the Communists had beaten the world's foremost democracy into space. How had this happened? As part of the soul-searching that took place after the Soviet triumph, the United States decided that the scientific establishment—including public school science instruction—was seriously in need of an overhaul. The newly established National Science Foundation (NSF) beefed up funding for basic scientific research and instituted directorates for education that would fund research to improve science education.

One goal was to improve the content and pedagogy of high school textbooks. It was an ambitious effort: scientists and master teachers were assembled to prepare textbooks in the disciplines of physics, chemistry, earth science, and biology. When university-level scientists began working with the NSF-funded Biological Sciences Curriculum Study (BSCS), they were shocked to discover the poor quality of extant textbooks. Evolution, the foundation of biology, was absent from almost all of them. They decided that in addition to improving the pedagogical approach to learning ("to get away from the 'parade-of-the-plant-and-animal-kingdoms' approach, to stress concepts and experimental science, and to encourage the personal involvement of students in their learning" [Moore 2002: 165]), the new BSCS books would treat evolution as it was treated in college-level texts: as an indispensable component of the biological sciences that students must understand to understand biology fully. In 1963, the first three BSCS textbooks were released, and all of them included evolution as a prominent theme (Grobman 1998).

The new BSCS approach brought about a revolution in textbooks. Partly because these new textbooks carried the stamp of approval of the NSF, but also because they were so much more interesting and up-to-date than extant books, school boards and textbook selection committees were eager to adopt them. Once the BSCS books began selling, commercial publishers began to try to produce books in the same mold (Skoog 1978: 24). As one of the BSCS writers described it, "Subsequent events showed that nearly every objecting school board ended up adopting the books—evolution, sex, and all. Word was spreading the BSCS biology was the 'new thing,' and there were community pressures on school boards to be up to date, even if a little wicked, rather than behind the times and fully virtuous. Once this situation was understood, nearly every newly published biology book included an explicit discussion of evolution" (Moore 1976: 192–193).

After the decline of evolution in textbooks and science curricula after the Scopes trial, antievolutionists had not been very active; they had not needed to be. But the resurgence in the 1960s of evolution in textbooks generated new resistance to the teaching of evolution in the public schools. Although since the 1700s some supporters of a literal interpretation of the Bible had argued that scientific evidence existed to support their views, such arguments had diminished considerably after Darwin's *Origin of Species*. Now, in the mid-twentieth century, such views were being revived, partly in response to the increasing presence of evolution in textbooks and in curricula. Much as the Scopes trial in 1925 had been a response to the post-Darwinian appearance of evolution in the curriculum, so did the return of evolution to the curriculum in the 1960s spark a reaction from religious conservatives (Numbers 1992). But the Scopes-era reaction to evolution was almost entirely centered on religious objections: evolution should not be taught to children, they argued, because it was unbiblical and would lead children away from faith. By the mid-twentieth century, however, science was a far more powerful cultural force than it had been earlier, and antievolutionists sought to exploit its authority (Larson 2003).

If students were going to be taught evolution, antievolutionists argued, students also should be exposed to a biblical view. The frank advocacy of a religious view such as creationism in the public schools would of course be unconstitutional, but creationists reasoned that if creationism could be presented as an alternative scientific view—creation science—then it would deserve a place in the science curriculm. No one was more important in shaping this approach than the late Henry M. Morris.

Creation Science and Henry M. Morris

The Genesis of Creation Science. Henry M. Morris is widely considered the father of the twentieth-century movement known as creation science. Morris was trained as a hydraulic engineer and began his career as a creationist with the publication in 1946 of his first book, *That You Might Believe*, written while he was an instructor at Rice University during World War II. In graduate school, he revised the book, which was then issued as *The Bible and Modern Science* (1951). In these early efforts, Morris, a self-proclaimed biblical literalist, promoted a recent six-day (twenty-four hours per day) creation, and a literal, historical flood, but he additionally claimed that special creationism can be supported by the facts and theories of science. Although both of these books are still in print and continue to sell, the modern creation science movement crystallized in 1961 with the publication of Morris's book *The Genesis Flood*, written with the theologian John Whitcomb.

Like Morris's previous works, *The Genesis Flood* argued that Noah's Flood could explain most modern geological features, a view that had originally been popularized by the early twentieth-century Seventh-Day Adventist geologist George McCready Price (Numbers 2006). Termed *flood geology*, this view became the core of the new movement called creation science.

Morris provided the scientific references and Whitcomb provided the theological arguments. The book's mix of theology and science is characteristic of creation science, and it continues to be widely read in evangelical and fundamentalist circles. *The Genesis Flood* proposed that there is scientific evidence that Earth is less than ten thousand years old, and that evolution was therefore impossible. This view became known as young-Earth creationism. Fundamentalists were eager to claim scientific support for their religious views and use it to "balance" the teaching of evolution.

Morris worked tirelessly to strengthen the evangelical antievolutionist movement. To promote scientific research supporting the young age of Earth and universe, the special creation of all living things, and Noah's Flood, he worked with a group of conservative Christian scientists to found the Creation Research Society (CRS) in 1963, soon after the publication of *The Genesis Flood*. *Creation Research Society Quarterly* (CRSQ) began publishing shortly thereafter, in 1964. Although in the early days, the board included some non–flood geology proponents, the CRS eventually evolved into a young-Earth organization. The CRS requires all voting members to sign a statement of belief; this reveals the essentially religious orientation of the organization, as scientific societies do not hold members to faith statements regarding salvation. Reflecting the special creationist and YEC orientation of Morris and other influential founders, the statement includes the following provisions (emphases in the original):

1. The *Bible* is the written Word of God, and because it is inspired throughout, all its assertions are historically and scientifically true in the original autographs. To the student of nature this means that the account of origins in *Genesis* is a factual presentation of simple historical truths.
2. All basic types of living things, including man, were made by direct creative acts of God during the Creation Week described in *Genesis*. Whatever biological changes have occurred since Creation Week have accomplished only changes within the original created kinds.
3. The great flood described in *Genesis*, commonly referred to as the Noachian Flood, was an historic event worldwide in its extent and effect.
4. We are an organization of Christian men and women of science who accept Jesus Christ as our Lord and Savior. The account of the special creation of Adam and Eve as one man and one woman and their subsequent fall into sin is the basis for our belief in the necessity of a Savior for all mankind. Therefore, salvation can come only through accepting Jesus Christ as our Savior.

Tenets of Creation Science. Special creationism is a religious view accepted in whole or in part by many American Christians (see chapter 3). Creation science reflects special creationism in that it professes that the universe came into being in its present form relatively suddenly, over a period of days rather than billions of years. The galaxies, Earth, and living things on Earth appeared during six twenty-four-hour days of creation, according to this view. Creation science, as outlined by Henry M. Morris, also proposes that the universe is young, with its age reckoned in the thousands rather than the billions or even millions of years. Creation science includes these ideas derived from special creationism but adds that this account of creation can be supported with scientific data and theory: its proponents do not consider creation science to be limited to a religious view.

Creation science argues that there are only two views, special creationism and evolution; thus, arguments against evolution are arguments in favor of creationism.

Literature supporting creation science thus centers on alleged examples of evidence "against" evolution, which are considered not only disproof of evolution but also positive evidence for creationism. As you will read later, evidence against evolution comprises the bulk of the creationist canon, it being difficult to amass empirical data in support of special creation.

Creation Science Expands

To counter the BSCS and other evolution-based textbooks, in 1970 the CRS published its own high school biology textbook, *Biology: A Search for Order in Complexity*, which by its title revealed its orientation of seeking divine design in nature (Moore and Slusher 1970). The CRS textbook did not sell many copies, however, and ran into legal difficulty in several states because of its frankly religious orientation. In 1974, Henry Morris published a textbook of his own, *Scientific Creationism*. To try to avoid the criticisms that such books were Christian apologetics masquerading as science books, Morris published a Christian schools edition containing an extra chapter of biblical references and a general edition that did not. Claims that religious references were not present in the general edition were not persuasive, however, when textbook selection committees encountered statements referring to such biblical events as Noah's Flood and the Tower of Babel: "The origin of civilization would be located somewhere in the Middle East, near the site of Mount Ararat (where historical tradition indicates the survivors of the antediluvian population emerged from the great cataclysm) or near Babylon (where tradition indicates the confusion of languages took place)" (Morris 1974: 188).

The Institute for Creation Research. In 1972, Henry Morris and others founded the Institute for Creation Research (ICR) as the research division of the Bible-based Christian Heritage College, which Morris had founded two years earlier with the evangelist Timothy LaHaye. The ICR became an independent institution in 1980, moving from the San Diego suburb of El Cajon to nearby Santee, California, where, until 2007, it had its headquarters in two large buildings. In that year, the main offices moved to an expanded office site in Dallas at the four-acre Henry M. Morris Center for Christian Leadership. After the turn of the century, Morris's son John D. Morris, a geologist, gradually assumed more responsibility for running ICR as his father became more elderly. The "torch was passed" in May 2002 (Rasche 2002:1); the elder Morris continued to write almost until his death in February 2006 at the age of eighty-seven.

The ICR has grown steadily since its inception in the early 1970s, taking pride in always ending the year in the black and never borrowing money for its building projects. To promote creation science, ICR conducts extensive outreach to churches and individuals. In any given week, ICR staff may be found around the country leading workshops, lecturing, or occasionally debating evolution with scientists. The ICR's popular Back to Genesis program, begun in 1988, consists of two days of lectures, movies, and workshops for adults and children. Other programs, such as the Good Science Workshops, are aimed at school-age children, parents, and teachers, and focus on creation science education. The ICR's foremost debater, Duane Gish, trained as a biochemist, holds workshops on how to debate evolutionists. One of ICR's radio programs, *Science, Scripture and Salvation*, airs on approximately 700 stations, and the other, a one-minute filler by director John D. Morris, *Back to Genesis*, has 860 outlets (www.icr.org/radio/rad-hist.htm).

Each month, ICR mails literature to 200,000 or more recipients. For thirty-six years, the mailing contained a small-format newsletter, *Acts and Facts*, and one or more pamphlets, among them the *Impact* series, in which scientific issues were discussed, and the much more evangelical *Back to Genesis* series. Often an advertising circular promoting books, videos, CDs, or other media was included, as well as a letter from the president of ICR and a request for financial support. In August 2007, ICR dramatically altered its communications format, shifting to a full-color magazine that includes news (as did *Acts and Facts*) as well as articles promoting creation science and attacking evolution (as did *Impact*). Also in 2007, ICR launched a new research journal, the *International Journal for Creation Research*, a competitor to the CRSC (Holden 2007).

Early in ICR's history, Morris helped found Creation-Life Publishers, which through its Master Books division maintains an extensive catalog of antievolution books. The division is promoted not only by ICR but also by virtually every creationist organization large enough to sell merchandise. The catalog includes more than 150 books. The ICR also maintains the Museum of Creation and Earth History at its Santee headquarters, which an estimated 25,000 individuals visited during its first year (Institute for Creation Research, 1993b). Remodeled in 1992, the museum currently reaches thousands of schoolchildren each year, most of them who are homeschooled or attend Christian schools. Because of the religious orientation of the museum, few local public school teachers take their students to the ICR museum. The museum presents a journey through the seven days of creation, mixing biblical and scientific references. True to Morris's concern with flood geology, there is a Noah's ark diorama that presents calculations of how many animals could have been housed on the ark.

To promote the establishment of creation science, ICR supports a graduate school that offers master's degrees in science education, biology, geology, and astrogeophysics. Until recently, the school was accredited not by the Western Association of Colleges and Schools, the accrediting agency for most other California institutions of higher learning, but by the TransNational Association of Christian Schools (TRACS). None other than Henry M. Morris founded TRACS, and he served as its president for many years. The purpose of TRACS is to accredit Bible-based institutions that pledge to promote creation science. When the ICR moved to Texas, which does not recognize TRACS accreditation, it failed to gain state certification for its graduate school—a decision that it appealed in 2008.

Most of the graduate school courses are taught during the summer. There are also annual trips down the Grand Canyon, where ICR geologist Steve Austin explains how the many layers of the canyon were formed by the receding waters of Noah's Flood.

Ken Ham and Answers in Genesis. In January 1987, a young Australian evangelist, Ken Ham, came to work for the ICR (Anonymous 1986). Inspired by Henry Morris's creation science ministry, Ham had co-founded the Australian Creation Science

Foundation in 1978 and had built it into a successful young-Earth ministry. In April 1988, ICR announced the institution of the Back to Genesis program, consisting of two-and-a-half-day public meetings built around creation science, but being more explicitly evangelical and religious than other ICR programs (Institute for Creation Research, 1988a, 1988b). Although other staff members were necessarily involved, the Back to Genesis program relied heavily on Ham. The more evangelical focus of its meetings may have been because Ham lacks a background in science, unlike most other ICR professional staff. By all accounts, the former teacher was a popular and successful evangelist, and the Back to Genesis programs began to play a larger role in ICR activities.

Ham also wrote the new *Back to Genesis* evangelical pamphlets that, beginning in January 1989, accompanied the ICR newsletter *Acts and Facts*, and Ham-led Back to Genesis revivals soon were held at least once a month throughout the United States. By August 1993, the Back to Genesis program had apparently expanded to the limits of ICR staff capabilities, and an article appeared in *Acts and Facts* encouraging churches to sign up for other, smaller ICR programs such as the Case for Creation seminars, "as well as speakers for pulpit supply, parent/teacher science workshops, school assemblies, conventions, campus conferences, and other types of meetings, even field trips are possible, especially with graduate student [sic], wherever a creationist message is in demand and can be scheduled. Fees are very reasonable compared to those of other types of specialty speakers" (Institute for Creation Research, 1993a: 5).

In January 1994, Ham moved to adjunct faculty status at the ICR's graduate school and left for Florence, Kentucky, to establish a branch of the Australian Creation Science Foundation. Ham first called his organization Creation Science Ministries, but soon changed its name to Answers in Genesis (AIG), and a few years later (in November 1997) the Australian Creation Science Foundation also became Answers in Genesis. The Australian and U.S. organizations formed the core of an international movement with other branches in Canada, New Zealand, and Great Britain. The international YEC establishment began to unravel, however, and in 2004, the Australian, Canadian, and New Zealand branches renamed themselves Creation Ministries International (CMI), and in 2005 they split from Answers in Genesis (Ham 2008). Up until 2007, AIG had sent its 40,000 members the Australian glossy magazine *Creation* as a membership premium, but because of the falling-out with CMI, Ham began publishing his own magazine, *Answers*. A lawsuit ensued over whether Ham had the right to use the *Creation* mailing list—reputed to be worth $250,000 to the Australian creationist organization (McKenna 2007).

Ham's claim that AIG is the largest YEC organization in the nation appears to be accurate. According to Internal Revenue Service (IRS) records (available for all nonprofit organizations), AIG's income in 2005 was slightly in excess of $13 million. (In the same year, ICR's income was about $7.8 million. In case you are curious, the organization I work for, the National Center for Science Education, had an income in 2005 of about $1.2 million). The AIG sponsors lectures and workshops led by Ham and other employees, and his monthly newsletter, *Answers Update*, offers books, videos, tapes, and other resources promoting YEC, in addition to the full-color magazine, *Creation*. A $27 million museum of creationism occupying 60,000 feet of AIG's headquarters in northern Kentucky opened in May 2007, to great fanfare. With the move in 2007 of ICR to its enlarged, four-acre headquarters in Dallas and the expansive and well-funded AIG museum, it appears that young-Earth creationism is expanding its potential for influence toward the end of the first decade of the twenty-first century.

Other YEC Ministries. In addition to the ICR, other national YEC organizations have heeded the creation science message of Henry Morris. The Bible-Science Association (BSA), founded in 1964, focused on spreading the message of creation science to the general public rather than on publishing scientific research. There were, in fact, some tensions between the BSA and the CRS. The BSA published the *Bible Science Newsletter* until the late 1990s, and then, falling on lean times, cut back publications and activities to focus on the creationist radio program (first aired in 1987) *Creation Moments.* The BSA itself became Creation Moments Inc. in 1997, and it now concentrates wholly on producing short radio programs.

Creation science is communicated to the public primarily through ICR and AIG publications, and through the less widely distributed *Creation Research Society Quarterly*, published by CRS. In 1978, Students for Origins Research began publishing the newsletter *Origins Research* from Santa Barbara, California. Although more moderate than the BSA, the students nonetheless usually promoted a YEC orientation. In 1996, Origins Research changed title and format and emerged as Origins and Design, and Students for Origins Research morphed into Access Research Network, shedding its overt young Earthism to promote intelligent design creationism (see chapter 6).

There also are regional and local organizations that promote the YEC views of Henry Morris, such as the Paulden, Arizona–based Van Andel Research Center (named after Jay Van Andel, founder of the Amway company), headed by an adjunct professor of biology at ICR, John R. Meyer. In Grand Junction, Colorado, Dave and Mary Jo Nutting (ICR graduate school alumni) operate the Alpha Omega Institute, which provides creationist geology and natural history tours of the Rockies and surrounding areas, as well as school assemblies for Christian and public schools. A St. Louis–based organization, the Creation-Science Association of Mid-America, was instrumental in the late 1990s in providing information to Kansas board of education members wishing to promote creation science. The Pittsburgh-based Creation Science Fellowship Inc. has been promoting young-Earth creationism since 1980, and it sponsors periodic international conferences on creationism. Several independent evangelists focus on creationism, including Carl Baugh of Glen Rose, Texas, and Walter Brown, of the Center for Scientific Creation in Phoenix. Perhaps the most successful of the creation science ministries was that of "Dr. Dino," or Kent Hovind, of Pensacola, Florida. However, Hovind ran afoul of the IRS in 2007 and was imprisoned for failing to pay withholding taxes for his employees (Stewart 2007).

Some national televangelists such as Hank Hanegraaff (Rancho Santa Margarita, California) and John Ankerberg (Ankerberg Theological Research Institute, Chattanooga, Tennessee) regularly present programs criticizing evolution, and they rely on and promote creationist views. Before his death in September 2007, D. James Kennedy's Coral Ridge Ministries' broadcasts regularly bashed evolution and promoted both young-Earth creationism and IDC. The nationwide Maranatha Campus

Ministries also promotes creation science. A sizable corpus of antievolutionary material consisting of books, videos, CDs, filmstrips, tape recordings, posters, and curricula promoting young-Earth creationism is publicly available from these organizations and individuals and their Web sites.

Equal Time for Creation and Evolution

Undoing Scopes. By the early 1960s, evolution was returning to science textbooks and classrooms after largely having been absent since the 1930s. It is not coincidental that Whitcomb and Morris's *The Genesis Flood* was published in 1961 and that the ICR was founded a few years later: the increased exposure of public school students to evolution was a cause for alarm among conservative Christians. In Arizona, opposition to the use of BSCS books in the Phoenix school district stimulated state legislators to introduce legislation that would require "equal time and emphasis to the presentation of the doctrine of divine creation, where such schools conduct a course which teaches the theory of evolution" (the legislation did not pass) (Larson 2003: 97). But the renewed textbook emphasis on evolution generated conflicts in states with antievolution laws for teachers who wished to teach modern science but who would thereby break the law.

In 1965, Arkansas was one of the few remaining states with Scopes-era antievolution laws still on the books (the others were Tennessee, Louisiana, and Mississippi [Larson 1997]). In that year, the Arkansas Education Association (AEA) decided to challenge the state's antievolution law, partly because the presence of evolution in textbooks put teachers on a collision course with the law. Rather than a Scopes-style teacher defendant who would be prosecuted for breaking the law, the AEA instead challenged the law itself with a teacher-plaintiff who sought to legally teach evolution (Moore 1998). Arkansas teacher Susan Epperson argued that the Arkansas antievolution law was unconstitutional because it violated her freedom of speech, and a co-plaintiff, a father of a student, argued for the right of the student to learn the banned subject. The trial itself was very short, taking only about two hours; the judge ruled that the antievolution law was unconstitutional. To the surprise of Epperson and the AEA, the Arkansas Supreme Court reversed the lower court in a two-sentence decision in 1967.

The case was appealed to the U.S. Supreme Court, which ruled in 1968 in *Epperson v. Arkansas* that the antievolution law was unconstitutional because it "selects from the body of knowledge a particular segment which it proscribes for the sole reason that it is deemed to conflict with a particular religious doctrine." The First Amendment requires schools to be neutral toward religion; to ban a subject (evolution) because a religious view (fundamentalism) finds it objectionable violates the Establishment Clause of the First Amendment. Finally, in 1968, forty-three years after the Scopes trial, it was unlawful to ban the teaching of evolution.

Epperson had more of a psychological effect on antievolutionism than an actual one, as the Arkansas and other antievolution laws had hardly ever been enforced (Larson 2003). But if evolution could not be banned, how could children be protected from it? Keeping evolution out of the classroom was obviously not possible, as evolution was widely included in textbooks. Teaching the Bible along with evolution was one solution.

"Neither Advances nor Inhibits Religion". In 1963, the Supreme Court struck down laws requiring prayer in public schools (*Abington School District v. Schempp*). The First Amendment of the U.S. Constitution sets forth freedoms of religion, speech, and assembly. The Religion Clause reads, "Congress shall make no law respecting the establishment of religion, nor inhibiting the free exercise thereof." The Establishment Clause prohibits the state from promoting religion, and the Free Exercise Clause prohibits the state from inhibiting or restricting religion. In *Schempp*, the justices clearly stated the requirement for religious neutrality in the public schools, stating that "to withstand the strictures of the Establishment Clause there must be a secular legislative purpose and a primary effect that neither advances nor inhibits religion."

William Jennings Bryan had argued that neutrality consisted of teaching neither evolution nor creationism in the schools: antievolution laws removed evolution from the curriculum so that students would not be exposed to what some considered an antireligious doctrine. As evolution returned to textbooks and to the curriculum, creationists protested that the classroom was no longer neutral. To restore neutrality, they argued, both evolution and creationism should be taught. Even before the *Epperson* decision struck down antievolution laws, parents Nell Segraves and Jean Sumrall petitioned the California Board of Education in 1963 to restore neutrality to the classroom by adding creationism to the curriculum if evolution were taught.

A Movement Builds. Henry Morris's original approach to promoting creation science was to reach out to the scientific and educational communities: Morris conceived of the ICR as a research and educational institute that eventually would persuade the academic community of the value of creation science. He believed that after scientists, educators, and the public understood creation science, the subject would trickle down to school science curricula. Other creationist organizations (such as Nell and Kelly Segraves's Creation Science Research Center) sought to promote creationism through political action; Morris, perhaps because of his background as a former college professor, preferred to work through education (Numbers 2006).

Scientists and educators, however, ignored creation science. The top-down approach wasn't working, so ICR shifted its strategy toward the grass roots. Although never embracing the Creation Science Research Center's approach of filing lawsuits to force the teaching of creationism, the ICR nonetheless encouraged citizens to take an active role in promoting creation science at the local level. In ICR's publication *Impact*, the lawyer Wendell Bird encouraged local citizens to present school boards with "resolutions" encouraging the teaching of creation science in science curricula (Bird 1979).

The model resolution laid out the definition of creation science: "special creation from a strictly scientific standpoint is hereinafter referred to as 'scientific creationism'" (Bird 1979: ii). It claimed that the presentation of only evolution in the classroom "without any alternative theory of origins" was unconstitutional "because it undermines their [students'] religious convictions," it would require students to attest to course materials they did not believe in, and it "hinders religious training by parents" (Bird 1979: ii). The resolution claimed that evolution-only teaching would promote belief systems such as "religious Liberalism, Humanism, and other religious faiths." It claimed that the "theory of special creation is an alternative model of origins at least as satisfactory as the theory of evolution and that theory of special creation can be presented from a strictly scientific standpoint without reference to religious doctrine" (Bird 1979: ii). School districts were then urged to give "balanced treatment to the theory of scientific creationism and the theory of evolution" in all aspects of the curriculum, including classroom time, textbook contents, and library materials.

Even before the ICR model resolution appeared, the conservative Christian layman Paul Ellwanger had submitted his own resolution to the Anderson, South Carolina, school district, proposing a "balanced treatment of evolution and creation in all courses and library materials dealing in any way with the subject of origins" (Institute for Creation Research, 1979: ii). Feedback between the Ellwanger and the ICR resolutions resulted in Ellwanger preparing sample legislation for districts or states to pass.

The legislation presented two alternative—and allegedly scientifically equivalent—views of "origins": evolution science and creation science, both of which should be taught to maintain a "balanced" curriculum. If evolution were taught, schools would be required also to teach creation science. Inspired by his efforts, a movement began to introduce Ellwanger bills in state legislatures. The late 1970s campaign to promote equal time for creation science legislation was truly a grassroots effort in the classic American tradition. The campaign spread largely by word of mouth and did not yet have the blessing or resources of national religious denominations or religiously oriented political organizations such as the Moral Majority. Although legislators in Ellwanger's home state of South Carolina failed to pass an Ellwanger bill, legislation soon began appearing in other states.

By the early 1980s, equal time legislation had been introduced in at least twenty-seven states, including Alabama, Colorado, Florida, Illinois, Indiana, Iowa, Louisiana, Missouri, Nebraska, Oklahoma, Oregon, South Carolina, Texas, and Washington (Moyer 1981: 2), Georgia, Kentucky, Minnesota, New York, Ohio, Tennessee, West Virginia, and Wisconsin (American Humanist Association 1981: back cover), Maryland (Weinberg 1981b: 1), Arkansas (Weinberg, 1981a), Mississippi, Arizona, and Kansas (Weinberg 1982: 1). All died in committee, except for those in Arkansas and Louisiana. Many scientists and educators were involved in campaigns to prevent the passage of equal time legislation. Creation science finally was receiving attention from scientists, though not the kind Henry M. Morris had desired.

McLean v. Arkansas. That Arkansas in 1981 was the first state to pass a creation and evolution equal time bill has an ironic twist: as discussed earlier in this chapter, in 1968, Arkansas had been the site of the Supreme Court case that struck down Scopes-era antievolution laws. Now it was to be in the spotlight again as the site of the first challenge to equal-time legislation. Arkansas Act 590 proposed "balanced treatment" for "evolution-science" and "creation-science," defining creation science as follows:

1. Sudden creation of the universe, energy, and life from nothing;
2. The insufficiency of mutation and natural selection in bringing about development of all living kinds from a single organism;

3. Changes only within fixed limits of originally created kinds of plants and animals;
4. Separate ancestry for man and apes;
5. Explanation of the earth's geology by catastrophism including the occurrence of a worldwide flood; and
6. A relatively recent inception of the earth and living kinds.

The act defined evolution science as follows:

1. Emergence by naturalistic processes of the universe from disordered matter and emergence of life from nonlife;
2. The sufficiency of mutation and natural selection in bringing about the development of present living kinds from simple earlier kinds;
3. Emergence by mutation and natural selection of present living kinds from simple earlier kinds;
4. Emergence of man from a common ancestor with apes;
5. Explanation of the earth's geology and the evolutionary sequence by uniformitarianism; and
6. An inception several billion years ago of the earth and somewhat later of life.

According to Act 590, to present only evolution in the schools would create a hostile climate for religious students, undermining "their religious convictions and moral or philosophical values" and violating "protections of freedom of religious exercise and of freedom of belief and speech for students and parents" (Anonymous 1983: 18). Teaching only evolution was held to be a violation of academic freedom "because it denies students a choice between scientific models and instead indoctrinates them in evolution-science alone" (Anonymous 1983: 18). Creation science was presented as a "strictly scientific" view.

Upon passage, Governor Frank White signed Act 590, and the Arkansas ACLU immediately challenged the bill. Plaintiffs in the lawsuit included religious leaders, science education organizations, civil liberty organizations, and several individual parents. Methodist clergyman William McLean was the lead plaintiff, joined by the bishops or other spokespeople for the Arkansas Episcopal Church, the United Methodists, Roman Catholics, African Methodist Episcopalians, Presbyterians, and Southern Baptists. Also joining the suit were the Arkansas Education Association, the National Association of Biology Teachers, the American Jewish Congress, the Union of American Hebrew Congregations, the American Jewish Committee, and the National Coalition for Public Education and Religious Liberty. The presence of so many religious plaintiffs helped to defuse the argument that opposition to the bill equated to opposition to religion.

McLean v. Arkansas was tried in federal district court. The Arkansas ACLU received considerable assistance from a large New York law firm—Skadden, Arps, Slate, Meagher, and Flom—which offered its services pro bono. The Arkansas ACLU would argue that because creation science was inherently a religious idea, its advocacy as required by Act 590 would violate the Establishment Clause of the U.S. Constitution. Furthermore, because creation science was not scientific, there was no secular purpose for its teaching (Herlihy 1983). The state, defending the law, had to argue the opposite: that creation science was scientific, and thus its advocacy would have a secular purpose. The state ignored the issue of whether creation science was religious. Each side brought in witnesses to testify in favor of its position. Much time was spent in the trial over the definition of science and whether creation science fulfilled that definition.

The state was reluctant to put Henry Morris or any other ICR spokesperson on the stand, notwithstanding the prominence of these people in the creationism movement (Larson 2003: 162). Because of the Christian apologetic nature of so much of Morris's writings, the defense was unwilling to have Morris cross-examined: it would be apparent that creation science was primarily a religious view, and the state's case would be lost from the beginning. Witnesses for the defense, therefore, consisted of other less-well-known creationists and some noncreationist scientists who questioned some aspects of evolution. An example of the latter was the British astrophysicist Chandra Wickramasinghe, who argued against the standard chemical origin-of-life model. His explanation for the origin of life on Earth was not special creation, however, but a natural explanation, having to do with the seeding of life on Earth with organic molecules from comets. When questioned about creation science, he stated that "no rational scientist could believe that the earth was less than one million years old" (Holtzman and Klasfeld 1983: 95). In the aftermath of the case, creationists blamed the state attorney general for not using the "strongest advocates" of creation science, although such a move would have been legal suicide.

The plaintiffs assembled a cast of eminent scholars—scientists, theologians, philosophers of science, sociologists, and educators—to make the case that creation science was not science but a form of sectarian religion. Included were three members of the National Academy of Sciences, the nation's most prestigious scientific organization. Press accounts attest to the articulateness and depth of knowledge of these witnesses, and to the superiority of the Skadden, Arps lawyers in cross-examination and in the general presentation of the case. The consensus was that the defense was simply outgunned. It was so apparent to the plaintiffs that their case would be successful that lawyers and witnesses had their victory party on the third day of the trial (Ruse 1984: 338).

Indeed, when the judge issued his decision, it was in favor of the plaintiffs; Act 590 was declared unconstitutional. In a strongly worded decision (*McLean v. Arkansas*, 529 F. Supp. 1255), Judge William Overton relied on a 1971 Supreme Court decision, *Lemon v. Kurtzman*, which had established three tests to determine whether a law or practice violated the Establishment Clause. The three prongs of *Lemon* are the purpose, effect, and entanglement rules.

Lemon (as did the earlier *Schempp* case) requires that the "statute must have a secular legislative purpose"; if the legislature's purpose in passing the law was to advance religion, then the law fails (*Lemon v. Kurtzman*, 403 U.S. 603 at 612–613). Judge Overton ruled that the legislative history of the law clearly demonstrated that the legislators intended to promote a religious view.

"Second, its principal or primary effect must be one that neither advances nor inhibits religion" (*Lemon* at 612). This effect prong was likewise judged to be violated by Act 590; Judge Overton decided that requiring creation science to be taught would promote a sectarian religious view, because creation science was a religious view, not a science. Much of the legal decision, in fact, was devoted to showing how creation science did not meet a general definition of science accepted by practitioners.

Lemon also states that "the activity must not foster 'an excessive government entanglement with religion'" (*Lemon* at 613). Judge Overton ruled that because the classroom must not be a place for religious proselytization, the administration would have to monitor teachers and instructional material to guard against willing or unwitting advancement of religion.

Because the *McLean* decision declared so strongly that Act 590 was unconstitutional, the state declined to appeal the case to the court of appeals. Equal time for creation science and evolution had failed in Arkansas, but a law very similar to the Arkansas law had been introduced into neighboring Louisiana only a few months before the *McLean* decision.

The Louisiana Equal-Time Law. The law that the Louisiana legislature passed in the spring of 1981 was another Ellwanger clone, with a few modifications intended to make it more likely to pass constitutional muster. The plaintiffs in McLean had been able to show that Act 590's definition of creation science paralleled biblical literalist creationism, a similarity that figured into Judge Overton's decision to strike it down. The framers of the Louisiana law, "Balanced Treatment for Creation Science and Evolution Science in Public School Instruction," sought to mount a stronger case by not defining creation science in recognizably religious terms. Again the ACLU challenged the law in federal district court, but because proponents of the law also requested an injunction, courts had to sort out jurisdictional issues, and both cases slogged through the courts for several years. Finally the federal district court heard the case. Rather than hold a full trial, as in Arkansas, the district court tried the case by summary judgment: the judge accepted written statements from both sides and decided the outcome of the case on the basis of these documents.

In 1985, the federal district court decided that the law was unconstitutional because it advanced a religious view by prohibiting the teaching of evolution unless creationism—a religious view—was also taught. The court of appeals agreed, and finally the case made its way to the Supreme Court in 1987. The highest court concurred with the lower courts: "The preeminent purpose of the Louisiana Legislature was clearly to advance the religious viewpoint that a supernatural being created humankind. . . . The Louisiana Creationism Act advances a religious doctrine by requiring either the banishment of the theory of evolution from public school classrooms or the presentation of a religious viewpoint that rejects evolution in its entirety" (*Edwards v. Aguillard*, 482 U.S. 578 at 591).

Equal time for creation science was no longer a legal option in the schools of the United States. Shortly after the filing of the *Edwards* decision, however, the creationist attorney Wendell Bird wrote an *Impact* pamphlet for the ICR in which he proposed the next strategy of antievolutionism: the repackaging of creation science so that it might survive such Establishment Clause challenges as had doomed it in Arkansas and Louisiana. The next stage of American antievolutionism, neocreationism, was beginning to evolve.

LEGAL CASES

Epperson v. Arkansas, 393 U.S. 97 (1968).
Lemon v. Kurtzman, 403 U.S. 602 (1971).
McLean v. Arkansas, 529 F. Supp. 1255 (1982).
Edwards v. Aguillard, 482 U.S. 578 (1987).

REFERENCES

American Humanist Association. 1981. Update on creation bills and resolutions. *Creation/Evolution* 2 (1): 1–44.

Anonymous. 1983. Act 590 of 1981. In *Creationism, science, and the law: The Arkansas case*, ed. M. C. La Follette, 15–19. Cambridge, MA: MIT Press.

Armstrong, Karen. 2000. *The battle for God: A history of fundamentalism*. New York: Ballantine Books.

Bird, Wendell R. 1979. Resolution for balanced presentation of evolution and scientific creationism. *Impact* (May): 4.

Carnegie, Andrew. 1889. Wealth. *North American Review* (June): 653–664.

Grabiner, Judith V., and Peter D. Miller. 1974. Effects of the Scopes trial. *Science* 185 (4154): 832–837.

Grobman, Arnold B. 1998. National standards. *American Biology Teacher* (October): 562.

Ham, K. 2008. The history of AIG-U.S. to the middle of 2008. Retrieved October 27, 2008 from http://www.answersingenesis.org/home/area/about/history.asp.

Herlihy, Mark E. 1983. Trying creation: Scientific disputes and legal strategies. In *Creationism, science and the law: The Arkansas case*, ed. M. C. La Follette, 96–103. Cambridge, MA: MIT Press.

Hileary, William, and Oren Metzger. 1990. *The world's most famous court trial: Tennessee evolution case*, 2nd reprint ed. Dayton, TN: Bryan College.

Holden, C. (2007). A journal is created. *Science* 316 (5827): 961.

Holmes, S. J. 1927. Proposed laws against the teaching of evolution. *Bulletin of the American Association of University Professors* 13 (8): 549–554.

Holtzman, Eric, and David Klasfeld. 1983. The Arkansas creationism trial: An overview of the legal and scientific issues. In *Creationism, science, and the law: The Arkansas case*, ed. M. C. La Follette, 86–96. Cambridge, MA: MIT Press.

Institute for Creation Research. 1979. Creation science and the local school district. *ICR Impact* (January): 4.

Institute for Creation Research. 1986. Ken Ham joins ICR staff. *Acts and Facts* (December): 1–4.

Institute for Creation Research. 1988a. "Back to Genesis" programs have a successful start. *Acts and Facts* (April): 2.

Institute for Creation Research. 1988b. First "Back-to-Genesis" program a resounding success; over 2000 in New England states hear creation message. *Acts and Facts* (July): 1, 5.

Institute for Creation Research. 1993a. Creationist speakers available. *Acts and Facts* (August): 5.

Institute for Creation Research. 1993b. ICR museum visited by 25,000 in first year. *Acts and Facts* (November): 1.

Larson, Edward J. 1997. *Summer for the gods: The Scopes trial and America's continuing debate over science and religion*. New York: Basic Books.

Larson, Edward J. 2003. *Trial and error: The American controversy over creation and evolution*, rev. ed. New York: Oxford University Press.

Lawrence, Jerome, and Robert E. Lee. 1960. *Inherit the wind*. New York: Bantam Books.

Marsden, George M. 1980. *Fundamentalism and American culture: The shaping of twentieth-century evangelicalism, 1870–1925*. New York: Oxford University Press.

McKenna, M. 2007. Lord of the ring. *The Australian*, June 5. Retrieved October 13, 2008 from http://www.theaustralian.news.com.au/story/0,20867,21848726-28737,00.html.

Moore, John A. 1976. Creationism in California. In *Science and its public: The changing relationship*, ed. Gerald Horton and William A. Blanpied, 191–208. Dordrecht: D. Reidel.

Moore, John A. 2002. *From Genesis to genetics: The case of evolution and creationism*. Berkeley: University of California Press.

Moore, John N., and Harold S. Slusher, eds. 1974. *Biology: A search for order in complexity*, rev. ed. Grand Rapids, MI: Zondervan.

Moore, Randy. 1998. Thanking Susan Epperson. *American Biology Teacher* (November/December) 60(9): 642–646.

Morris, H. M. 1946. *That you might believe*. Chicago: Good Books.

Morris, H. M. 1951. *The Bible and modern science*. Chicago: Moody Press.

Morris, Henry M., ed. 1974. *Scientific creationism*, public school ed. San Diego: Creation-Life Publishers.

Moyer, Wayne. 1981. Legislative initiatives. *Scientific Integrity* (June): 1–4.

Numbers, Ronald. 2006. *The creationists*. Cambridge: Harvard University Press.

Rasche, M. D. 2002. To God be the glory! *Acts and Facts* 31 (8): 1–2.

Ruse, Michael. 1984. A philosopher's day in court. In *Science and creationism*, ed. A. Montagu, 311–342. New York: Oxford University Press.

Skoog, Gerald. 1978. Does creationism belong in the biology curriculum? *American Biology Teacher* 40 (1): 23–29.

Skoog, Gerald. 1979. The topic of evolution in secondary school biology textbooks: 1900–1979. *Science Education* 63: 621–640.

Stewart, M. 2007. A decade for "Dr. Dino"; Kent Hovind gets 10 years for violating federal tax law. *Pensacola News-Journal*. January 20, p. A1.

Weinberg, Stanley. 1981a. *Memorandum to liaisons for committees of correspondence*, 1 (April): 1–4.

Weinberg, Stanley. 1981b. *Memorandum to liaisons for committees of correspondence*, 1 (November): 1–4.

Weinberg, Stanley. 1982. *Memorandum to liaisons for committees of correspondence*, 2 (February): 1–4.

Whitcomb, John C. Jr., and Henry M. Morris 1961. *The Genesis Flood: The biblical record and its scientific implications*. Phillipsburg, NJ: Presbyterian and Reformed.

CHAPTER 6

• • • • • • • • • • • •

Neocreationism

In 1987, the Supreme Court decision *Edwards v. Aguillard* struck down a Louisiana law requiring equal time for creationism and evolution. Creationism is a religious idea, said the Court, and the First Amendment prohibits the government from promoting religion. Antievolution strategies subsequently were developed that avoided the use of any form of the words *creation*, *creator*, and *creationism*. In effect, proponents shifted their strategy from proposing to balance evolution with creation science to proposing to balance evolution with creation science in other guises. Antievolutionists proposed the teaching of "scientific alternatives" to evolution or evidence against evolution— avoiding referring to such purported disciplines as creationism. A school district in Louisville, Ohio, which had an equal time for evolution and creation science regulation in place before *Edwards*, rewrote the science curriculum to "avoid mention of creationism in its curriculum guide, calling it alternative theories to evolution and adding it to the science classes" (Kennedy 1992). The avoidance of creation science terminology and the development of creation science–like alternatives to evolution, plus the renaming of the content of creation science as evidence against evolution, constitute what I call neocreationism, which continues into the twenty-first century. These approaches were encouraged by creationist interpretations of the *Edwards v. Aguillard* decision.

THE EDWARDS DECISION AND NEOCREATIONISM

Justice William Brennan wrote the *Edwards v. Aguillard* Supreme Court decision striking down Louisiana's balanced-treatment law. Seven justices signed the decision; Justice Antonin Scalia wrote a dissent joined by Chief Justice William Rehnquist. *Edwards v. Aguillard* was argued more narrowly than the earlier *McLean v. Arkansas* decision, which had struck down Arkansas's equal-time law after finding that it violated all three prongs of *Lemon*. *Edwards* declared that the Louisiana equal-time law violated the first prong (purpose) of *Lemon* and did not extend the argument to the other prongs (as had *McLean*). Another difference between the two cases appeared in the treatment of creation science as science: *McLean*—after a full trial—declared that creation science failed as science. *Edwards*—decided on a summary judgment—did not take a stand on whether creation science qualified as science. Instead, *Edwards* ignored whether creation science was science and went straight to the Establishment Clause's requirement that schools be religiously neutral. Because creation science was a form of creationism, the court declared that it was unconstitutional to advocate it in the public schools:

> The Act impermissibly endorses religion by advancing the religious belief that a supernatural being created humankind. The legislative history demonstrates that the term "creation science," as contemplated by the state legislature, embraces this religious teaching. The Act's primary purpose was to change the public school science curriculum to provide persuasive advantage to a particular religious doctrine that rejects the factual basis of evolution in its entirety. Thus, the Act is designed either to promote the theory of creation science that embodies a particular religious tenet or to prohibit the teaching of a scientific theory disfavored by certain religious sects. In either case, the Act violates the First Amendment. (*Edwards v. Aguillard* 482 U.S. 578 (1987) at 579)

The *Edwards* decision, however, suggested loopholes that creationists could, and did, seize on. One was a statement recognizing the extant ability of teachers to "supplant the present science curriculum with the presentation of theories, besides evolution, about the origin of life." Such theories, however, had to be secular and not religious: "We do not imply that a legislature could never require that scientific critiques of prevailing scientific theories be taught. ... In a similar way, teaching a variety of scientific theories about the origins of humankind to schoolchildren might be validly done with the clear secular intent of enhancing the effectiveness of science instruction" (*Edwards* at 593–594).

This wording encouraged antievolutionists to argue for the teaching of scientific alternatives to evolution.

SCIENTIFIC ALTERNATIVES TO EVOLUTION: ABRUPT APPEARANCE THEORY

Creation science is of course the original scientific alternative to evolution, but it had been identified as a religious view in both *McLean* and *Edwards* and could not constitutionally be advocated in the public schools. The lawyer Wendell Bird, who had advised the Institute for Creation Research on legal matters and who was appointed to argue Louisiana's position before the Supreme Court in the *Edwards* case, proposed a new scientific alternative to evolution that he claimed was distinct from creation science. His view, which he dubbed "abrupt appearance theory," was, however, indistinguishable in content from creation science.

While a graduate student at Yale University in the mid-1970s, Bird had written an article for the *Yale Law Journal* arguing for the constitutionality of teaching creation science. It was this article and Bird's later work as staff attorney for ICR that shaped the argument that creation science was a legal alternative to evolution, which, as a supposedly purely scientific position, could be taught without violating the Establishment Clause. Although this argument was unpersuasive to judges in both the *McLean* and the *Edwards* cases, both the district and state supreme courts recognized that it is indeed legal to teach a secular, nonreligious, truly scientific alternative to evolution. Although neither courts nor scientists have recognized such an alternative, Bird's abrupt appearance theory was the creationist's first public post-*Edwards* attempt to formulate such an alternative.

The phrase *abrupt appearance* in fact was part of the creationist's definition of creation science in the *Edwards v. Aguillard* case. Creation science was defined in *Edwards* as including "the scientific evidences for creation and inferences from those scientific evidences" (at 578), but also including "origin through abrupt appearance in complex form" (at 579). Bird reworked his brief for the *Edwards* case into a two-volume publication, *The Origin of Species Revisited*, published in 1987. Abrupt appearance theory was held to be the scientific evidence for the sudden appearance of all living things—in fact, the entire universe—in essentially its present form. No material or transcendent agent was identified as causing this event; Bird was meticulous in avoiding any references that could be interpreted as religious and would therefore expose abrupt appearance theory to the same First Amendment challenges as creation science.

Consciously attempting to distance himself from religious creationism, Bird identified two scientific alternatives to explain "origins": evolution and abrupt appearance theory. Evolution was defined broadly as encompassing cosmological (stellar) evolution, biochemical evolution (the origin of life), and biological evolution (the common ancestry of living things) (Bird 1987a: 17). Abrupt appearance theory contrasts sharply with the continuous unfolding of the universe expressed in evolutionary theory: "The theory of abrupt appearance involves the scientific evidence that natural groups of plants and animals appeared abruptly but discontinuously in complex form, and also that the first life and the universe appeared abruptly but discontinuously in complex form" (Bird 1987a: 13).

The essence of abrupt appearance theory, therefore, is discontinuity: stars and galaxies appear abruptly, and life and groups of living things appear abruptly, much as in the religious view of special creation. Abrupt appearance theory thus encompasses creation science and other religious views—though it is claimed to have a "totally empirical basis" (Bird 1987a: 13): "This theory of abrupt appearance is different from the theories of creation, vitalism, panspermia, and similar concepts. Discontinuous abrupt appearance is a more general theory and a more scientific approach than scientific views of creation, vitalism, or panspermia, although they can be formulated as submodels of abrupt appearance" (Bird 1987a: 20).

Although mammoth in its scope (its two volumes purport to summarize scientific, pedagogical, philosophical, and legal aspects of the creationism versus evolution debate) and prodigious in the number of citations from both the scientific and the creationist literature, *The Origin of Species Revisited* is rarely cited today in creationist literature. It was, and remains, ignored in the scientific literature, and after the mid-1990s it virtually disappeared from the political realm as well. It has been supplanted by another "alternative" to evolution that was evolving parallel to it and that expresses some of the same ideas.

SCIENTIFIC ALTERNATIVES TO EVOLUTION:
INTELLIGENT DESIGN

Intelligent design (ID) is a movement that began a few years before the *Edwards* decision and solidified in the few years after it. Like creation science and abrupt appearance theory, ID is presented as a scientific alternative to evolution, and it has been more successful than creation science in appealing to Christians who are not biblical literalists.

The Origin of Intelligent Design

Intelligent design creationism dates from the publication of *The Mystery of Life's Origin* (Thaxton, Bradley, and Olsen 1984). Thaxton, Bradley, and Olsen proposed that the origin of life not only was currently unexplained through natural causes but also could not be explained through natural causes. As the biologist Dean H. Kenyon wrote in the introduction, "It is fundamentally implausible that unassisted matter and energy organized themselves into living systems" (Thaxton et al. 1984: viii). The essential scientific claim of ID was made clear from the very beginning: some things in biology are categorically unexplainable through natural causes.

Encouragement for *The Mystery of Life's Origin* came from Jon Buell, a former campus minister who became president of the Dallas-based conservative Christian organization the Foundation for Thought and Ethics (FTE). He recruited the historian and chemist Charles Thaxton, the engineer Walter Bradley, and the geochemist Roger Olsen to write a document on scientific difficulties concerning the origin of life, which became *The Mystery of Life's Origin*. Buell, Thaxton, Bradley, Olsen, and others, many of whom were associated with the FTE, proposed a new form of creationism that did not rely directly on the Bible: there were no references to a universal flood, to the special creation of Adam and Eve or any other creature, or to a young Earth. But paralleling creation science, *Mystery* emphasized supposedly scientific problems of evolution. *Mystery* mostly stuck to science, with only brief references in an epilogue to the necessity of intelligence being involved in the origin of life. Much as had Bird in proposing abrupt appearance theory, the authors were agnostic on the identity of the creative agent. They offered the suggestion of Hoyle and Wickramasinghe (1979) that life on Earth possibly was produced by extraterrestrials of high intelligence, although the authors expressed their preference for creation by God.

The next major ID product to emerge was again from FTE: the 1989 high school biology supplementary textbook *Of Pandas and People*, written by the biologists Percival Davis and Dean Kenyon. Originally titled *Biology and Origins*, the book was submitted to secular publishers for more than two years before one was found—a small Texas press that specialized in agricultural materials (Scott 1989). By this time, the nascent movement had settled on the phrase *intelligent design* for its position and this term appeared in *Pandas*.

Although *Pandas* soon was proposed for adoption as an approved (and thus purchasable using state funds) textbook in at least two states (Idaho and Alabama), its supporters were unsuccessful. Publishers have claimed that it was in use in several school districts, but it has not been possible to verify this claim. In general, though, it would be safe to say that *Pandas* made little splash in the world of science education. The ID movement remained largely unnoticed until the publication in 1991 of *Darwin on Trial* by University of California, Berkeley, law professor Phillip Johnson.

Because previously the antievolution movement had been based in small, nonacademic, nonprofit organizations such as ICR and FTE, the publication of an antievolution book by a tenured professor at a major secular university came as a surprise to the educated public. Although the scientific community ignores books by Henry Morris, a few scientists reviewed *Darwin on Trial* in popular publications such as *Scientific American*, and discussions of this new form of antievolutionism appeared in the popular press. Scientists uniformly criticized what they considered uninformed science in Johnson's book. On the other hand, educated conservative Christians, for whom creation science was unacceptable because of its often-outlandish scientific claims, found Johnson's message very attractive indeed. Largely because of its more respectable academic associations, ID obtained considerably more coverage in the popular media than did creation science—though the latter then as well as now boasts many more organizations and activists than does the ID movement.

What, in detail, are the claims of ID? As is the case with creation science, ID combines a scholarly focus with cultural renewal—an effort to promote a sectarian religious view.

The Scholarly Focus of Intelligent Design

Intelligent design proponents posit that the universe, or at least components of it, have been designed by an "intelligence." They also claim that they can empirically distinguish intelligent design from design produced through natural processes (e.g., natural selection). This is done through the application of two complementary ideas, one promoted by a biochemist and the other by a philosopher-mathematician.

Irreducible Complexity. The biochemist Michael Behe contends that intelligence is required to produce irreducibly complex cellular structures (ones that couldn't function if a single part were removed) because such structures could not have been produced by the incremental additions of natural selection (Behe 1996).

Critics of Behe have pointed out that it is not clear that irreducibly complex structures actually exist—except perhaps by definition. Critics have also argued that the examples Behe gives of irreducibly complex structures can often be reduced and still be functional. Behe commonly uses a mousetrap as his example of an irreducibly complex structure, claiming that if any one of the five basic parts of a mousetrap (platform, hammer, spring, catch, and hold-down bar) is removed, it can no longer catch mice. Scientists gleefully set about producing four-part, three-part, two-part, and even one-part mousetraps to demonstrate the reducibility of Behe's prime example of an irreducibly complex structure.

Similarly, supposedly irreducibly complex biochemical structures such as a bacteria's flagellum can function with fewer parts than Behe originally claimed in *Darwin's Black Box*. Ultimately, of course, it is possible to reduce a structure to so few parts that, indeed, removal of any one part will make the structure cease functioning. More important than whether irreducibly complex structures actually occur other than by

Figure 6.1
An Irreducibly Complex Mousetrap. A mousetrap
has five parts. Behe contends that all five pieces
need to be assembled or it is impossible to catch
mice, hence the mousetrap is irreducibly complex.
Courtesy of Sarina Bromberg.

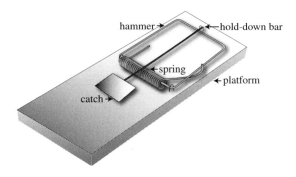

definition, however, is the critical question of whether they can be produced by natural mechanisms.

Behe answers no, claiming that natural selection, the main mechanism of evolutionary change, is inadequate to the task. He views natural selection as assembling a complex structure by stringing together components one at a time, with each addition requiring a selective advantage. Behe's view is that for a structure like the bacterial flagellum, consisting of more than forty proteins and enzymes, it is extraordinarily unlikely that so many elements could be assembled—by chance—one by one, and even more unlikely that there would be selective advantage to each addition. This piece-by-piece assemblage of the flagellum, one enzyme at a time, one after another, is also envisioned by William Dembski (2001), who claims that the probability against this occurring is astronomical; a bacterial flagellum, Dembski and Behe agree, cannot be produced through natural causes.

Critics have noted that Behe presents an incomplete picture of how natural selection operates: it is not the case that components of a complex structure must be added one after another, piece by piece, like stringing beads. It is clear from the study of components of a cell that a great deal of borrowing and swapping of bits and pieces takes place: each cellular structure is not composed of unique proteins and enzymes, or even of wholly unique combinations of proteins, and this is also true of the proteins composing the bacterial flagellum. In a recent analysis of the structure and origin of the flagellum, the authors note, "Three modular molecular devices are at the heart of the bacterial flagellum: the rotorstator that powers flagellar rotation, the chemotaxis apparatus that mediates changes in the direction of motion and the T3SS that mediates export of the axial components of the flagellum. *In each module, the apparatus is fashioned from recycled parts that occur elsewhere in nature*" (Pallen and Matzke 2006; emphasis added).

The cross-linking proteins of flagella, for example, have other functions elsewhere in the cell. Somewhat fewer than half of the proteins found in a bacteria's flagellum are the same as or very similar to those found among other bacteria in a structure called the type-III secretory apparatus, which performs some of the functions of a flagellum. An adaptive advantage for a structural element may exist and cause it to be selected for—but for a different purpose and perhaps in a different cell component than that of the final, supposedly irreducibly complex structure under discussion. Natural selection thus can produce complex structures without having to separately string protein after protein together, in which each addition requires a separate action of natural selection. Behe's critics thus have argued that some of the components of an irreducibly complex structure could be assembled separately for some purposes and then combined for other functions. This is not through a random or chance process: natural selection is intimately involved during all stages.

There is another way in which natural selection can be more flexible than Behe appears to allow. So-called irreducibly complex structures may indeed have been assembled piece by piece, but the pathway of assembly may not be obvious because of a process evolutionary biologists have called scaffolding.

Consider an arch made of stone. The keystone, the stone at the top of the arch, must be in place or the arch will fall down; an arch is an irreducibly complex structure. To build it, stonemasons will often build a scaffolding to support the sides of the arch as they build toward the center. When the keystone is laid and the arch is thereafter stable, the scaffolding is removed, leaving the irreducibly complex arch.

An irreducibly complex biochemical or molecular structure may be built in a similar way, in the sense that at an earlier time in the history of the structure, there might have been components that supported the function of the structure, much as a scaffolding supports the sides of an arch until the keystone is in place. These supporting biochemical components may be made redundant by the addition of more efficient components, much as the arch's scaffolding is made redundant for holding up the sides of the arch once the keystone is laid. The now-superfluous components can be removed by natural selection. Without knowing the entire history of the structure, it might seem that all the parts of the structure appeared all at once, fully formed and functional in their final configuration, with no history of earlier, simpler structural predecessors. But just as an arch attains irreducible complexity only at the end of construction, so too do these supposedly irreducibly complex biochemical structures: they actually had a history, though the exact events may not yet have been traced. Evolutionary biologists have proposed a number of similar ways in which a complex structure could be formed (Pond 2006).

But even if natural selection were unable to explain the construction of irreducibly complex structures, does this mean that we must now infer that intelligence is required to produce such structures? Only if there are no other natural causes—known or unknown—that could produce such a structure. Given our current knowledge of the mechanisms of evolution, there is no reason natural selection cannot explain the assembly of an irreducibly complex structure—but it is also the case that a future researcher might come up with an additional mechanism or mechanisms that can explain irreducibly complex structures by some other natural process.

Some scientists have described Behe's approach as an "argument from ignorance" (Blackstone 1997: 446) because the intelligent creator is used as an explanation when a natural explanation is lacking. This is reminiscent of the God-of-the-gaps argument, in which God's direct action is called on to explain something that science has not yet explained. Both theologians and scientists reject God-of-the-gaps arguments. To scientists, using God to explain natural phenomena of any kind violates the practice of methodological naturalism, in which scientific explanations are limited only to natural causes. To theologians, the God-of-the-gaps approach creates theological problems of the irrelevance or diminution of God when natural explanations for natural events ultimately replace the direct hand of God. Intelligent design proponents, however, claim that the issue is not current ignorance of a discoverable natural cause but the impossibility of a natural cause.

Behe's idea of irreducible complexity was anticipated in creation science; much as in Paley's conception, creation science proponents hold that structures too complex to have occurred "by chance" require special creation (Scott and Matzke 2007). Behe, following ID convention, doesn't mention God directly, but the logical consequence of the irreducible complexity argument is that irreducibly complex structures—unable to be produced by natural causes—are evidence for God's direct action. As such, ID verges on being a variety of progressive creationism in which God intervenes at intervals to create irreducibly complex structures like DNA, bacterial flagellum, the blood-clotting cascade, and so on. Although many ID proponents find the progressive creationist position attractive, ID is not necessarily wedded to the progressive creation position. Some of its proponents suggest that design could have been prearranged (or "front-loaded"). In *Darwin's Black Box*, Behe suggests that perhaps all the irreducibly complex structures of all living things were somehow present in the first living cell, and then appeared through time as various organisms evolved. "Suppose that nearly four billion years ago the designer made the first cell, already containing all of the irreducibly complex biochemical systems discussed here and many others. (One can postulate that the designs for systems that were to be used later, such as blood clotting, were present but not 'turned on.' In present-day organisms plenty of genes are turned off for a while, sometimes for generations, to be turned on at a later time.)" (Behe 1996: 228).

As noted by the cell biologist Kenneth Miller, such an übercell would somehow have to avoid the mutational drift of genes controlling such structures for billions of years until it was "time" for such structures to appear (Miller 1996: 40). The probability of genes for, say, the bacterial flagellum remaining intact for so long violates much of what we know about the behavior of genes in the absence of natural selection. Such genes tend to accumulate mutations that make the gene nonfunctional.

William Dembski has expanded this concept and proposes that God might have front-loaded everything in the universe in the Big Bang—all the irreducibly complex structures are merely unfolding like so many homunculi as time passes (Dembski 2001). In this view, God would not be progressively creating but would have acted only once.

Complex Specified Information: The Design Inference. Dembski's design inference takes a probability theory approach to distinguish those phenomena in nature that are designed by intelligence from those that are the result of natural causes or chance. Although arguments against evolution based on probability have long been a mainstay in creation science (Gish 1976; Morris 1974; Perloff 1999), Dembski's design inference is at least superficially more impressive, couched as it is in a mathematical idiom. In proposing an explanatory filter decision tree, Dembski contends that there are three ways to explain phenomena on the basis of their frequency of occurrence (see Figure 6.2).

Things that occur commonly or with high predictability can be attributed to the unfolding of natural laws. That the moon goes through phases every month can be explained by the passage of the moon around Earth and the changing angle between the moon and sun, as we see here on Earth; it is not necessary to attribute design to this phenomenon. Phenomena that occur at intermediate probability can be attributed to chance—even very low-probability events will occur some of the time, just by chance alone. But some kinds of low-probability phenomena—Dembski refers to them as specified low-probability events—that are not due to law or to chance compose the class of phenomena that must be attributed to intelligent design. Dembski proposes that complex specified information distinguishes intelligently designed phenomena.

Specification is a sort of side information that we add about a phenomenon or event. Consider the explanation for finding an arrow in a bull's-eye. If we see an arrow in a bull's-eye, we might consider that the archer got lucky, but if we see ten arrows in ten bull's-eyes, we attribute this to an archer with a high level of skill. On the other hand, if we knew that the archer shot the arrows first, and then drew the targets around them, we would not attribute the perfect shots to skill. Knowing that the targets were present before the arrows is a specification or additional information that allows us to attribute the arrows in the bull's-eyes to design rather than chance (or cheating).

Dembski's filter (Figure 6.2) allows the assignment of the causes of some phenomena to natural law, chance, or design, using the combination of probability and specification. The natural law filter stops high-probability events; medium or low, unspecified-probability events are attributed to chance, and only low, specified-probability events are attributed to intelligent design. Dembski's filter is therefore an elimination algorithm: something is explained by design when it is not explained by law or chance. But this approach allows false positives where something is attributed to design because of missing or unknown information at the first, natural law level.

For example, let's say that while walking through the forest, we come upon a circle of toadstools that has sprung up overnight.[1] The ring wasn't there yesterday, and in a few days it will largely be in tatters: fungi are fragile things. If this walk were taking place in the ninth century in Europe, as peasants we would recognize the circle of toadstools as a fairy ring, a location indicating the presence of fairies the night before. Applying Dembski's filter, we would conclude that the sudden and random appearance of the circle, and of course its symmetrical shape, certainly were not the result of natural processes: rings of toadstools crop up with no warning, unlike the phases of the moon. So in the year 800, a fairy ring would pass through the first (regularity) filter. It would also not be attributed to chance, as the likelihood of a fairy ring occurring at a given place is very improbable. However, this low-probability event has a specification, its circular shape. Therefore, following Dembski's filter, we would attribute the appearance of a fairy ring to ID; European peasants of the year 800 knew that fairy rings were the remains of midnight revels held by tiny fairies in the woods.

Perhaps because no one ever found tiny beer cans next to the toadstools, eventually a natural explanation was found for fairy rings: they are the result of one of the ways toadstools reproduce themselves. These fungi send out underground, threadlike mycelia from a central point, and when circumstances of moisture and temperature

Figure 6.2
Dembski's Explanatory Filter. William Dembski proposes that design can be detected by eliminating regularity and chance. Small probability events that are specified are the result of design, according to Dembski (1998: 37). Courtesy of Alan Gishlick.

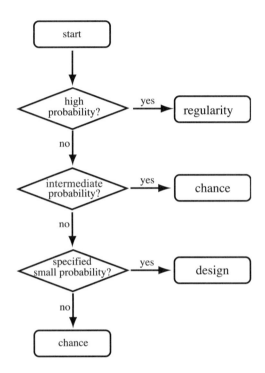

are suitable—and when two mycelia mate—toadstools form aboveground. These toadstools produce spores that are carried by the wind and eventually land and start growing new fungi. Because the mycelia radiate from a center, circles of toadstools are formed. With additional information, it can be seen that fairy rings actually are not improbable, though many variables are involved in their appearance, and they actually do not appear randomly but in specific environments. In the twenty-first century, we recognize that fairy rings have a natural explanation; in the ninth century, the circles were explained by design. Because Dembski's filter depends on the extent of scientific knowledge of the time, it thus fails to be a reliable predictor of design by intelligence (Wilkins and Elsberry 2001).

The Problem of Natural Intelligent Cause. As shown in Table 6.1, ID proponents contend that design can be produced both by natural causes (e.g., natural selection has some limited ability to shape organisms to meet some environmental pressures) and by intelligent causes. An intelligently designed phenomenon could be the product of transcendent intelligence such as a creator God, or it could be the product of material

Table 6.1
Natural and Intelligent Causes: The Intelligent Design View

Natural causes	Intelligent causes
Natural selection	Transcendent agent (God)
	Natural (material) agents: Humans, Higher primates, Extraterrestrials

agents such as extremely intelligent extraterrestrials—an argument first made, in fact, in the original ID book *The Mystery of Life's Origin.*

Unfortunately, the dichotomy between natural and intelligent is artificial, because some of the agents on the intelligent side are actually natural. To a scientist, anything that is the result of matter, energy, and their interactions is a natural phenomenon, whether nonliving phenomena such as stars and rocks or living phenomena such as plants and animals. Material agents such as humans, higher primates, or extraterrestrials (if such beings exist) are therefore natural, as are their behaviors. No one disputes that the behavior of humans and animals can be studied and understood through the application of scientific principles; such behavior is the subject matter of physiology and psychology. We study bird or whale communication, for example, and attempt to explain it by using general theories about regularities of these behaviors. Behavior is the product of natural entities and is thus itself natural.

To answer whether the *intelligent* behavior of material beings can similarly be studied and explained requires a definition of intelligent behavior—which is not as easy as it seems. Psychologists define intelligence broadly, as having elements of problem solving and some degree of abstraction. But problem solving is also a broad category: bees solve the problem of communicating the location of nectar through their waggle dance, which surely is a complex behavior having elements of abstraction (the dance indicates the direction and distance from the hive to the food source), but is this behavior intelligent? Intelligent behavior is usually conceived of as having some element of choice involved rather than as the result of largely uncontrolled or genetically wired causes—yet clearly choice is a continuum. A bee may be largely hardwired to return to the hive when a source of pollen is discovered and to perform the waggle dance: her genes make it extremely likely that she will respond to a food source by returning directly to the hive. But what if, on the way home, the bee finds a larger source of nectar? Can she make a choice not to report on the original source but to bring back a message about the second one?

In other instances of behavior, choices clearly are made. A chimpanzee attains a high or low social status through complex interactions with many individuals over a long period of time. Such actions are not genetically hardwired like the waggle dance of the bee, and in fact even involve examples of conscious manipulation of other group members, including efforts to deceive one other (especially over food sources). When primatologists and psychologists study such intelligent behaviors, they attempt to explain them through theoretical principles—in other words, they study them scientifically. Certainly economists, psychologists, and political scientists also study the intelligent (and sometimes unintelligent!) behavior of human beings and attribute

Table 6.2
An Alternative View of Natural and Intelligent Causes

Natural causes	Transcendent causes
Natural selection	Transcendent agent (God)
Natural (material) agents: Humans, Higher primates, Extraterrestrials	

all or part of it to patterns. The intelligent behavior of material creatures is therefore natural and an appropriate subject for scientific investigation.

Therefore, all of the natural intelligent agents on the right side of Table 6.1 should be moved to the left side. If intelligence produced by material beings is moved to the natural side of the equation, as in Table 6.2, the ID dichotomy of natural and intelligent must be restructured. What ID proponents wish to label "intelligent" reduces to one agent: God. The natural-intelligent dichotomy is in reality a natural-transcendent dichotomy. This is not only more empirically accurate but also a more logically satisfying relationship: if there is a transcendent, omniscient, and omnipotent agent such as God, then such an agent by definition could not be explained by natural causes and more properly would form a dichotomy with natural cause.

Intelligent design supporters cannot accept this, however, because appealing to transcendent causes is, of course, a form of religion. They are well aware that the First Amendment does not allow the advocacy of creationism in public schools. Hence it better suits the ID movement to try to combine all forms of intelligent cause into one heterogeneous list—whether or not such a division is empirically or logically defensible.

Intelligent design supporters are hostile to methodological materialism and propose a new kind of science: theistic science. This is an alleged subclass of science concerned with those scientific problems that deal with origins (i.e., origins science), which are unrepeatable. Such phenomena as the origin of life and the evolution of living things (unspecified) constitute origins science. Although the majority of science may be performed in a methodologically materialistic fashion, explaining only with reference to natural causes, origins science allows (indeed, requires) the occasional intervention of intelligence—by which is meant the direct hand of God. Theistic science, then, is a proposal to radically change how we do science by abandoning methodological materialism in favor of allowing explanation by supernatural causes—and still call the process *science*. It is not a position that either philosophers of science or scientists have embraced (Pennock 1999). Theistic science leads to the second focus of the ID movement, an effort to promote a sectarian religious view.

ID's Cultural Renewal Focus

According to Dembski, "Two animating principles drive intelligent design. The more popular by far takes intelligent design as a tool for liberation from ideologies that suffocate the human spirit, such as reductionism and materialism. The other animating principle, less popular but intellectually more compelling, takes intelligent design as the key to opening up fresh insights into nature" (Dembski 2002: 1).

Whether or not ID actually has opened fresh insights into nature, there is no doubting the popularity of what ID proponents call cultural renewal. In this focus of ID, the movement seeks to replace the alleged philosophical materialism of American society with a theistic (especially Christian) religious orientation. Perhaps the most vocal proponent of the cultural renewal focus of ID is the retired law professor Phillip Johnson. Although his first antievolution book, *Darwin on Trial*, made only a few references to the purported evils of materialism in American society, subsequent books have been much more evangelical in tone and have strongly and clearly promoted the ID vision for a society with more theistic sensibilities. Conferences (such as "Mere Creation" in 1996) have also promoted sectarian Christian views. Under Johnson's guidance—and taking advantage of his prominence and connections as a professor holding an endowed chair at a leading secular university—the ID movement sought to find acceptance first and foremost from the secular academic community. Within a few years of the publication of *Darwin on Trial*, the rapidly expanding ID movement found a new institutional locus beyond the FTE at the conservative think tank Discovery Institute in Seattle. Perhaps proponents believed that the new ID movement would have more credibility with academics if it were housed in a more neutral institution than the FTE, which has long been associated with evangelical Christianity and thus with creation science. The Discovery Institute rapidly replaced the FTE as the hub for ID activities during the 1990s.

The Discovery Institute was founded by the politician Bruce Chapman in 1991 and "promotes ideas in the common sense tradition of representative government, the free market and individual liberty" (Discovery Institute mission statement at http://www.discovery.org/about.php). In 2008, Discovery Institute programs included Technology and Democracy, Cascadia (regional transportation), Bioethics, Russia, and Science and Culture. Minor programs touched on education, the environment, religion liberty and public life, and C. S. Lewis and public life.

The ID-promoting Center for Renewal of Science and Culture (CRSC) originally was announced in a 1996 press release: "For over a century, Western science has been influenced by the idea that God is either dead or irrelevant. Two foundations recently awarded Discovery Institute nearly a million dollars in grants to examine and confront this materialistic bias in science, law, and the humanities. The grants will be used to establish the Center for the Renewal of Science and Culture at Discovery, which will award research fellowships to scholars, hold conferences, and disseminate research findings among opinion makers and the general public" (Chapman 1996a).

In the Discovery Institute's members' newsletter, Chapman further described the CRSC as having specific religious goals: "The more you read about the program—and there will be about six books to read from this center in the next four years—the more you will realize the radical assault it makes on the tired and depressing materialist culture and politics of our times, as well as the science behind them. Then, when you start to ponder what society and politics might become under a sounder scientific dispensation, you will become truly inspired" (Chapman 1996b).

The goals of the CRSC have been identified as explicitly religious in other Discovery Institute publications as well: "To defeat scientific materialism and its destructive moral, cultural and political legacies. To replace materialistic explanations with the theistic understanding that nature and human beings are created by God" (Discovery Institute 2003). Also, "Accordingly, our Center for the Renewal of Science and

Culture seeks to show that science supports the concept of design and meaning in the universe—and that that design points to a knowable moral order" (Chapman 1998: 3).

Until August 2002, the cultural renewal focus was reflected in the name of the Center for Renewal of Science and Culture. In that month, the word *renewal* was dropped from all Web pages, and the CRSC became the Center for Science and Culture (CSC). One may speculate that *cultural renewal* may have been too reminiscent of the goals of twentieth-century creation science, distracting attention from the scholarly focus: scientific and other scholarly organizations do not typically have as their goal the renewal of culture (Holden 2002).

Although ID proclaims itself a scholarly movement, its cultural renewal focus is fundamentally incompatible with the openness and flexibility that a scientific theoretical perspective demands. Enamored of an ideological, political, or social goal, it is all too easy to misrepresent or ignore empirical data when they do not support the goal; certainly creation science is infamous for doing so (Scott 1993). A few ID proponents appear to be aware that the scholarly aspect of ID has taken a backseat to the political and the ideological. Bruce Gordon (2001: 9) has been especially eloquent on this issue: "design-theoretic research has been hijacked as part of a larger cultural and political movement. In particular, the theory has been prematurely drawn into discussions of public science education, where it has no business making an appearance without broad recognition from the scientific community that it is making a worthwhile contribution to our understanding of the natural world."

Gordon (2001: 9) also commented, "If design theory is to make a contribution in science, it must be worth pursing on the basis of its own merits, not as an exercise in Christian 'cultural renewal,' the weight of which it cannot bear."

And indeed, the scientific component of ID seems to have taken a backseat to the cultural renewal component, resulting in a dearth of actual models and theories in ID, recognized even by proponents. The Discovery Institute fellow Paul Nelson, a philosopher of science, has commented: "Easily the biggest challenge facing the ID community is to develop a full-fledged theory of biological design. We don't have such a theory right now, and that's a problem. Without a theory, it's very hard to know where to direct your research focus. Right now, we've got a bag of powerful intuitions, and a handful of notions such as 'irreducible complexity' and 'specified complexity'—but, as yet, no general theory of biological design" (quoted in *Touchstone* 2004: 64–65).

Intelligent Design's Content Problem

Intelligent design is criticized not only for a lack of theory but also for a lack of empirical content. This objection is presented both by scientists and by young-Earth creationists, noting that ID proponents seem reluctant to commit to claims about what happened in the history of life. As detailed by the YEC Carl Wieland, on the *Answers in Genesis* Web site: "They generally refuse to be drawn on the sequence of events, or the exact history of life on Earth or its duration, apart from saying, in effect, that it 'doesn't matter.' However, this is seen by the average evolutionist as either absurd or disingenuously evasive—the arena in which they are seeking to be regarded as full players is one which directly involves historical issues. In other words, if the origins debate is not about a 'story of the past,' what is it about?" (Wieland 2002).

Most ID proponents accept an ancient age of the universe and Earth, but there are some prominent ID supporters who are YECs, such as Paul Nelson and John Mark Reynolds. These creation science adherents reject evolution altogether, whereas some ID supporters such as Michael Behe have gone so far as to accept common ancestry of humans and apes (Behe 2007). The range of scientific opinion within the ID camp, therefore, runs from young-Earth creationism to mild forms of theistic evolution, although Dembski and others have declared theistic evolution to be incompatible with ID (Dembski 1995). The ID movement surely is a proverbial big tent, though it remains to be seen whether the differences among the tent's occupants will be reconcilable if ID takes any specific empirical positions on what Wieland has called the "story of the past" (Scott 2001).

Darwinism

However, regardless of their stand on issues such as the age of Earth or common ancestry of living things, common to all ID proponents is the rejection of Darwinism. In ID literature, *Darwinist* becomes an epithet, though it is not always clear in any given passage exactly what is meant by *Darwinism*. In evolutionary biology, *Darwinism* may refer to the general idea of evolution by natural selection or it may specifically refer to the ideas held by Darwin in the nineteenth century. Usually the term is not used for modern evolutionary theory, which, because it goes well beyond Darwin to include subsequent discoveries and understandings, is more frequently referred to as neo-Darwinism or just evolutionary theory. Evolutionary biologists hardly ever use *Darwinism* as a synonym for evolution, though historians and philosophers of science occasionally use it this way. In ID literature, however, *Darwinism* can mean Darwin's ideas, natural selection, neo-Darwinism, post-neo-Darwinian evolutionary theory, evolution itself, or materialist ideology inspired by Godless evolution.

The public, on the other hand, is unlikely to make these distinctions, instead simply equating Darwinism with evolution (common descent). For decades, creation science proponents have cited the controversies among scientists over *how* evolution occurred—including the specific role of natural selection—in their attempts to persuade the public that evolution itself, the thesis of common ancestry, was not accepted by scientists, or at least was in dispute. Within the scientific community, of course, there are lively controversies, for example over how much of evolution is explained by natural selection and how much by additional mechanisms such as those being discovered in evolutionary developmental biology ("evo-devo"). No one says natural selection is unimportant; no one says that additional mechanisms are categorically ruled out. But these technical arguments go well beyond the understanding of laypeople and are easily used to promote confusion over whether evolution occurred.[2] Intelligent design proponents exploit this public confusion about Darwinism to promote doubt about evolution.

The rejection of Darwinism, however, is not merely the glue that holds the disparate ID proponents together; it is also central to their movement. The natural selection mechanism of evolutionary change has long vexed conservative Christians who have difficulty reconciling it with the concept of a loving, all-good creator who is personally involved with creation. Concern with theodicy (the problem of evil) in Christian theology of course predated Darwin's discoveries, but there is no escaping that natural selection has implications for certain Christian views. There are many ways that Christian theologians have integrated the natural selection mechanism into different views of God (Peters and Hewlett 2003), though these compromises, along with theistic evolution, are rejected by ID proponents. Natural selection is acceptable—in fact, undeniable—on the level of a population of organisms: neither ID nor creation science proponents deny the ability of natural selection to lengthen bird beaks or produce antibiotic-resistant bacteria or pesticide-resistant insects. But for God to use the wasteful and cruel mechanism of natural selection to produce the diversity of living things today is theologically unacceptable to many in the ID camp. In presentations to the public, however, they focus more on the alleged scientific failings of natural selection, which they believe lacks the creative power to produce new body plans and bring about significant evolutionary changes among groups.

Is Intelligent Design Creationist?

Intelligent design proponents do not refer to themselves as creationists, associating that term, as many do, with the followers of Henry M. Morris. Indeed, most ID proponents do not embrace the young-Earth, flood geology, and sudden creation tenets associated with YEC. Yet by Phillip Johnson's definition, ID proponents arguably are creationists: "'Creationism' means belief in creation in a more general sense. Persons who believe that the earth is billions of years old, and that simple forms of life evolved gradually to become more complex forms including humans, are 'creationists' if they believe that a supernatural Creator not only initiated this process but in some meaningful sense *controls* it in furtherance of a purpose" (Johnson 1991: 4).

Phillip Johnson contends that the scientific data and theory supporting evolution are weak, and that evolution persists as a scientific idea only because it reinforces philosophical materialism. To him and most other ID proponents, the most important issue in the creation/evolution controversy is whether the universe came to its present state "through purposeless, natural processes known to science" (Johnson 1990: 30) or whether God had meaningful involvement with the process. Intelligent design proponents clearly believe that God is an active participant in creation, though they are divided as to whether this activity takes the form of front-loading all outcomes at the Big Bang, episodic intervention of the progressive creationism form, or other, less well-articulated possibilities. Theistic evolution, however, is ruled out or at best viewed as an ill-advised accommodation.

EVIDENCE AGAINST EVOLUTION

The *Edwards* decision, as mentioned, rejected equal time for creationism and evolution but allowed secular, scientific alternatives to evolution legally to be taught. Antievolutionists generated abrupt appearance theory and ID because scientific alternatives to evolution were not found in the scientific community. Creationists looking for an alternative to the now-unconstitutional creation science had another option suggested to them in the dissent to *Edwards* written by Justice Antonin Scalia, who wrote, "The people of Louisiana, including those who are Christian fundamentalists, are quite entitled, as a secular matter, to have *whatever scientific evidence there may be against evolution presented in their schools*, just as Mr. Scopes was entitled to present whatever scientific evidence there was for it" (Scalia, dissenting, *Edwards*, 482 U.S. 578, 634 (emphasis added)).

A month after the *Edwards* decision was published, the attorney Wendell Bird, who had argued the creationist side before the Supreme Court in *Edwards*, analyzed the decision in a joint paper published with ICR staff. The ICR staff seized on teaching evidence against evolution as a potential legal strategy for creationists—as creation science was no longer legal to teach. The article said: "In the meantime, school boards and teachers should be strongly encouraged at least to stress the scientific evidences and arguments against evolution in their classes (not just arguments against some proposed evolutionary mechanism, but against evolution per se), even if they don't wish to recognize these as evidences and arguments for creation (not necessarily as arguments for a particular date of creation, but for creation per se)" (Bird 1987: 4).

Teaching evidence against evolution (EAE) thus was viewed as a way of teaching creationism on the sly. Given the two models mind-set of young-Earth creationism, this made perfect sense: evidence against evolution is considered evidence for creationism. Creationists believe (probably correctly) that students think in the same dichotomous way: if students learn that evolution is weak or invalid science, they automatically will conclude even without urging from a teacher that special creationism is the true explanation for nature. Given this reasoning, denigrating evolution by teaching the evidence against or a critical analysis of evolution becomes a backdoor way of teaching creationism.

A dissent, however, is not legally binding, and so there is no legal directive to teach EAE, though this is how creationists often present Scalia's dissent to the public.

After *Edwards* (1987), there were a number of efforts by creationists to pass legislation requiring not equal time for creationism and evolution, but equal time for evolution and the alleged evidence against evolution. A series of bills with similar wording got their start with a 1996 Ohio bill resulting from the grassroots efforts of a retired Wisconsin teacher. John Hansen founded Operation T.E.A.C.H.E.S. (Teach Evolution Accurately, Comprehensively, Honestly, Equitably, Scientifically) and traveled around the country between 1995 and 2000 trying to persuade state legislators in Wisconsin, Minnesota, Indiana, Iowa, Ohio, Kentucky, Alaska, Georgia, and New Mexico to sponsor his model bill (Hansen 1997, 1999, 2000). Ohio State Representative Ron Hood formatted Hansen's idea as legislation (Trevas 1996) and introduced it (without success) in 1996 and 2000. A Georgia legislator also introduced the Hood bill, and when Hansen retired to Arizona, he persuaded an Arizona legislator to introduce the bill as well. None of these bills passed. Hansen claimed in his newsletter that legislators in Alaska, New Mexico, and Kentucky also introduced his legislation, but I could find no record of such bills.

The Ohio legislation submitted by Representative Hood (HB 62, submitted in 1996, and HB 679, submitted in 2000) was virtually identical to the Georgia and Arizona bills: "Whenever a theory of the origin of humans or other living things that might commonly be referred to as 'evolution' is included in the instructional program provided by any school district or educational service center, both scientific evidence and related arguments supporting or consistent with the theory and scientific evidence and related arguments problematic for, inconsistent with, or not supporting the theory shall be included."

A 2001 Arkansas bill (HB 2548) forbade educational agencies to use public funds to purchase textbooks or other instructional materials lacking antievolution arguments from ID and traditional creation science sources: "No state agency, city, county, school district or political subdivision shall use any public funds to provide instruction or purchase books, documents or other written material which it knows or should have known contain descriptions, conclusions, or pictures designed to promote the false evidences set forth in subsection (d) of this section."

Subsection (d) of the bill listed supposed examples of evolutionary fraud taken from ID proponent Jonathan Wells's book, *Icons of Evolution*: Haeckel's embryos, the Miller-Urey experiment, *Archaeopteryx* (the ancient bird), and the peppered moth example of natural selection. From traditional creation science (Chick 2000) the bill's author took such staples as Piltdown man, Nebraska man, and Neanderthal man—all claimed to be fraudulent, though only the first actually was. Elsewhere in the bill were references to other creation science claims such as gaps in the fossil record, falsity of the geological column, and flaws in radiometric dating. Even if one were not familiar with creation science literature, the reference to *evidences*—a term from Christian apologetics rather than science—would reveal the inspiration for this bill.

In addition to assuming that scientific evidence against evolution exists, such bills—like the equal time bills they supplanted—appeal to the American public's appreciation of fairness; the third "pillar of creationism." The American political tradition of local decision making (e.g., by town councils or local school boards) encourages a wide variety of voices to contend for influence and authority. Part of the American political and cultural tradition is for all voices to have an opportunity to be heard, even if later rejected. This is enshrined in the First Amendment's Free Speech and Assembly clauses and manifests even in journalistic traditions in which the reporter is expected to present both views of a controversy. As will be discussed elsewhere (chapter 11), the fairness approach, though culturally very powerful, is misapplied in the realm of science, which actually is highly discriminating—against those views that fail to accurately explain nature.

Scientific knowledge grows because ideas are considered, weighted against the evidence, and provisionally accepted or rejected depending on how well they fare. In the initial stages of the consideration of a scientific explanation, a variety of positions are likely to be entertained, but as any scientist will be quick to admit, most explanations eventually end up on the cutting room (or perhaps laboratory) floor, or are seriously reworked. Once rejected, however, there must be a compelling reason for discarded explanations to be taken seriously again. Scientific claims for the world and its inhabitants suddenly coming into being, at one time, in their present form, have not been taken seriously since the end of the eighteenth century, and it is unfair to pretend to students that this view is a viable scientific option in the twenty-first century. On reflection, the American cultural tradition of fairness is most appropriately applied in matters of opinion, rather than in matters of fact and logic. The 1897 attempts by an Indiana legislator to pass a law setting the value of pi to 3.0 (Mikkelson 2007) are viewed as comical: we would respond the same way to an effort by an enthusiast of the

Old South to require textbooks to report that General Grant surrendered to General Lee at the end of the Civil War. There are some things that one's preferences simply cannot change.

Evidence against evolution is emerging as a popular antievolution approach, especially after the failure of ID to survive a constitutional challenge in the federal district court case *Kitzmiller v. Dover* (see chapter 7). It is attractive to legal specialists among the antievolutionists because it appears to avoid the Establishment Clause of the First Amendment by not obviously promoting religion. It remains to be seen whether this strategy will be effective; as discussed in chapter 7, in a small number of cases, judges have alluded to the importance of looking at the historical context of policies promoting evidence against evolution and have declared them, in effect, creationism in disguise.

There are many phrases that express the underlying idea of EAE—that evolution is weak science that warrants careful student examination. One approach is to require students to critically analyze evolution—meaning that students should criticize it. Another phrase used is "strengths and weaknesses of evolution"; yet another is presenting evolution as "theory not fact," meaning to present evolution as a theory in the popular rather than the scientific sense as a guess or hunch. Frequently these theory-not-fact policies take the form of disclaimers that are to be read to students or pasted into textbooks.

Critically Analyzing Evolution

One EAE variant was promulgated by ID supporters on the Ohio State Board of Education during a controversy over the content of state science education standards in 2002. Lacking enough votes to have ID included in the standards, pro-ID board members arranged for a public hearing in March 2002 that would include ID proponents and opponents. Stephen C. Meyer, the director of the Center for Science and Culture at the Discovery Institute, testified that as a compromise, the board members who had been pushing to include ID in the curriculum should instead encourage teachers to "teach the controversy" about evolution (Miller 2002:6). He contended that there was a vigorous debate going on within the scientific community over the validity of evolution. Jonathan Wells presented examples from his book *Icons of Evolution* illustrating the kinds of problems with evolution that students supposedly should be taught. The anti-ID testifiers, biologist Kenneth R. Miller and physicist Lawrence Krauss, strongly discouraged adding ID to the standards, and also rebutted the claim that—at least among scientists—there was a controversy over whether evolution had occurred.

After much wrangling over wording, in October 2002 the board finally approved the standards, including one referring to evolution that read, "Describe how scientists continue to investigate and critically analyze aspects of evolutionary theory." The wording illustrated how political the issue had become, as both the proevolution and antievolution factions could (and did) claim victory. Supporters of evolution education claimed that the standard required students to critically analyze different ideas within evolutionary theory, emphasizing the word *aspects*: "'What we're essentially saying here is evolution is a very strong theory, and students can learn from it by analyzing evidence as it is accumulated over time,' Tom McClain, a board member and co-chairman of the Ohio Board of Education's academic standards committee, told the Associated Press" (Olsen 2002).

Intelligent design supporter Phillip Johnson, on the other hand, interpreted the standard as requiring students not to critically analyze but to criticize evolutionary theory: "The recent decision of the Ohio Science Standards Committee of the State School Board has been a big breakthrough. [Critics] are calling it a compromise, but it isn't. It's our position. *It allows teachers to present evidence against the theory of evolution.* This evidence includes the facts that the drawings of embryos in the textbooks are fraudulent and that the peppered moth experiment was botched if not an outright hoax" (Staub 2002; emphasis added).

A few years later, in 2006, in the wake of the *Kitzmiller v. Dover* decision, the Ohio Board of Education dropped the "critically analyze" standard, as well as a model antievolution classroom lesson plan that accompanied it. But *critical analysis of evolution* has proved popular wording for a basic EAE approach; between 2002 and 2007, the National Center for Science Education recorded sixteen state or local policies promoting this approach in thirteen different states.

Strengths and Weaknesses of Evolution

Another approach creationists use is to propose that when evolution is taught, both "strengths" and "weaknesses" of the subject should be taught; sometimes the language used calls for teaching evidence for and evidence against evolution. In all cases, the content presented is the familiar creation science and ID claims that there are gaps in the fossil record, that natural selection cannot produce big changes like body plan differences, that the overwhelming complexity of even the simplest cell cannot be explained by natural processes but requires special creation, and so on.

State science education standards are often divided into process skills and content sections. Process skills include information students should know about science as a way of knowing—that science includes observation, experimentation, testing, and so on. The content sections outline the concepts and facts students should know within any discipline. Physics content standards, for example, usually include the requirement that students understand concepts of mass and density. Texas state science standards are known as the Texas Essential Knowledge and Skills, or TEKS. They include in each discipline (physics, chemistry, biology, and other fields) a process skill that states the following:

> (3) Scientific processes. The student uses critical thinking and scientific problem solving to make informed decisions. The student is expected to (A) analyze, review, and critique scientific explanations, including hypotheses and theories, as to their strengths and weaknesses using scientific evidence and information.

Because process skill 3A accompanies the TEKS for each field of science, one might infer that the intent of the writers was that students should learn to be critical thinkers, a worthy pedagogical goal. A teacher could choose any of a number of scientific explanations to assist students in learning these skills. However, in 2002–2003, when biology textbooks were being considered for adoption, creationists on the state board of education lobbied to require the textbooks submitted by the publishers to include strengths and weaknesses of evolution—but no other scientific theory. Publishers were loath to rewrite their books to include a lot of bad science, but fortunately after much wrangling, the majority of the board voted to adopt the books largely as they were (Stutz 2003). Calling for strengths and weaknesses of evolution was again a contentious issue when the TEKS came up for revision in 2008 (Scharrer 2008).

During the 2003 hearings, creationists urged textbook publishers to include examples of alleged strengths and weaknesses of evolution from the Discovery Institute Fellow Jonathan Wells's book *Icons of Evolution* (Wells 2000). Wells himself spoke at the hearings about the various failings of the textbooks submitted for adoption. Without specifically mentioning intelligent design, *Icons* instead vigorously attacks evolution—the idea of common ancestry, as well as natural selection as a mechanism of evolution. Identifying commonly used textbook illustrations of evolution or of natural selection as "icons," Wells lambastes (among other examples) the Miller-Urey "sparking" experiments, which produced organic molecules from inorganic molecules; the concept of homology; the nineteenth-century embryologist Ernst Haeckel's drawings of vertebrate embryos; the peppered moth natural selection experiments; and the idea of humans evolving from apes. Ostensibly a critique of high school and college textbooks, the book uses the presentation of the alleged icons in textbooks as an excuse to attack the validity of evolution by natural selection itself. The book has been widely panned by scientists (Coyne 2002; Padian and Gishlick 2002; Scott 2001b), but it forms a template for the EAE approach. (A rebuttal to the claim of the supposed fraudulence of the peppered moth natural selection experiments is presented in the readings in chapter 11.)

A new book from the Discovery Institute, *Explore Evolution*, repeats many of these claims and is apparently intended for use as a textbook promoting the "critical analysis of evolution" or "strengths and weaknesses of evolution" approaches (Meyer, Minnich, Moneymaker, Nelson, and Seelke 2007).

Explore Evolution's chapters are organized into "Arguments For" and "Arguments Against" sections. Unfortunately, the "Arguments For" are strawman presentations of evolutionary biology, from which students will learn little about standard science. The "Arguments Against" are familiar creationist claims. About half of the book is devoted to challenging the common ancestry of living things, arguing instead for the barely disguised alternative of special creation to explain similarities and differences ordinarily explained by evolution. Because the goal is to have *Explore Evolution* used in the public schools, obvious creationist language is avoided. As is typical in ID publications, natural selection comes in for special attack as being inadequate to explain the diversity of living things. Again typical of ID publications, evolution is presented as an active scientific controversy, despite statements from a wide range of scientific associations that, on the contrary, evolution is considered mainstream science (Sager 2008).

"Just a Theory" Disclaimers

Another EAE approach is to denigrate evolution by requiring that it be distinguished from all other scientific explanations as a theory, by which they mean a guess or a hunch. Often such efforts are coupled with requirements that disclaimers ("evolution is just a theory") be included in textbooks or be read to students. As discussed in chapter 1, scientific theories are far from guesses: there are many explanations in science, and the best ones are elevated to theories. When school boards or state legislatures attempt to single out evolution as just a theory, it is clear that they are not using this term in its scientific sense. But such disclaimers and policies have the net effect of drawing attention to evolution as a particularly controversial subject, which makes it less likely that evolution will be taught.

Efforts to require disclaimers for evolution began in Texas, when in 1974 the state board of education required that all biology textbooks bought in the state treat evolution as a theory and not factually verifiable. "Furthermore, each textbook must carry a statement on an introductory page that any material on evolution included in the book is clearly presented as theory rather than verified" (Mattox, Green, Richards, and Gilpin 1984: 1). Although in 1984 the Texas attorney general opined that the Texas disclaimer was illegal (see chapter 10), other states and communities have regularly proposed and passed such evolution-only disclaimers.

The vast majority of theory, not fact, policies and disclaimers do not pass, but the publicity given to them contributes to the general perception that evolution is somehow less valid than other scientific subjects. A disclaimer that was passed by the board of education in Tangipahoa, Louisiana, in 1994 singled out evolution for special treatment. Teachers were directed to read the disclaimer to students before discussing evolution or assigning readings. The disclaimer read in part:

> It is hereby recognized by the Tangipahoa Parish Board of Education, that the lesson to be presented, regarding the origin of life and matter, is known as the Scientific Theory of Evolution and should be presented to inform students of the scientific concept and not intended to influence or dissuade the Biblical version of Creation or any other concept.
>
> It is further recognized by the Board of Education that it is the basic right and privilege of each student to form his/her own opinion or maintain beliefs taught by parents on this very important matter of the origin of life and matter. Students are urged to exercise critical thinking and gather all information possible and closely examine each alternative toward forming an opinion.

The Tangipahoa disclaimer was challenged in federal district court, which ruled in *Freiler v. Tangipahoa* (1997) that the purpose of the regulation was to promote religion and that the preceding paragraph's attempt to present the disclaimer as having the purpose of promoting critical thinking was a sham. This determination was made on the basis of the facts of the case, in which it was apparent from minutes and other reports of the board of education that the policy was intended to promote specific sectarian (biblical Christian) views. In 1999, the Fifth Circuit Court of Appeals upheld the decision, which noted that it was possible that some form of disclaimer could be constitutional—although the Tangipahoa disclaimer, with its specific mention of the Bible, was not.

In June 2000, the Supreme Court let the appeals court decision stand by refusing to hear the case. But as you will read in the next chapter, the Freiler case did not stop the effort to disclaim evolution.

NOTES

1 The example of fairy rings as a refutation of the design influence was originally suggested by Flietstra (1998).

2 Note that none of the active proponents of either ID or creation science are contributing to the scientific discussion of these points: instead of debating evolution theory at professional scientific conferences or in journals, they perform no research on evolutionary biology but merely report on the work of other scientists (e.g., Johnson 1991; Wells 2000), often distorting it severely in the process (Branch 2002; Coyne 1996, 2002; Gould 1992; Padian and Gishlick 2002; Scott 2001b; Scott and Sager 1992). They prepare their articles and books for the general reader rather than the scientific public.

LEGAL CASES

Edwards v. Aguillard 482 U.S. 578 (1987).
McLean v. Arkansas Board of Education, 529 F. Supp. 1255 (E.D. La. 1982).
Freiler v. Tangipahoa Parish Board of Education, 975 F. Supp. 819 (E.D. La. 1997).

REFERENCES

Behe, Michael. 1996. *Darwin's black box: The biochemical challenge to evolution.* New York: Free Press.

Behe, M. J. 2007. *The edge of evolution: The search for the limits of Darwinism.* New York: Free Press.

Bird, Wendell. 1987. *The* Origin of Species *revisited: The theories of evolution and abrupt appearance*, vol. 1. New York: Philosophical Library.

Bird, W. R. 1987. The Supreme Court decision and its meaning. *ICR Impact* (170): 1–4. With "ICR Evaluation," 3–4.

Blackstone, Neil W. 1997. Argumentum ad ignorantiam: A review of "Darwin's Black Box: The Biochemical Challenge to Evolution." *Quarterly Review of Biology* 72 (4): 445–447.

Branch, Glenn. 2002. Analysis of the Discovery Institute bibliography. *Reports of the National Center for Science Education* 22 (4): 12–18, 23–24.

Chapman, Bruce. 1996a. [Press release.] Seattle: Discovery Institute, October 10.

Chapman, Bruce. 1996b. Ideas whose time is come. *Discovery Institute Journal* (August), p. 1.

Chapman, Bruce. 1998. Letter from the president. Seattle: Discovery Institute.

Chick, Jack T. 2000. *Big daddy* (pamphlet). Ontario, CA: Chick Publications.

Coyne, Jerry A. 1996. God in the details. *Nature* 383: 227–228.

Coyne, Jerry A. 2002. Creationism by stealth. *Nature* 410: 745–746.

Dembski, William. 1995. What every theologian should know about creation, evolution, and design. *Center for Interdisciplinary Studies Transactions* 3 (2): 1–8.

Dembski, W. 1998. *The design inference: Eliminating chance through small probabilities.* New York: Cambridge University Press.

Dembski, William. 2001. *No free lunch: Why specified complexity cannot be purchased without intelligence*. Lanham, MD: Rowman & Littlefield.

Dembski, W. 2002. Becoming a disciplined science: Prospects, pitfalls, and reality check for ID. Retrieved October 14, 2008, from http://www.iscid.org/papers/Dembski_Disciplinedscience_102802.pdf.

Discovery Institute. 2003. The "Wedge document": So what? Retrieved September 6, 2008, from http://www.discovery.org/scripts/viewDB/filesDB-download.php?id=349.

Flietstra, Rebecca J. 1998. The design debate. *Christianity Today* 4 (5): 34.

Gish, Duane T. 1976. *Origin of life: Critique of early stage chemical evolution theories*. San Diego: Institute for Creation Research.

Gordon, Bruce. 2001. Intelligent design movement struggles with identity crisis. *Research News and Opportunities in Science and Theology* (January): 9.

Gould, Stephen Jay. 1992. Impeaching a self-appointed judge. *Scientific American* (July): 118–121.

Hansen, John. 1997. *Operation T.E.A.C.H.E.S. 1997 progress report*. Williams Bay, WI: Operation T.E.A.C.H.E.S.

Hansen, John. 1999. *Operation T.E.A.C.H.E.S. 1998 progress report*. Williams Bay, WI: Operation T.E.A.C.H.E.S.

Hansen, John. 2000. *Operation T.E.A.C.H.E.S. 2000 progress report*. Williams Bay, WI: Operation T.E.A.C.H.E.S.

Holden, Constance. 2002. Design's evolving image. *Science* 297(5589): 1991.

Hoyle, Fred, and Chandra Wickramasinghe. 1979. *Diseases from space*. New York: Harper & Row.

Johnson, Phillip E. 1990. *Evolution as dogma: The establishment of naturalism*. Dallas, TX: Haughton.

Johnson, Phillip E. 1991. *Darwin on trial*. Washington, D.C.: Regnery Gateway.

Kennedy, J. 1992. Teaching of creationism splits Louisville parents. *The Repository*, September 30, B4.

Mattox, Jim, Tom Green, David R. Richards, and Rick Gilpin. 1984. Opinion No. JM-134. Austin: Office of the Attorney General, State of Texas. Retrieved October 29, 2008 from http://www.oag.state.tx.us/opinions/47mattox/op/1984/htm/jm0134.htm.

Meyer, Stephen C., Scott Minnich, Jonathan Moneymaker, Paul A. Nelson, and Ralph Seelke 2007. *Explore evolution*. Melbourne: Hill House Publishers.

Mikkelson, Barbara. 2007. Alabama's slice of pi. July 16. Retrieved April 6, 2008, from http://www.snopes.com/religion/pi.asp.

Miller, Kenneth R. 1996. Review of "Darwin's Black Box: The Biochemical Challenge to Evolution." *Creation/Evolution* 16 (2): 36–40.

Miller, Kenneth R. 2002. Goodbye, Columbus. *Reports of the NCSE* 22(2): 6–8.

Morris, Henry M., ed. 1974. *Scientific creationism*. San Diego: Creation-Life Publishers.

Olsen, Ted. 2002. Ohio science standards don't mandate intelligent design, but may open door. October 15, 2002. Retrieved October 14, 2007, from http://www.christianitytoday.com/ct/2002/140/22.0.html.

Padian, Kevin, and Alan D. Gishlick. 2002. The talented Mr. Wells. *Quarterly Review of Biology* 77 (1): 33–37.

Pallen, M. J., and N. J. Matzke (2006). From *The Origin of Species* to the origin of bacterial flagella. *Nature Reviews Microbiology* 4 (10): 784–790.

Pennock, Robert T. 1999. *Tower of Babel: The evidence against the new creationism*. Cambridge, MA: MIT Press.

Perloff, James. 1999. *Tornado in a junkyard: The relentless myth of Darwinism*. Arlington, MA: Refuge Books.

Peters, Ted, and Martinez Hewlett. 2003. *Evolution from creation to new creation*. Nashville, TN: Abingdon Press.

Pond, F. 2006. The evolution of biological complexity. *Reports of the National Center for Science Education* 26 (3): 22, 27–31.

Sager, Carrie E., ed. 2008. *Voices for evolution 3rd ed.* Oakland, CA: National Center for Science Education.

Scharrer, Gary. 2008. Debate over biology is brewing. *San Antonio Express-News*. May 31.

Scott, Eugenie C. 1989. New creationist book on the way. *NCSE Reports* 9 (2): 21.

Scott, Eugenie C. 1993. The social context of pseudoscience. In *The natural history of paradigms*, ed. J. H. Langdon and M. E. McGann, 338–354. Indianapolis: University of Indiana Press.

Scott, Eugenie C. 2001a. The big tent and the camel's nose. *Reports of the National Center for Science Education* 21 (2): 39. Retrieved October 29, 2008 from http://ncseweb.org/rncse/21/1-2/big-tent-camels-nose.

Scott, Eugenie C. 2001b. Fatally flawed iconoclasm. *Science* 292 (5525): 2257–2258.

Scott, Eugenie C., and Thomas C. Sager. 1992. "Darwin on Trial": A review. *Creation/Evolution* 12 (2): 47–56.

Scott, Eugenie C., and Nicholas J. Matzke. 2007. Biological design in science classrooms and the public arena. *Proceedings of the National Academies of Science* 104 (Suppl. 1): 8669–8676.

Staub, Dick. 2002. The Dick Staub Interview: Phillip Johnson. December 3, 2002. Retrieved September 22, 2007, from http://www.christianitytoday.com/ct/2002/decemberweb-only/12-2-22.0.html.

Stutz, Terrence. 2003. Board backs adoption of biology books. *Dallas Morning News*. Nov. 7, p. 1

Thaxton, Charles B., Walter L. Bradley, and Roger L. Olsen. 1984. *The mystery of life's origin: Reassessing current theories*. New York: Philosophical Library.

Touchstone. 2004. The measure of design. *Touchstone* 17 (6): 60–65.

Trevas, D. 1996. Rep: teach evolution as a theory, not fact. *Vindicator* (Youngstown, OH) April 28.

Weiland, Carl. 2002. AIG's views on the intelligent design movement. Answers in Genesis. Retrieved September 4, 2002, from http://www.answersingenesis.org/docs2002/0830_IDM.asp.

Wells, Jonathan. 2000. *Icons of evolution: Science or myth?* Washington, DC: Regnery.

Wilkins, J., and W. Elsberry. 2001. The advantages of theft over toil: The design inference and arguing from ignorance. *Biology and Philosophy* 16: 711–724.

CHAPTER 7

• • • • • • • • • • • • • • • •

Testing Intelligent Design and Evidence against Evolution in the Courts

The first decade of the twenty-first century has seen a flurry of legal decisions dealing with the two neocreationist approaches, scientific alternatives to evolution and evidence against evolution, and their variants. To put these legal decisions in context, recall the brief discussion in chapter 5 about the First Amendment of the U.S. Constitution. The Establishment and the Free Exercise clauses, taken together, require schools and other government agencies to be religiously neutral. In fact, the courts have been "particularly vigilant in monitoring compliance with the Establishment Clause in elementary and secondary schools" (*Edwards v. Aguillard*, 482 U.S. 578 at 583–584) because students in a classroom are a captive audience. Among other things, freedom of religion means that parents rather than the government have the right to instruct children in religious views.

Also in chapter 5, you read about *Lemon v. Kurtzman*—the three-prong standard that courts use to decide whether a government action violates the Establishment Clause (*Lemon v. Kurtzman*, 403 U.S. 602).

To be constitutional, a law or policy must have a legitimate secular purpose and must not, when implemented, have the primary effect of either advancing or inhibiting religion. The third prong of *Lemon* is that a law or policy must not cause undue entanglement of state and religion. In addition, the Supreme Court has added a test that augments the effect prong of *Lemon*, the endorsement test. This test, most clearly articulated in *Santa Fe v. Doe* (*Santa Fe Independent School District v. Doe*, 530 U.S. 290), asks whether a policy or law would be viewed by an informed observer from the community as endorsing a religious view or religion in general. The endorsement test also considers whether individuals who do not adhere to the religious view being presented would be made to feel like outsiders in the community while people who profess the view would be perceived as insiders or as more favored members of the community. As we will see, the *Lemon* and endorsement tests have played key roles in recent lawsuits concerning creationism and evolution.

TESTING INTELLIGENT DESIGN

An Uncompromising School Board

The small central Pennsylvania community of Dover had for several years feuded over the teaching of evolution. From at least 2001 on, some school board members had made public comments derogatory of evolution or in favor of teaching creationism. In 2002, a four-foot by sixteen-foot student-painted mural depicting a line of progressively more human "ape-men" was removed from the wall of the science classroom and burned by a school district custodian, allegedly while a school board member looked on. The custodian considered the naked figures obscene and irreligious (Lebo 2005).

Fueling the fire was that, in 2001, the state of Pennsylvania adopted science education standards that required the teaching of evolution. In 2003, when it was time for Dover to select a new biology textbook, teachers chose a textbook that included a conventional treatment of this subject: a standard commercial textbook published by Prentice Hall, *Biology* by Kenneth R. Miller and Joseph Levine.

This choice did not sit well with some of the school board members, who delayed the purchase of the book for more than a year. At a school board meeting in June 2004, board members contended that a new book should be chosen that included both creationism and evolution. Teachers argued that this would be bad educational policy and would unconstitutionally promote religion. Board members also urged teachers to use an intelligent design (ID) video, *Icons of Evolution*. Teachers dutifully reviewed it but judged it unsuitable for the classroom.

One board member, William Buckingham, sought advice from the Thomas More Law Center (TMLC), a Michigan-based organization that describes itself as "the sword and shield for people of faith," and was told of a supplemental textbook, *Of Pandas and People*, that presented ID. The TMLC had, in fact, been searching for a school district willing to mount a test case of the legality of teaching ID (Goodstein 2005). Buckingham proposed to the board that *Pandas* could be used to counter the evolution presented in the Prentice Hall book. (*Pandas* was discussed in chapter 6: produced by the Foundation for Thought and Ethics (FTE), it is the first book to use the phrase *intelligent design* in its modern context.) Teachers examined *Pandas* and rejected it as not matching the curriculum for high school students and as scientifically inaccurate. They also criticized its old-fashioned pedagogical approach. School board members, led by Buckingham, persisted in holding up the textbook adoption and refused to vote to approve the Prentice Hall book unless *Pandas* also was approved. Finally, at a school board meeting in August 2004, enough board members voted to approve the new textbooks. Teachers resisted using *Pandas* as a supplementary textbook alongside the regular textbook, but as a compromise, partly in fear of losing their jobs, they agreed to place *Pandas* in the classroom as a reference book (*Kitzmiller v. Dover Area School District*, 400 F.Supp.2d 707 at 755). Because some community members were raising objections to the use of public money to buy a creationist book, Buckingham requested donations from his church, and raised $850 to purchase sixty copies of *Pandas* for donation to the school district. Church members believed that they were supporting the teaching of creationism.

But the teachers left the books in the packing boxes and showed no inclination to use them. Furthermore, at a meeting in early October 2004, the district superintendent clarified that because *Pandas* was only a reference book, teachers would not be required to use it. In response, board members decided that an antievolution policy was necessary, and in mid-October 2004, passed a resolution requiring, "Students will be made aware of gaps/problems in Darwin's theory and of other theories of evolution including, but not limited to, intelligent design. Note: Origins of Life is not taught."

Although *origins of life* usually refers to the appearance of the first living things from nonliving chemicals, to the school board members most actively opposing evolution, the phrase instead meant common ancestry (*Kitzmiller*, at 749). These school board members thought, therefore, that the policy would forbid the teaching of evolution (in the sense of common ancestry) and promote the teaching of ID. The "gaps/problems in Darwin's theory" and intelligent design were to be taught in lecture form, and *Pandas* was to be used for readings.

The policy was controversial, and two board members resigned over their colleagues' action. At noisy school board meetings, many parents tried to persuade the school board not to bring what they considered creationism into the science classroom; other parents applauded the board's action for doing precisely that. Some members of the community began talking about a lawsuit, and in November 2004, board members appeared to back off slightly from their earlier enthusiasm for ID and composed a disclaimer for teachers to read to students before teaching evolution. The policy would go into effect at the beginning of the January 2005 school term. This statement was more detailed than the October resolution, proclaiming:

> The Pennsylvania Academic Standards require students to learn about Darwin's Theory of Evolution and eventually to take a standardized test of which evolution is a part.
>
> Because Darwin's Theory is a theory, it continues to be tested as new evidence is discovered. The Theory is not a fact. Gaps in the Theory exist for which there is no evidence. A theory is defined as a well-tested explanation that unifies a broad range of observations.
>
> Intelligent Design is an explanation of the origin of life that differs from Darwin's view. The reference book, *Of Pandas and People*, is available for students who might be interested in gaining an understanding of what Intelligent Design actually involves.
>
> With respect to any theory, students are encouraged to keep an open mind. The school leaves the discussion of the Origins of Life to individual students and their families. As a Standards-driven district, class instruction focuses upon preparing students to achieve proficiency on Standards-based assessments.

The science teachers unanimously refused to read the statement to their classes; when the policy was implemented in January 2005, administrators, rather than teachers, went from class to class to read the board-passed statement. Several teachers, in fact, joined in late fall with other Dover parents to request that the American Civil Liberties Union (ACLU) represent them in a lawsuit against the school district. A complaint was filed in federal district court in December, naming parent Tammy Kitzmiller as the lead plaintiff. *Kitzmiller v. Dover* thus became the first legal test of ID.

Legal Teams Square Off

The plaintiffs' legal team included two civil liberties organizations, the Pennsylvania affiliate of the ACLU and Americans United for Separation of Church and State (AU). It also included the large Philadelphia-based law firm of Pepper Hamilton LLP, and a consultant, the National Center for Science Education (NCSE). The school district was defended by the Thomas More Law Center (TMLC); its regular attorney had warned it not to adopt the antievolution policy. The Discovery Institute (DI), the leading ID organization in the country, began corresponding with Buckingham and another board member, Alan Bonsell, in June 2004, and members of its staff and DI fellows were involved early in the case as expert witnesses. But later, the DI and the TMLC parted ways—according to the director of the TMLC, Richard Thompson—because personnel associated with the DI insisted on having their own attorneys present at pretrial depositions (NCSE 2005).

The claim of the plaintiffs was that the board's policies requiring the teaching of ID violated the First Amendment ban on the promotion of religion in the public schools, because ID was an inherently religious doctrine. In defense, the district's attorneys had to show that the policies were passed not to promote religion but to improve science education. The defense would argue that large numbers of scientists were questioning evolution, and that students should be able to think critically about its so-called gaps and problems. The defense would contend that any religious implications of ID were incidental to ID as a valid science—the claimed secular reason for teaching ID. Demonstrating that ID was an up-and-coming scientific field thus formed a major component of the defense's strategy.

The legality of the policy would ultimately stand or fall on whether ID was primarily or secondarily religious: was ID valid science, as the defense claimed, or merely the most recent variant of creation science, as the plaintiffs claimed? Since the inception of ID, its proponents have assiduously tried to avoid the creationist label; creationism had previously been judged to be unconstitutional by the Supreme Court in *Edwards v. Aguillard*. It was essential to the defense that ID be judged as valid science, and it was just as essential to plaintiffs that it be judged either nonscience or inaccurate science (or both), because if either were true, there would be no valid secular, pedagogical reason to teach it. Both sides, therefore, organized their cases at least partly around the scientific status of ID and consequently requested a ruling by the judge on this issue. This necessarily would require a ruling on the nature of science, as well as on whether ID fulfilled the definition of science (Jones 2007).

Plaintiffs' lawyers prepared to attack as a sham the defense's claim that teaching ID would improve students' science education: on the contrary, they would claim, teaching ID would miseducate students. First, ID does not follow the established approach universally used by scientists of restricting scientific explanation to natural causes: the intelligent agent was God. Second, the (few) fact claims ID makes, such as the impossibility of the evolution of an irreducibly complex structure, were simply wrong. They would further argue that ID relies on arguments (e.g., irreducible complexity) wherein evolution is denigrated as a way of supporting ID. This, they would contend, is merely a variant of creation science's two-model approach, which denigrates evolution to promote special creationism. In reference to the gaps and problems aspect of the Dover policy, plaintiffs' attorneys again would point out the history of the denigration of evolution as a creationist strategy. Because evolution is sound science, teaching students that evolution is weak or unreliable science would miseducate them about a central scientific concept. Because there was no real pedagogical purpose or effect of teaching ID and/or denigrating evolution, the only purpose and effect of the policy would be to advance religion, and the policy should therefore be struck down. It was also necessary for plaintiffs to show that ID was a religious view: plaintiffs' attorneys would try to convince the judge that the history of ID indicated a direct ancestral relationship to the unconstitutional creation science, both in personnel and content.

Expert witnesses for the plaintiffs were cell biologist Kenneth R. Miller (the coauthor of the textbook used in Dover's schools), paleontologist Kevin Padian, philosophers Robert Pennock and Barbara Forrest, theologian John Haught, and professor of education Brian Alters. Mathematician Jeffrey Shallit was listed and deposed as a rebuttal witness (a deposition is a questioning of a witness by the opposing attorneys in the fact-gathering period before the trial itself). Expert witnesses for the defense included biochemist Michael Behe, microbiologist Scott Minnich, communications professor John Angus Campbell, professor of education Dick M. Carpenter II, theologian, philosopher, and mathematician William A. Dembski, and philosopher Warren A. Nord. Sociologist Steve Fuller and philosopher Stephen Meyer were listed as rebuttal witnesses. Of the defense witnesses, only Behe, Minnich, and Fuller actually testified, however; others—Campbell, Dembski, and Meyer, all DI fellows or employees—were withdrawn, and Nord and Carpenter mysteriously were not called as witnesses. Both sides also called plaintiffs, defendants, and other citizens to testify as to the facts of the case.

The trial began on September 26, 2005, and stretched over six weeks, ending on November 4. In all, court was in session for twenty-one days—a long trial. The federal district court judge John E. Jones III presided.

All of the plaintiffs' expert witnesses spoke to the question of the nature of science, and all defined it as restricted to explaining nature through natural causes. Scientist expert witnesses Miller and Padian testified on the soundness of evolution as science, and on the invalidity of the fact claims of ID (such as the unevolvability of irreducible complexity and the inaccuracy of statements about genetics and paleontology in *Of Pandas and People*). Theologian Haught testified that ID was a religious position with a long history in Christian theology. Philosopher of science Robert Pennock testified on the nature of science, and as part of a team of scholars researching the computer modeling of evolutionary processes, he also spoke to the invalidity of ID's claims that natural selection could not produce significant changes in an evolving population. Educational pedagogy specialist Brian Alters evaluated the policies of the Dover board from an educational standpoint and found them to foreclose rather than broaden students' understanding.

The most dramatic testimony came from philosopher Barbara Forrest, coauthor of a vigorous history and critique of ID, *Creationism's Trojan Horse: The Wedge of Intelligent Design* (Forrest and Gross 2004). During the pretrial wrangling, the defense had filed a legal challenge to her credentials to be an expert witness, saying "she is little more than a conspiracy theorist and a Web-surfing, 'cyber-stalker' of the Discovery Institute" (Muise 2005). After examining Forrest's academic credentials and scholarly accomplishments, the judge dismissed its motion and accepted Forrest as an expert witness on the history of ID.

Forrest's testimony traced the history of ID as an outgrowth of the earlier creation science movement. She identified creation science proponents who morphed into ID proponents, such as Dean Kenyon, the coauthor of *Of Pandas and People*. Kenyon had

Figure 7.1
The lines represent the number of times the words *creationism* or *creationist*
(top line) or the phrase *intelligent design* (bottom line) occurred in each of the
manuscripts associated with *Of Pandas and People.* In the early manuscripts,
creationist and *creationism* occur frequently and the phrase *intelligent design*
is rare. In 1987, the frequencies reverse, with creationist wording becoming
almost extinct, replaced by intelligent design. The Supreme Court case *Edwards
v. Aguillard,* striking down the teaching of creation science in public schools,
was delivered in 1987.

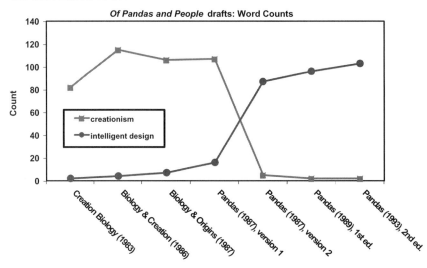

been scheduled to testify in *McLean v. Arkansas* on behalf of the defense, supporting
the legality of teaching creation science along with evolution. He also had prepared
an affidavit for the later *Edwards v. Aguillard* decision, in which he described creation
science in terms very much like modern-day ID proponents describe ID.

But perhaps the most striking evidence—the judge in his decision later called it
"astonishing"—was the deliberate change from creationist language to ID language in
early drafts of the FTE manuscripts for the book that became *Of Pandas and People.*
During discovery (the preparation period before the trial), the plaintiffs' consultant
NCSE located newspaper articles and FTE correspondence in its archives suggesting
the possibility that earlier drafts of *Pandas* had very creationist-sounding titles and
content. Plaintiffs' lawyers subpoenaed any early drafts of the manuscript from FTE.
After some legal wrangling, FTE delivered them to the court. Plaintiffs' consultants
analyzed them for content, finding that the number of times the terms *creation, creationist,* and their cognates appeared in the texts fell off dramatically in 1987—the
date of the *Edwards v. Aguillard* Supreme Court decision. Between two 1987 drafts,
the terms were replaced with other terms like *intelligent design* and *design proponents,*
demonstrating that intelligent design really was just creationism (Figure 7.1).

As further proof that ID was equivalent to creationism in the minds of the authors, a
crucial passage defining the topic of *Pandas* was compared. In the earlier manuscripts,
the definition was as follows: "*Creation* means that the various forms of life began
abruptly through the agency of an *intelligent creator* with their distinctive features

already intact—fish with fins and scales, birds with feathers, beaks, and wings, etc." (emphasis added).

In the second 1987 and subsequent published versions of *Pandas*, the same words are used to define ID: "*Intelligent design* means that the various forms of life began abruptly through an *intelligent agency*, with their distinctive features already intact—fish with fins and scales, birds with feathers, beaks, and wings, etc." (emphasis added).

Here, too, the change from the creationist to the ID terminology took place in 1987: the same year that the Supreme Court issued its decision striking down laws requiring equal time for creation science.

When the defense took the stand, the lawyers presented expert witness testimony by scientists Michael Behe and Scott Minnich that evolution had lots of gaps and problems, and that ID was a valid, cutting-edge science that students would benefit from learning. The testimony of a sociologist of science, Steve Fuller, was intended to support the idea that methodological naturalism was not really necessary in science, that ID fell under a broadened definition of science, and that it was pedagogically valuable for students to learn it. Although the defense had announced that it would call two other expert witnesses, they were never called, and the defense of ID and the arguments regarding the nature of science rested on Behe, Minnich, and Fuller.

Testimony from the expert witnesses was not the only testimony heard, of course. The judge also heard from fact witnesses: plaintiffs, school board members, and even from a few newspaper reporters. Although school board members denied having religious motivations for their actions, it was clear from testimony and evidence that key school board members vigorously opposed evolution for religious reasons and believed that teaching ID would bring creationism into the classroom. In some instances, school board members appeared to have lied under oath, damaging their overall credibility, including the credibility of claims that they had no religious motivation for their actions.

"Breathtaking Inanity"

Judge Jones did not find the arguments of the defense expert or fact witnesses to be persuasive. The decision in *Kitzmiller v. Dover* was handed down on December 20, 2005, and it was a complete victory for the plaintiffs, who won on every one of their points. Judge Jones declared Dover's educational policies regarding evolution and ID to be unconstitutional. The judge was unpersuaded by the claims of secular purpose for the Dover policies, writing:

> Although as noted Defendants have consistently asserted that the ID Policy was enacted for the secular purposes of improving science education and encouraging students to exercise critical thinking skills, the Board took none of the steps that school officials would take if these stated goals had truly been their objective. The Board consulted no scientific materials. The Board contacted no scientists or scientific organizations. The Board failed to consider the views of the District's science teachers. The Board relied solely on legal advice from two organizations with demonstrably religious, cultural, and legal missions, the Discovery Institute and the TMLC. Moreover, Defendants' asserted secular purpose of improving science education is belied by the fact that most if not all

of the Board members who voted in favor of the biology curriculum change conceded
that they still do not know, nor have they ever known, precisely what ID is. To assert a
secular purpose against this backdrop is ludicrous. (*Kitzmiller*, at 763)

Jones also noted that several of the most actively antievolutionist school board
members had lied under oath during deposition and on the witness stand. Such
behaviors, he said, further devalued any claims they might have had for a secular
purpose for teaching ID. He laid the blame for the expensive and lengthy trial squarely
at the feet of a religiously motivated school board, goaded on by the TMLC:

Those who disagree with our holding will likely mark it as the product of an activist judge.
If so, they will have erred as this is manifestly not an activist Court. Rather, this case
came to us as the result of the activism of an ill-informed faction on a school board, aided
by a national public interest law firm eager to find a constitutional test case on ID, who
in combination drove the Board to adopt an imprudent and ultimately unconstitutional
policy. The breathtaking inanity of the Board's decision is evident when considered
against the factual backdrop which has now been fully revealed through this trial. The
students, parents, and teachers of the Dover Area School District deserved better than
to be dragged into this legal maelstrom, with its resulting utter waste of monetary and
personal resources. (*Kitzmiller*, at 765–766)

The judge was clear in his view that ID did not qualify as science for a number of
reasons:

Finally, we will offer our conclusion on whether ID is science not just because it is
essential to our holding that an Establishment Clause violation has occurred in this case,
but also in the hope that it may prevent the obvious waste of judicial and other resources
which would be occasioned by a subsequent trial involving the precise question which is
before us. . . .

We find that ID fails on three different levels, any one of which is sufficient to preclude
a determination that ID is science. They are (1) ID violates the centuries-old ground
rules of science by invoking and permitting supernatural causation; (2) the argument of
irreducible complexity, central to ID, employs the same flawed and illogical contrived
dualism that doomed Creation Science in the 1980s; and (3) ID's negative attacks on
evolution have been refuted by the scientific community. . . . it is additionally important
to note that ID has failed to gain acceptance in the scientific community, it has not
generated peer-reviewed publications, nor has it been the subject of testing and research.
(*Kitzmiller*, at 735)

In its first legal outing, then, ID failed to defend itself as a valid science, or even as
science at all. It is doubtful, however, that the *Kitzmiller* decision will completely stop
efforts to teach ID. The *Kitzmiller* case was not appealed, hence the judge's decision
is precedent only in the Middle Federal District of Pennsylvania. It will, however,
be highly influential in discouraging the teaching of ID because the trial record was
so long and complete, and because the decision was so thorough. Much as *McLean
v. Arkansas* was an opportunity for creation science proponents to demonstrate that

theirs was a valid science, so was *Kitzmiller* the opportunity for ID proponents to demonstrate its scientific validity. Complaints of creation science proponents after *McLean* that the "best" creation scientists did not testify cannot be repeated for *Kitzmiller*, as Behe, a tenured professor, is arguably the most highly qualified scientist who is a leading promoter of ID. But just as creation science proponents continued to promote their views in the public schools even after the Supreme Court declared its teaching unconstitutional in the 1987 case *Edwards v. Aguillard*, so it is likely that ID proponents similarly will not abandon their efforts to promote ID.

However, even before the *Dover* trial, the most prominent ID-supporting organization, the Discovery Institute, had already pulled back from earlier efforts to try to bring ID into the classroom. Instead, in about 2002, it began to propose (as it currently proposes) that ID should not be mandated; rather, teachers should teach the alleged strengths and weaknesses of evolution. It argues that administrators ought not explicitly require ID to be presented as an alternative, though teachers should be allowed to teach it without penalty if they wish. But the focus has moved away from encouraging the teaching of ID to teaching that evolution is weak science.

EVIDENCE-AGAINST-EVOLUTION SUBSTITUTES FOR CREATIONISM

As discussed in chapter 6, the current manifestation of the old creation science two-model approach is for creationists to propose that evolution be "balanced" by the teaching of alleged evidenced against evolution (EAE). Believing that EAE is evidence for creationism, creationists presume that students taught that evolution is weak or inadequate theory, even in the absence of direct instruction by a teacher, will conclude that therefore God must have specially created living things. This perspective is found both in creation science and in ID; the logical basis of ID is that evolution is inadequate to explain complexity, hence an intelligent designer (God) by default must have specially created complex natural phenomena.

After ID failed to survive its first court challenge, EAE has become the most popular manifestations of creationism. This approach has not yet been systematically dealt with in the courts, but as EAE policies become more popular, there will be more opposition to them in the future. One case in which EAE was a component was that of a high school biology teacher in Minnesota.

Rodney LeVake and Arguments against Evolution

The EAE approach was first tested in the courts in Faribault, Minnesota, in 1998, where the teacher Rodney LeVake was to begin his first year as a high school biology teacher. Colleagues learned that LeVake was omitting evolution from his course, because he thought it was "impossible" (Moore 2004: 327). His administrators requested that LeVake clarify his approach by preparing an essay describing what he would teach. The document he submitted consisted of a list of typical arguments against evolution, including a mixture of creation science arguments and ones popularized by ID. Concerned that students in LeVake's class would not be taught a standard curriculum, his principal and superintendent reassigned LeVake to teach another class in which evolution would not be part of the curriculum. There was no change in his salary, rank, or seniority.

With support from a conservative legal foundation, the American Center for Law and Justice, LeVake sued the district for his free speech right to critically analyze evolution, and he claimed religious discrimination. The Minnesota State Court decided in favor of the district, citing considerable case law holding that a district is within its legal right to direct the teacher on class content. The courts have generally held that a teacher who signs a contract with a district is agreeing to teach the curriculum of that district. Case law does not recognize much academic freedom for teachers at the K–12 level; the situation is very different from that at the college level. LeVake appealed but did not succeed in getting a rehearing.

Another manifestation of EAE, the disclaiming of evolution by presenting it as theory, not fact, had its first court test in the Tangipahoa, Louisiana, case (chapter 6). A carefully worded district court decision struck down the Tangipahoa disclaimer largely on the grounds that it was too overtly religious. A second court case took on a textbook disclaimer that was more carefully worded.

Evolution as Theory, Not Fact: The Cobb County Textbook Sticker

A final EAE approach is related to a misunderstanding of the terminology of science: this is the attempt to require that evolution be taught as theory, not fact. Many of the theory, not fact, policies involve disclaimers to be read to students or inserted into textbooks. In readings presented in chapter 10 there are examples of disclaimers going back to the 1970s, but here I will discuss a more recent case of a textbook disclaimer in Cobb County, Georgia.

In 2001, the Cobb County district began the process to choose new biology textbooks, and as often happens, controversy emerged over how the candidate books treated evolution. By 2001, the Georgia Board of Education had passed science education standards that called for instruction in evolution in the high school biology curriculum. The Cobb County district had, since at least 1979, singled out evolution for special treatment in a series of policies and resolutions. A policy passed in 1995, for example, set up a set of regulations around the teaching of human evolution and removed the topic as a graduation requirement. In 1996, the district requested that a publisher remove a chapter in a fourth-grade science book that discussed a natural origin of the universe and the solar system after parents protested that it ignored creationist teachings (the publisher complied). As a result of the 1995 policy, pages discussing evolution were regularly cut out of textbooks. This was not a district that took casually the teaching of evolution.

The textbook that teachers selected was, as it happens, the same Prentice Hall textbook authored by Kenneth Miller and Joseph Levine that a few years later offended the Dover Area School Board, leading ultimately to *Kitzmiller v. Dover*. The textbook committee was concerned that district policies conflicted with the state standards, so the board promised to review the policies. The Prentice Hall book was thereafter adopted. Some parents, however, objected to the adoption of the book, and one parent, Marjorie Rogers, collected about 2,300 signatures on a petition requesting that alternate views to evolution be presented, and that a "statement [be] placed

Figure 7.2
The Cobb County, Georgia, 2002 textbook disclaimer.

This textbook contains material on evolution. Evolution
is a theory, not a fact, regarding the origin of living things.
This material should be approached with an open mind,
studied carefully, and critically considered.

Approved by
Cobb County Board of Education
Thursday, March 28, 2002

prominently at the beginning of the text that warns the students that some of the information contained in the book is not factual but rather theory, and that there are other theories regarding these matters which are accepted by other scientists."

Shortly thereafter, the Cobb County Board of Education required that the new books would have a sticker inserted to inform students that evolution was theory, not a fact (see Figure 7.2).

In August 2002, a group of parents sued the district on the grounds that the disclaimer sticker was unconstitutional because it favored the beliefs of fundamentalist Christians by denigrating evolution. By expending public funds on the sticker and its maintenance, the district was unconstitutionally promoting religion. The school district argued in defense that its intent was to strengthen the teaching of evolution, as required by the state standards, and that having a textbook disclaimer sticker would assuage some of the parental and community opposition to this controversial topic. It claimed that it never intended to promote religion, and that the revised policies it instituted after the sticker was decided on stated that religious neutrality was to be maintained in the classroom. It took a long time before the case was scheduled for trial, and finally the sides met in early November 2004 before the district court judge Clarence Cooper. After a four-day trial, the judge filed his opinion in January 2005: the disclaimers were unconstitutional and the stickers must be removed.

The *Lemon* test, the commonly used litmus test for the constitutionality of creationist policies, was again applied. Cooper believed the school board members when they claimed that they had no religious purpose in passing the disclaimer requirement: they stated that they had intended to strengthen the teaching of evolution in the district and, because evolution had long been a controversial subject in Cobb County, they required the textbooks to have disclaimers to assuage the concerns of some parents. But on the effect prong of *Lemon*, as modified by the endorsement test, Cooper decided that a reasonable observer in the community would recognize the close ties between disclaiming or criticizing evolution and certain Christian religious views, and would conclude that adherents of these views were being politically favored. He wrote, "The Court's review of pertinent law review articles affirms that encouraging the teaching of evolution as a theory rather than as a fact is one of the latest strategies to dilute evolution instruction employed by antievolutionists with religious motivations" (*Selman v. Cobb County School District*, 390 F.Supp.2d 1286 at 1309). Pursuant to a court order, the district had the stickers removed from the books over the summer.

The case dragged on, however, because the district appealed the ruling to the Third Circuit Court of Appeals, which took more than a year to issue its decision. Finally, in May 2006, the three-judge panel vacated the district court's decision. Because there had been irregularities in the handling of evidence (e.g., Marjorie Rogers's 2,300-person petition was not among the court exhibits), and confusion in the court record as to the specific order of events, the appeals court declared that it was unable to judge the case on its merits and returned the case to Judge Cooper, who had the choice of either retrying the case or correcting the trial record.

Back before Judge Cooper, plaintiffs were prepared to retry the case, bringing in a new legal team that included Eric Rothschild from Pepper Hamilton and Richard Katskee from Americans United for Separation of Church and State—two members of the team that successfully had argued *Kitzmiller v. Dover*. They asked for and won permission from the judge to bring in expert witnesses and reopen discovery. Before long, the defense had settled. The settlement agreement stipulated that neither antievolution nor pro-creationism or ID disclaimers of any kind, oral or written, would be allowed in the district, and the district was directed to follow the state curriculum regarding the teaching of evolution. Mindful of the district's previous policy of cutting references to evolution out of textbooks, the settlement also forbid "excising or redacting materials on evolution in students' science textbooks." The first court trial of a theory, not fact, disclaimer policy, part of the EAE arsenal, had ended in defeat for creationism.

POST-DOVER PREDICTIONS

Prediction is very difficult, especially if it's about the future (as the saying goes). But anyone interested in the creationism/evolution controversy is doubtless curious about what the future will bring, now that ID has had its first (unsuccessful) go-round in the courts. Will this be the end of ID, and will antievolutionists concentrate on evidence against evolution and similar strategies? This last section of the chapter speculates about what we might anticipate in the creationism/evolution controversy in the next five years or so.

ID over in Dover?

Intelligent design proponents indeed suffered a major defeat in *Kitzmiller v. Dover*. As was the case in *McLean v. Arkansas*, a full trial provided the ID side an opportunity to make its best case that ID was a valid scientific alternative to evolution. Judge Jones's decision was long, detailed, and devastating to that contention: on the basis of the testimony and other submitted materials, he ruled that ID was a form of creationism and, at best, a failed science that proponents could not argue was pedagogically appropriate to teach. The decision was not appealed; therefore, regardless of how strong the decision was, it is precedent only in the Middle Federal District of Pennsylvania. In another district, with another fact situation (perhaps with a less obviously religious school board) a policy promoting ID might be proposed and fare better. Although Judge Jones is a conservative Republican appointed by President George W. Bush, an even more conservative judge with perhaps less respect for precedent might be more open to ID arguments. Still, any new venue would have to overcome the strong evidence presented in *Kitzmiller* regarding the creationist history of ID.

But it is not easy to pick a venue for a trial. There might be a federal district court or even a state court that interprets Establishment Clause jurisprudence in a way friendly to ID—but can ID proponents convince a school board in that district to pass a policy? According to news reports, TMLC unsuccessfully tried for four years to interest a school district in passing a pro-ID policy before it found the Dover school board as its test case (Goodstein 2005). And for there to be a trial, there must be someone willing to bring suit against the policy: there must be one or more plaintiffs. And they may be hard to find in any school district where there is solid support for a pro-ID policy. Thus, finding the perfect constellation of facts, so to speak, might be difficult for ID proponents. But because *Kitzmiller* is only advisory rather than precedential outside of its district, a new *Dover*-like trial elsewhere remains a possibility.

There still remains the difficulty for ID proponents of demonstrating that ID is valid science, which is key to its survival in the courts. Thus far, ID has failed to produce any research that supports, much less explains, biological design. As Jones noted in his decision, a negative argument that evolution is inadequate theory does not demonstrate ID. To do so requires at minimum an attempt to present a mechanism: what did the designer do and when did the designer do it? Thus far, ID has made many promises of research breakthroughs just around the corner, but such promises remain to be fulfilled. Without them, ID will likely remain unpersuasive in convincing a judge that it is a valid science.

Relabeling

A predictable strategy for antievolutionists will be to continue to relabel creationism (or ID) so that the legal objections to it are reduced. Proponents of creation science attempted to reduce its legal liability by dropping the word *creation*, although the phrase *intelligent design* itself had shortcomings. *Intelligent design* implies an agent—a designer. A judge would be inclined to ask, "Who is the designer?" Even though the standard ID position is that the identity of the agent is unimportant, and that the agent doesn't have to be God, it takes little digging to discern that a transcendent designer is really what proponents have in mind. As a sectarian religious view, ID would find no place in the public schools. To avoid this problem, ID proponents may attempt to relabel their movement with a term or phrase that does not evoke an agent. In the draft of the FTE's book edited by William Dembski and Jonathan Wells, *The Design of Life*, submitted as evidence in the *Kitzmiller* trial, the phrase *sudden emergence* abruptly appears in place of what, in *Of Pandas and People*, had been the definition of ID. In *Pandas*, the sentence read: "*Intelligent design* means that the various forms of life began abruptly through an *intelligent agency*, with their distinctive features already intact—fish with fins and scales, birds with feathers, beaks, and wings, etc." (Davis and Kenyon 1993, pp. 99–100; emphasis added).

In the manuscript for *The Design of Life*, the paragraph redefines ID and omits any reference to agency: "*Sudden emergence* holds that various forms of life began with their distinctive feature already intact, fish with fins and scales, birds with feathers and wings, animals with fur and mammary glands" (Dembski n.d., 28; emphasis added).

Wendell Bird's abrupt appearance theory similarly was an agentless form of creationism, but this phrase is too closely associated with creation science. Similarly, *sudden emergence theory* may be too closely tied to ID to survive. But given the history of creationism, a new term for the movement may indeed be right around the corner, and this time, it will omit reference to any agent that could be interpreted as God.

There is another type of relabeling that might take place in the post-*Dover* era. On the heels of the *Kitzmiller* decision, Sharon Lemburg, a teacher in one school district proposed teaching ID in social studies. In early January 2006, a teacher in the southern California El Tejon Unified School District began to teach a four-week elective intersession (between semesters) course: Philosophy of Intelligent Design. A professional geologist, Kenneth Hurst, and other parents had protested the course when it was first suggested in the early part of December because of both its negative effect on science education and its promotion of religion. Science teachers in the district also protested the course, contending that it would undermine the science curriculum.

In truth, the course as originally conceived had little to do with ID and consisted almost entirely of videos promoting creation science. The course description read:

> The class will take a close look at evolution as a theory and will discuss the scientific, biological, and Biblical aspects that suggest why Darwin's philosophy is not rock solid. This class will discuss Intelligent Design as an alternative response to evolution. Topics that will be covered are the age of the earth, a world wide flood, dinosaurs, pre-human fossils, dating methods, DNA, radioisotopes, and geological evidence. Physical and chemical evidence will be presented suggesting the earth is thousands of years old, not billions. The class will include lecture discussions, guest speaker, and videos. The class grade will be based on a position paper in which students will support or refute the theory of evolution.

In addition to being confused about the distinction between ID and creation science, Lemburg, a special education teacher, had no credentials for teaching science, a significant omission considering the large number of science topics—from physics and radioisotopic dating to biology—that would be included in this intersession course. Guest speakers for both sides were listed, including local creation science proponents and a minister, and two guest speakers for evolution. One was listed as "Francis Krich," apparently Francis Crick, a Nobel laureate who had died a year and a half before the intersession. During December 2005, the curriculum was revised to remove the more egregious creation science elements, but it remained a pro-ID and antievolution curriculum.

Hurst and other parents, unable to persuade the school board to drop the course, sued the district on the grounds that teaching creation science was an unconstitutional advancement of religion. They were represented by Americans United for Separation of Church and State. The district settled out of court, ending the course early, and promising not to teach "the course entitled 'Philosophy of Design' or 'Philosophy of Intelligent Design' or any other course that promotes or endorses creationism, Creation Science, or intelligent design" (Stipulated Order of Dismissal, *Hurst v. Newman*, No. 1:06-CV-00036, at 2).

It is not constitutionally permissible for a public school teacher to advocate creationism, creation science, or ID. This proscription holds whether the advocacy is taking place in science class or in some other class. Teaching about any religious idea, of course, is not forbidden and is appropriate in many academic disciplines, such as history or sociology. But the courts have held that there is a significant difference between discussing religion in a comparative context and presenting religious views as factually correct, as occurs in the various forms of creationism. Relabeling creationism as philosophy, or some other nonscientific field, then, is not likely to be a successful strategy.

Embedding Evolution

The *Epperson* case struck down Scopes-type antievolution laws partly because religious bias was indicated by the singling out of evolution among all other scientific topics. Other cases dealing with the creationism issue have similarly noted that singling out evolution implies religious purpose or endorsement of religion; as a result, creationists are experimenting with "embedding" evolution in a list of scientific topics to be treated "critically." Then, when the policy is actually implemented, only evolution among the laundry list of scientific topics is singled out for this special treatment. For example, in Phoenixville, Pennsylvania, an antievolution policy failed to pass the school board in 2002. The fallback policy that the majority of school board members could agree on promoted critical thinking and was not specific to evolution—but discussions at the school board as well as letters to the editor made it clear that evolution was the policy that was most likely to be treated critically (Hardy 2002).

Permissive and Academic Freedom Strategies

Another creationist approach is to propose policies that *allow* rather than require the teaching of creationism or antievolutionism, and some of these bills and regulations attempt to protect teachers who may do so. An example is a sample policy that the Discovery Institute began circulating in 2002, and which it encouraged the Dover school board to adopt: "Teachers, in their discretion, may encourage students to consider both the scientific strengths and weaknesses of evolutionary theory in order to better understand the assigned curriculum. This policy does not call for the study of creationism, nor does it call for the study of intelligent design theory" (Cooper 2004).

Although the Dover school board did not adopt the policy, in 2004, the school board in Grantsburg, Wisconsin did, with slight modifications: "Students are expected to analyze, review, and critique scientific explanations, including hypotheses and theories, as to their strengths and weaknesses using scientific evidence and information. Students shall be able to explain the scientific strengths and weaknesses of evolutionary theory. This policy does not call for the teaching of creationism or intelligent design" (Quick 2004).

The Discovery Institute has argued since about 2002 that teachers should not be required to teach ID, but if they choose to teach it, they should be allowed to do so. This perspective may also be intended to make legal challenges more difficult to mount: a policy requiring the teaching of ID could be challenged on its face (i.e., a facial challenge), whereas an individual teacher who chose to teach ID in his or her classroom would require an "as-applied" legal challenge, which is a much more difficult undertaking. A teacher would have to be caught in the act of teaching creationism, which would require more monitoring than is usually possible in an American classroom. Permissive policies of this nature also have the advantage of appealing to a teacher's or the public's support of academic freedom.

This was the approach used in Union County, North Carolina, in 2006 when a citizens' group, the Fair Science Committee, unsuccessfully called for "an objective critique of the theory of evolution as currently being taught" and "academic freedom within the classroom" (Fair Science Committee, 2005). Although popular, calls for academic freedom at the precollege level, under present legal interpretations, fall on deaf ears: the law provides the K–12 teacher with little academic freedom. Courts have consistently held that a teacher has the responsibility to teach the curriculum of the district, as directed by administrators and any governing body. This has been demonstrated clearly in creationism/evolution cases such as *Peloza v. San Juan Capistrano*. In this California case, a teacher lost his suit claiming that his free speech and free exercise of religion were compromised when the district required him to teach evolution. The district's right also to tell teachers what not to teach was illustrated in *Webster v. New Lenox*, an Illinois case where a teacher sued for his claimed right to teach creationism even when the district had directed him not to. The court held that the district was within its right to restrict Webster's curriculum in this matter. On the other hand, if states were to pass legislation giving K–12 teachers more academic freedom, the strategy of using the academic freedom to teach creationism might gain some legal credibility.

Conservative legislators in Florida reacted to the inclusion of evolution in the 2008 Florida science education standards by introducing legislation they called the Academic Freedom Act. Modeled on a sample policy written by the Discovery Institute, the bill called for teachers to "objectively present scientific information relevant to the full range of scientific views regarding biological and chemical evolution," it protects a teacher from discrimination for doing so, and also provides that students will not be "penalized . . . for subscribing to a particular position on evolution.'" Although the bill contended that its implementation would not require any change in the Florida science standards, teachers and scientists disagreed, denouncing the bill as "a subterfuge for injecting the religious beliefs held by some into the science classroom" (Florida Citizens for Science 2008). This is because the bill called for the presentation of scientific information, and proponents of ID and creation science both promote their views as scientific. Teachers were also wary of the bill, fearing that it would leave the door open for students to promote creationist views in their assignments. Eventually, the two houses of the legislature were unable to reach agreement on two different versions of the bill, which therefore died when the legislature adjourned in May 2008.

Also in early 2008, the state of Louisiana enacted its own Academic Freedom Act, similar in spirit to the Discovery Institute's model legislation but based on a policy used in the northern Louisiana parish of Ouachita. The original House version of the bill called for strengths and weaknesses of evolution, but such language was dropped in the Senate version of the bill. This bill (SB 733) was couched in the familiar critical analysis language, calling for "critical thinking skills, logical analysis, and open and objective discussion of scientific theories being studied including, but not limited to, evolution, the origins of life, global warming, and human cloning."

The heart of the bill is the encouragement of the state board of education to "create and foster and environment within public elementary and secondary schools that promotes critical thinking skills, logical analysis, and open and objective discussion of scientific theories being studied including, but not limited to, evolution, the origins of life, global warming, and human cloning." Teachers are permitted to use "supplemental textbooks and other instructional materials to help students understand, analyze, critique, and objectively review [the] scientific theories being studied." Teachers are allowed to use additional instructional materials to critique evolution. This seems to allow the purchase and use of creationist materials, as the content of such materials is composed of critiques of evolution. Although the final version of the bill had been modified considerably from its origin, teachers and other critics of the bill contended that its passage was unnecessary: that teachers were not being stifled over the teaching of evolution.

Demonizing Darwinists

Antievolutionists have long associated evolution with negative historical figures and movements such as Hitler, Stalin, slavery, eugenics—and just about every ism one can imagine. Such demonization of evolution is not new, but in the first decade of the twenty-first century, such accusations seem to be increasing (Coral Ridge Ministries 2007; Ham and Ware 2007; Weikart 2004). In addition to the efforts of the Discovery Institute and the various YEC organizations, the Islamist creationist Harun Yahya has been particularly vociferous in several books published in the late 1990s and early 2000s about the alleged linkage between evolution and social evils like Nazism and communism (Yahya n.d.)

A common theme in such treatments is the familiar confusion of methodological naturalism with philosophical naturalism. Because Darwin (as all scientists) restricted himself to natural causes in explaining evolution, he is accused of promoting the philosophy of naturalism and therefore atheism. The belief is that without God, humankind will suffer moral degeneration and be capable of the kinds of inhuman brutality associated with Hitler and Stalin. The same view can be found in the new Answers in Genesis museum, which presents evolution as inspiring Hitler, Stalin, and Lenin.

There are serious flaws with all parts of the argument, of course: there is no necessary link between methodological naturalism and philosophical naturalism, evidenced by the many scientists who are people of faith. Second, there is no necessary link between religion and morality: there are moral and ethical systems that are not Christian, or even theistic, thus nonbelievers certainly can be ethical and moral (and believers can sometimes fail to live up to religiously based ethical standards). Further, the link among atheism and Hitler, Stalin, and other leaders rightly condemned for their brutality is weak: the origins of such leaders and the cultural, historical, economic, and political forces that bring them to and maintain their power are always complex. In general, historians have treated claims that evolution was a predominant or even serious component to events like the Holocaust as conceptually naive, mistaken in their history, and as better examples of polemics than of scholarship (e.g., Gliboff 2004). As Farber said in a review of Weikart's *From Darwin to Hitler*, "But it is a very long way from barnacles to the death camps" (2005: 390).[1]

It is of more than passing sociological interest that the choice of the demon with which to link evolution (i.e., Darwinism) varies through time and reflects cultural sensibilities. Ken Ham of Answers in Genesis roundly excoriates evolution as the source of racism, whereas creationists in the 1950s and 1960s, before the civil rights movement, were not nearly as concerned with linking racism and evolution. Hitler and the Nazis, however, are always good candidates for demonizing ideas, ideologies, or individuals.

Summary

The teaching of ID in the public schools, like creation science before it, was evaluated in a court of law for its constitutionality. In Dover, Pennsylvania, the attempt by a school board to teach ID resulted in a full trial in a federal district court. The judge, following the *Lemon* decision and the endorsement test, ruled that there was no secular reason for teaching ID, and, because it was a religious view, it could not legally be taught. *Kitzmiller v. Dover* was a landmark decision, and though not appealed to a higher court, will nonetheless be highly influential in any future trials involving ID.

Antievolutionism is thriving even in the absence of a legal warrant to teach ID. The most recent strategy involves denigrating evolution—a two-model approach in which denigrating evolution is seen as promoting creationism. Common phrases associated with this approach include *evidence against evolution, strengths and weaknesses of evolution, critical analysis of evolution*, and *teach the controversy*. Intelligent design proponents also have been encouraging a permissive approach by which teachers are not required to teach ID but supposedly will be protected from lawsuits or negative treatment by superiors if they do.

NOTE

1 Perhaps the most ambitious recent effort to demonize evolution (and evolutionists) is the 2008 documentary, *Expelled: No Intelligence Allowed*, produced by a Canadian production company, Premise Media, and starring the American comic actor and political conservative Ben Stein. In it, big science—represented by mainstream scientific organizations such as the American Association for the Advancement of Science, the National Academy of Sciences, and research scientists at colleges and universities around the country—is presented as unable to accept ID because of an alleged commitment to atheism. The scientific establishment is castigated for ridiculing, denying tenure to, and firing ID scientists "for the 'crime' of merely

believing that there might be evidence of 'design' in nature, and that perhaps life is not just the result of accidental, random chance" (see http://www.expelledthemovie.com). A major theme in the movie is that Darwinism is the source of Nazism and the Holocaust, and other reprehensible social movements such as racism and eugenics. A critique is at http://www.expelledexposed.com.

LEGAL CASES

Epperson v. Arkansas, 393 U.S. 97 (1968).

Lemon v. Kurtzman, 403 U.S. 602 (1971).

McLean v. Arkansas, 529 F. Supp. 1255 (E.D. Ark. 1982).

Edwards v. Aguillard, 482 U.S. 578 (1987).

Webster v. New Lenox, 917 F.2d 1004 (7th Cir. 1990).

Peloza v. San Juan Capistrano Unified School District, 37 F.3d 517 (9th Cir. 1994).

Freiler v. Tangipahoa, 185 F.3d 337 (5th Cir. 1999).

Santa Fe Independent School District v. Doe, 530 U.S. 290 (2000).

Kitzmiller v. Dover Area School District, 400 F.Supp.2d 707 (M.D. Pa. 2000).

Selman v. Cobb County School District, 390 F.Supp.2d 1286 (N.D. Ga. 2005).

Stipulated Order of Dismissal, *Hurst v. Newman*, No. 1:06-CV-00036 (E.D. Cal. Jan. 17, 2006).

REFERENCES

Cooper, Seth. 2004. E-mail to Dover board member Alan Bonsell, December 10. *Kitzmiller v. Dover Area School District*, 400 F.Supp.2d 707, Plaintiffs' Exhibit No. 57.

Coral Ridge Ministries. 2007. *Darwin's deadly legacy* (movie). Available from Coral Ridge Ministries, Fort Lauderdale, Florida.

Davis, Percival W., and Dean H. Kenyon. 1993. *Of pandas and people*. Dallas, TX: Haughton.

Dembski, William. n.d. Manuscript *The Design of Life*, *Kitzmiller v. Dover Area School District*, 400 F.Supp.2d 707. M.D. Pa. 2000, Plaintiffs' Exhibit No. 775.

Fair Science Committee. 2005. Petition for fair science in Union County schools. Retrieved September 6, 2008, from http://www.fairscience.org/Petition.htm.

Farber, Paul L. 2005. Review of Weikart, *From Darwin to Hitler*. *Journal of the History of Biology* 38: 390–391.

Florida Citizens for Science. 2008. Untitled press release. Retrieved May 14, 2008, from http://www.flascience.org/release17mar08.doc.

Forrest, Barbara. and Paul R. Gross. (2003). *Creationism's Trojan horse: The wedge of intelligent design*. New York: Oxford University Press.

Gliboff, Sander. 2004. "Review of Richard Weikart, *From Darwin to Hitler: Evolutionary Ethics, Eugenics, and Racism in Germany*." H-German, H-Net Reviews. Retrieved March 14, 2008, from http://www.h-net.org/reviews/showrev.cgi?path=37981105462766.

Goodstein, Laurie. 2005. In intelligent design case, a cause in search of a lawsuit. *The New York Times*, November 4.

Ham, Ken, and A. Charles Ware. 2007. *Darwin's plantation: Evolution's racist roots*. Green Forest, AR: Master Books.

Hardy, Dan. 2002. District is open to theories about life. *Philadelphia Inquirer* Dec. 3, p. B1

Jones, John E., III. 2007. Our Constitution's intelligent design. *Litigation* 33 (3): 3–6, 57.

Lebo, Lauri. 2005. Mural at issue. *Daily Record* (Dover, PA), September 30.

Miller, Kenneth R. 2002. Goodbye, Columbus. *Reports of the NCSE* 22 (2): 6–8.

Moore, Randy. 2004. Standing up for our profession: A talk with Ken Hubert. *American Biology Teacher* 66 (5): 325–327.

Muise, Robert J. 2005. Defendant's Brief in Support of Motion *in limine* to Exclude the Testimony of Barbara Forrest, Ph.D., *Kitzmiller et al. versus Dover Area School District,* In the United States District Court for the Middle District of Pennsylvania. Court document 159-1, Retrieved October 29, 2008 from http://www.ncseweb.org/webfm_send/470.

National Center for Science Education. 2005. Discovery Institute and Thomas More Law Center squabble in AEI forum. Retrieved October 29, 2008 from http://www.ncseweb.org/news/2005/10/discovery-institute-thomas-more-law-center-squabble-aei-form-00704.

Quick, Suzanne. 2004. Policy that reopened origins debate revised; Grantsburg board says creationism need not be taught. *Milwaukee Journal-Sentinel.* Dec. 10.

Weikart, Richard 2004. *From Darwin to Hitler: Evolutionary ethics, eugenics, and racism in Germany.* Basingstoke, UK: Palgrave Macmillan.

Yahya, Harun. n.d. *The disasters Darwinism brought to humanity,* http://www.harunyahya.com/disaster/.php.

PART III

• • • • • • • • • • •

Selections from the Literature

In part 3, I present selections from the antievolutionist literature and responses from the pro-evolution side. I do not present evidence for evolution; as emphasized throughout this book, evolution is the consensus view of the scientific community, supported by overwhelming evidence from a variety of scientific fields. The best way to begin to learn about the evidence for evolution is to take courses at the university level (they are rarely offered in high school) or to study popular—or better, scientific—sources that present the evidence and theory of this science, some of which are included as references in other chapters and in Further Reading. As discussed in chapter 2, evolution is included among a variety of sciences, including astronomy, geology, biochemistry, biology, and anthropology. Because this book concentrates on the creationism/evolution controversy, I have made antievolutionism the focus of selections from the literature.

Part 3 is organized topically, beginning with the physical sciences in chapter 8 and moving to biology in chapter 9 and legal issues in chapter 10. Educational issues are taken up in chapter 11, followed by religious issues in chapter 12 and topics relevant to the philosophy of science in chapter 13. In this second edition, a new chapter 14, "Creationism and Evolution in the Media and Public Opinion," has been added, which looks at the coverage of the controversy in the press and at surveys of opinions about evolution representing the general public, scientists, and teachers.

CHAPTER 8

• • • • • • • • • • • • • •

Cosmology, Astronomy, Geology

INTRODUCTION

Creationists claim that data from astronomy, geology, and cosmology (the study of the origins and development of the universe) support special creationism. The fields of chemistry and physics tend to be cited primarily through their relationship to cosmology: standard cosmological theory claims that atomic elements evolved in nuclear reactions in stars; special creationism contends that God created atoms and elements (and the rest of the physical universe) in their present form. The most frequently encountered creationist claim from the physical sciences is that the second law of thermodynamics makes evolution impossible; after a basic discussion of creation science, this will be the first scientific topic presented.

Much of the young-Earth creationist (YEC) literature relevant to the physical sciences concerns arguments for a young Earth—and most of these involve some sort of rate calculation. For example, Morris (1974) argues that Earth must be young because of the amount of helium currently in the atmosphere. The radioactive decay of many elements produces helium, and if Earth were indeed billions of years old, there would be far more helium in the atmosphere than has been measured. Another rate example is Morris and Parker's (1987) argument that the amount of oil seeping out of seafloor vents would result in oceans composed only of oil, if indeed Earth were billions of years old. Evidence refuting these and other rate arguments are conveniently presented in Isaak (2007) and online (see http://www.talkorigins.org/indexcc/list.html). These are, again, arguments against evolution rather than positive evidence for creationism.

To support the claim that Earth is young, creationists have long contended that radiometric dating is scientifically flawed, and that standard scientific claims for a 4-billion-year-old Earth are therefore unsupported. However, some creationists are abandoning many of the classic YEC arguments, such as Darwin's deathbed confession, NASA's discovery of a "missing day," women having one more rib than men, and others. In the readings countering creationist claims, I present criticisms by creationists themselves of some of these old chestnuts. Although many individual creationists hold these positions, Answers in Genesis, one of the major creationist organizations and from which these criticisms come, clearly does not promote them.

Intelligent design (ID), on the other hand, attempts to avoid the question of the age of Earth and, indeed, most other fact claims about the nature of the universe. The exception is the embrace by ID proponents of the anthropic principle, a cosmological concept. At heart, the anthropic principle is a design argument, proposing that God created a perfectly tuned universe in order that humankind would evolve. As such, it is embraced by proponents of both YEC and ID.

Creation science above all puts the revealed truth of the Bible before empirical observation, as shown by the first group of readings.

REFERENCES

Isaak, M. 2007. *The counter-creationism handbook*. Berkeley: University of California Press.
Morris, Henry M., ed. 1974. *Scientific creationism*. San Diego: Creation-Life Publishers.
Morris, H. M., and Gary E. Parker. 1987. *What is creation science?* El Cajon, CA: Master Books.

● ●

READING SUPPORTING CREATIONIST CLAIMS: GENERAL

The Tenets of Creationism

Creationism can be studied and taught in any of three basic forms, as follows:

1. Scientific creationism (no reliance on Biblical revelation, utilizing only scientific data to support and expound the creation model).
2. Biblical creationism (no reliance on scientific data, using only the Bible to expound and defend the creation model).
3. Scientific Biblical creationism (full reliance on Biblical revelation but also using scientific data to support and develop the creation model).

These are not contradictory systems, of course, but supplementary, each appropriate for certain applications. For example, creationists should not advocate that Biblical creationism be taught in public schools, both because of judicial restrictions against religion in such schools and also (more importantly) because teachers who do not believe the Bible should not be asked to teach the Bible. It is both legal and desirable, however, that scientific creationism be taught in public schools as a valid alternative to evolutionism.

In a Sunday School class, on the other hand, dedicated to teaching the Scriptures and "all the counsel of God," Biblical creationism should be strongly expounded and emphasized as the foundation of all other doctrine. In a Christian school or college, where the world of God is studied in light of the Word of God, it is appropriate and very important to demonstrate that Biblical creationism and scientific creationism are fully compatible, two sides of the same coin, as it were. The creation revelation in Scripture is thus supported by all true facts of nature; the combined study can properly be called scientific Biblical creationism. All three systems, of course, contrast sharply and explicitly with the evolution model.

The evolution and creation models, in their simplest forms, can be outlined as follows:[1]

Evolution Model	Creation Model
1. Continuing naturalistic origin	1. Completed supernaturalistic origin
2. Net present increase in complexity	2. Net present decrease in complexity
3. Earth history dominated by uniformitarianism	3. Earth history dominated by catastrophism

The evolution model, as outlined above, is in very general terms. It can be expanded and modified in a number of ways to correspond to particular types of evolutionism (atheistic evolution, theistic evolution, Lamarckianism [sic], neo-Darwinism, punctuated equilibrium, etc.).

The same is true of the creation model, with the Biblical record giving additional specific information, which could never be determined from science alone. The three key items in the creation model above are then modified as follows:

Biblical Creation Model

1. Creation completed by supernatural processes in six days
2. Creation in the bondage of decay because of sin and the curse
3. Earth history dominated by the great flood of Noah's day

Creationists, however, do not propose that the public schools teach six-day creation, the fall of man, and the Noachian flood. They do maintain, however, that they should teach the evidence for a complex completed creation, the universal principle of decay (in contrast to the evolutionary assumption of increasing organization), and the worldwide evidences of recent catastrophism. All of these are implicit in observable scientific data, and should certainly be included in public education.

Both the scientific creation model and the Biblical creation model can be considerably expanded to incorporate many key events of creation and earth history, in terms of both scientific observation on the one hand and Biblical doctrine on the other. These can, in fact, be developed as a series of formal tenets of scientific creationism and Biblical creationism, respectively, as listed below:

Tenets of Scientific Creationism

1. The physical universe of space, time, matter, and energy has not always existed, but was supernaturally created by a transcendent personal Creator who alone has existed from eternity.
2. The phenomenon of biological life did not develop by natural processes from inanimate systems but was specially and supernaturally created by the Creator.
3. Each of the major kinds of plants and animals was created functionally complete from the beginning and did not evolve from some other kind of organism. Changes in basic kinds since their first creation are limited to "horizontal" changes (variations) within the kinds, or "downward" changes (e.g., harmful mutations, extinctions).

4. The first human beings did not evolve from an animal ancestry, but were specially created in fully human form from the start. Furthermore, the "spiritual" nature of man (self-image, moral consciousness, abstract reasoning, language, will, religious nature, etc.) is itself a supernaturally created entity distinct from mere biological life.

5. Earth pre-history, as preserved especially in the crustal rocks and fossil deposits, is primarily a record of catastrophic intensities of natural processes, operating largely within uniform natural laws, rather than one of uniformitarian process rates. There is therefore no a priori reason for not considering the many scientific evidences for a relatively recent creation of the earth and the universe, in addition to the scientific evidences that most of the earth's fossiliferous sediments were formed in an even more recent global hydraulic cataclysm.

6. Processes today operate primarily within fixed natural laws and relatively uniform process rates. Since these were themselves originally created and are daily maintained by their Creator, however, there is always the possibility of miraculous intervention in these laws or processes by their Creator. Evidences for such intervention must be scrutinized critically, however, because there must be clear and adequate reason for any such action on the part of the Creator.

7. The universe and life have somehow been impaired since the completion of creation, so that imperfections in structure, disease, aging, extinctions and other such phenomena are the result of "negative" changes in properties and processes occurring in an originally perfect created order.

8. Since the universe and its primary components were created perfect for their purposes in the beginning by a competent and volitional Creator, and since the Creator does remain active in this now-decaying creation, there does exist ultimate purpose and meaning in the universe. Teleological considerations, therefore, are appropriate in scientific studies whenever they are consistent with the actual data of observation, and it is reasonable to assume that the creation presently awaits the consummation of the Creator's purpose.

9. Although people are finite and scientific data concerning origins are always circumstantial and incomplete, the human mind (if open to the possibility of creation) is able to explore the manifestation of that Creator rationally and scientifically, and to reach an intelligent decision regarding one's place in the Creator's plan. . . .

NOTES

1. See Henry M. Morris, ed., *Scientific creationism* (San Diego: Creation-Life Publishers, 1974), 12.

2. These tenets have recently been adopted by the staff of the Institute for Creation Research and incorporated permanently in its By-Laws.

Excerpted from Henry M. Morris, The tenets of creationism. *Impact*, July (1980): 1–4. Used with permission.

"Biblical Glasses"

Q: What are "biblical glasses"?

A: Christians need to understand that because the Bible is the revealed Word of God, and a true record of history—we need to look at the world through the Bible. In other words, we should always put on our biblical glasses in order to understand the world.

For example, if you took your children to the Grand Canyon and they asked you how the layers of rock and the canyon formed, what would you say? If you put on your biblical glasses, you could answer this way:

> Well, children, those layers contain fossils. The Bible teaches that there was no death before sin; therefore, these layers couldn't have been laid down millions of years ago—before Adam sinned. But the Bible tells us about a global Flood—this would have created layers burying lots of dead things. And the catastrophic runoff would have carved the canyon.

Here is another example regarding dinosaurs. When we put on our biblical glasses, we can say much about dinosaurs: they were created on the sixth day, so they coexisted with man; they ate plants before sin; they were on a huge boat that landed in the Middle East; and so on.

By building our thinking on the Bible—beginning with Genesis—we can put on biblical glasses so that we're always ready to give answers to a world that needs them.

Excerpted from Answers in Genesis, *Weekly News*, October 20, 2007, http://www. answersingenesis.org/e-mail/archive/answersupdate/2007/1020.asp.

● ●

READING OPPOSING CREATIONIST CLAIMS: GENERAL

Moving Forward: Arguments We Think Creationists Shouldn't Use

"Darwin Recanted on His Deathbed"

Many people use this story, originally from a Lady Hope. However, it is almost certainly not true, and there is no corroboration from those who were closest to him, even from Darwin's wife, Emma, who never liked evolutionary theory. Also, even if true, so what? If Ken Ham recanted Creation, would that disprove it? So there is no value to this argument whatever (Grigg 1995).

"Moon Dust Thickness Proves a Young Moon"

For a long time, creationists claimed that the dust layer on the moon was too thin if dust had truly been falling on it for billions of years. They based this claim on early estimates—by evolutionists—of the influx of moon dust, and worries that the moon landers would sink into this dust layer. But these early estimates were wrong, and by the time of the Apollo landings, most in NASA were not worried about sinking. So the dust layer thickness can't be used as proof of a young moon (or of an old one either).

"Women Have One More Rib Than Men"

AiG [Answers in Genesis] has long pointed out the fallacy of this statement. Dishonest skeptics wanting to caricature creation also use it, in reverse. The removal of a rib would not affect the genetic instructions passed on to the offspring, any more than a man who loses a finger will have sons with nine fingers. Note also that Adam wouldn't have had a permanent defect, because the rib is the one bone that can regrow if the surrounding membrane (periosteum) is left intact (Wieland 1999).

"NASA Computers, in Calculating the Positions of Planets, Found a Missing Day and 40 Minutes, Proving Joshua's 'Long Day' and Hezekiah's Sundial Movement of Joshua 10 and 2 Kings 20"

This is a hoax. Essentially the same story, now widely circulated on the Internet, appeared in the somewhat unreliable 1936 book *The Harmony of Science and Scripture* by Harry Rimmer. Evidently an unknown person embellished it with modern organization names and modern calculating devices.

Also, the whole story is mathematically impossible—it requires a fixed reference point before Joshua's long day. In fact we would need to cross-check between both astronomical and historical records to detect any missing day. And to detect a missing 40 minutes requires that these reference points be known to within an accuracy of a few minutes. It is certainly true that the timing of solar eclipses observable from a certain location can be known precisely. But the ancient records did not record time that precisely, so the required cross-check is simply not possible. Anyway, the earliest historically recorded eclipse occurred in 1217 B.C., nearly two centuries after Joshua. So there is no way the missing day could be detected by any computer.

Note that discrediting this myth doesn't mean that the events of Joshua 10 didn't happen. Features in the account support its reliability e.g., the moon was also slowed down. This was not necessary to prolong the day, but this would be observed from Earth's reference frame if God had accomplished this miracle by slowing Earth's rotation (Grigg 1997).

REFERENCES

Grigg, R. 1995. Did Darwin recant? *Creation* 18 (1): 36–37.
Grigg, R. 1997. Joshua's long day: Did it really happen—and how? *Creation* 19 (3): 35–37.
Wieland, C. 1999. Regenerating ribs: Adam and that "missing rib." *Creation* 21 (4): 46–47.

Excerpted from Jonathan Sarfati, Moving forward: Arguments we think creationists shouldn't use. *Creation* 24, no. 2 (2002): 20–24. Used with permission.

● ●

THE SECOND LAW OF THERMODYNAMICS PRECLUDES EVOLUTION: CREATIONIST CLAIMS

The Scientific Case Against Evolution: A Summary. Part II

The main scientific reason why there is no evidence for evolution in either the present or the past (except in the creative imagination of evolutionary scientists) is because one of the most fundamental laws of nature precludes it. The law of increasing entropy—also known as the second law of thermodynamics—stipulates that all systems in the real world tend to go "downhill," as it were, toward disorganization and decreased complexity.

This law of entropy is, by any measure, one of the most universal, best-proved laws of nature. It applies not only in physical and chemical systems, but also in biological and geological systems—in fact all systems, without exception.

> No exception to the second law of thermodynamics has ever been found—not even a tiny one. Like conservation of energy (the "first law"), the existence of a law so precise and so independent of details of models must have a logical foundation that is independent of the fact that matter is composed of interacting particles. (Lieb and Yngvason 2000)

The author of this quote is referring primarily to physics, but he [*sic*] does point out that the second law is "independent of details of models." Besides, practically all evolutionary biologists are reductionists—that is, they insist that there are no "vitalist" forces in living systems, and that all biological processes are explicable in terms of physics and chemistry. That being the case, biological processes also must operate in accordance with the laws of thermodynamics, and practically all biologists acknowledge this.

Evolutionists commonly insist, however, that evolution is a fact anyhow, and that the conflict is resolved by noting that the earth is an "open system," with the incoming energy from the sun able to sustain evolution throughout the geological ages in spite of the natural tendency of all systems to deteriorate toward disorganization. That is how an evolutionary entomologist has dismissed W. A. Dembski's impressive recent book, *Intelligent Design*. This scientist defends what he thinks is "natural processes' ability to increase complexity" by noting what he calls a "flaw" in "the arguments against evolution based on the second law of thermodynamics." And what is this flaw? Although the overall amount of disorder in a closed system cannot decrease, local order within a larger system can increase even without the actions of an intelligent agent. (Johnson 2000: 274)

This naive response to the entropy law is typical of evolutionary dissimulation. While it is true that local order can increase in an open system if certain conditions are met, the fact is that evolution does not meet those conditions. Simply saying that the earth is open to the energy from the sun says nothing about how that raw solar heat is converted into increased complexity in any system, open or closed. The fact is that the best known and most fundamental equation of thermodynamics says that the influx of heat into an open system will increase the entropy of that system, not decrease it. All known cases of decreased entropy (or increased organization) in open systems involve a guiding program of some sort and one or more energy conversion mechanisms.

Evolution has neither of these. Mutations are not "organizing" mechanisms, but disorganizing (in accord with the second law). They are commonly harmful, sometimes neutral, never beneficial (at least as far as observed mutations are concerned). Natural selection cannot generate order, but can only "sieve out" the disorganizing mutations presented to it, thereby conserving the existing order, but never generating new order. In principle, it may be barely conceivable that evolution could occur in open systems, in spite of the tendency of all systems to disintegrate sooner or later. But no one yet has been able to show that it actually has the ability to overcome this universal tendency, and that is the basic reason why there is still no bona fide proof of evolution, past or present.

From the statements of evolutionists themselves, therefore, we have learned that there is no real scientific evidence for real evolution. The only observable evidence is that of very limited horizontal (or downward) changes within strict limits. Evolution never occurred in the past, is not occurring at present, and could never happen at all. . . .

REFERENCES

Johnson, Norman A. 2000. Design flaw. *American Scientist* 88 (May/June): 274.
Lieb, E. H., and Jakob Yngvason. 2000. A fresh look at entropy and the second law of thermo-
        dynamics. *Physics Today* 53 (April): 32.

Excerpted from Henry M. Morris, The scientific case against evolution: A Summary, Part II.
        Impact 331 (2001): 1–4. Used with permission.

● ●

THE SECOND LAW OF THERMODYNAMICS PRECLUDES EVOLUTION: READING OPPOSING CREATIONIST CLAIMS

Entropy in Muffins: Why Evolution Does Not Violate the Second Law of Thermodynamics

The Second Law of Thermodynamics has to do with entropy—the entropy of the universe increases during any spontaneous process. A traditional way to understand this is that disorder increases in an isolated (closed) system. This is where some muffins come in handy.

1. Imagine you have 6 muffins hot from the oven and 6 frozen in the freezer. You place the dozen muffins in a special box alternating hot with cold muffins. You place a lid on the box, which will not allow any heat inside the box to escape or any outside temperature to affect the muffins. All heat in the muffins will remain in the box (a closed system).
2. Inside the box, your system is highly ordered: hot, cold, hot, cold. The average temperature in the box is obtained by averaging the temperature of all the muffins together. As time goes by, the heat from the hot muffins mixes with the cold from the frozen muffins to produce a situation where all muffins are the same temperature. Notice that the average temperature is still the same as it was when the muffins first went into the box; only the arrangement of the heat has changed. Entropy has increased; your system is no longer ordered.
3. To keep your system ordered, you would have to have some sort of action or intervention system that would continue to heat the hot muffins and cool the frozen ones. This energy would have to come from outside the system (as it does in the case of a refrigerator, which must be plugged into an external energy source). So you could keep the system ordered, but to do so you would have to have an open system (where energy can flow in).
4. Life is similar. You might have two human beings who seek to increase order by making the two human bodies into three. In a closed system, this increase in order would be impossible.

COSMOLOGY, ASTRONOMY, GEOLOGY

But humans exist in an open system where they take matter and energy in and can spin out additional humans at the rate of one every 9–12 months.

5. This is because the earth is not a closed system. Energy from the sun is like a giant generator powering life on earth. Plants increase the order and complexity in their own bodies as they grow from seed to flower (using the sun's light directly plus the minerals and water in the earth and the carbon from the atmosphere). Herbivores use the energy in plants, carnivores use herbivores, and so on. So a huge cascade of complexity is built on the very simple source of energy from the sun.

6. If the earth were a closed system, then every living organism on earth would be defying entropy on a daily basis. But. . .

7. The earth is not a closed system; thus, respiration, growth, reproduction, and evolution happen on earth on a daily basis without violating the Second Law of Thermodynamics.

8. Many physicists think the universe as a whole is a closed system. That is, not only will the sun burn out some day with the result that life on earth will no longer have the external energy source it needs (actually worse things will probably destroy life on earth before that, as the sun will probably expand and cook everything well before it burns out), but eventually all the energy in the universe—currently arranged like the muffins in the closed box—will even out to the point where no order will exist at all. When the muffins are all the same temperature, the game is over.

9. However, many physicists think that long before the universe falls into total entropy, other things will happen to the overall structure of the universe, so it hardly makes sense to talk about the entire universe as a closed system anyway.

Excerpted from Patricia Princehouse, Entropy in muffins: Why evolution does not violate the second law of thermodynamics, *Reports of the National Center for Science Education*, 25, no. 5–6 (2005): 27. Used with permission.

Biological Evolution and the Second Law

Consider how different the world would be if all systems became less energetic and less organized with time. There would be no puffy clouds, thunderstorms, or weather fronts. Their organization and energy would have dissipated long ago. There would be no trees or flowers. Their seeds would just decay. And we wouldn't be here either. Each of us would have died as a withering zygote that could not undergo development. Clearly the creationist implication that all systems tend toward decay and disorder is wrong. There are many systems besides evolution that tend toward greater order. Philip Morrison (1978), for example, has shown that spontaneous increases in order are common in our world. He points out that the second law really says that increases in order must be paid for in energy. Such increases are clearly not impossible except in closed systems lacking a source of energy. Where large amounts of energy are available, as in the sun-earth system, large increases in order are possible.

Creationists, of course, deny this while claiming that organisms contain some sort of God-given precoded plan and energy conversion system that allows them to escape the death and decay dictated by the second law. On the other hand, almost all scientists accept both the second law and evolution. We need to ask, therefore, just how the second law does affect living systems. A look at gene mutation should allow an answer to this question. A given normal gene will mutate to a nonfunctional version of itself with a characteristic frequency, often on the order of 1/1,000,000.

(For every 999,999 times this gene is transmitted correctly to the next generation, it is transmitted incorrectly one time.) We could call this type of mutation from functional to nonfunctional a "damaging" mutation.

It comes as a surprise to some people, but nonfunctional genes occasionally mutate back to the functional version. We could call this a "repair" mutation. If genes were likened to cars, this would be like saying that occasionally a dented car could be correctly fixed by being in a second accident! However, genes are not cars; chemical complexity is not the same thing as physical complexity. Even though an explosion in a print shop will not produce a dictionary, energy can change simple methane and ammonia into complex amino acids, as Stanley Miller and Harold Urey demonstrated in 1953. Similarly, even though a second collision probably will not undent a dented car, a second mutational event occasionally renders a gene functional again.

The effect of the second law is clearly seen when the repair mutation rate is measured. This repair rate is always less than the damaging mutation rate. In other words, it is easier to go from an ordered state (functional) to a disordered state (nonfunctional) than it is to go in the reverse direction. A typical rate for this repair type of mutation is on the order of 1/1,000,000,000. This is the most important consequence of the second law on living systems. Clearly, the second law does not prevent systems from going from disorder to order. All the law does in this case is to make such mutations rare compared to mutations going in the thermodynamically favored direction—toward disorder. If that's all there were to it, however, gene systems would still eventually all move to a disordered nonfunctional state. They obviously don't. Is this because of a mystical precoded plan, or is there another, nonsupernatural explanation?

Now we come to the essence of evolution: natural selection. All that any organism has to do to escape "degeneration in accord with the second law of thermodynamics" is to be able to produce more young than are needed to replace the parents. As long as that is true, the occasional mutants (almost all less fit than the original version) will usually reproduce poorly or even die without adversely affecting the population. Since the harmful mutations are underrepresented in succeeding generations, these mutations simply cannot build up to a level that threatens the well-being of the population. Thus, mutations are random changes, usually toward disorder, but the effect of natural selection is to remove the relatively common disordered genes and prevent the genetic system from degenerating.

In the same way, natural selection can replace genes with the rare mutant genes that represent an improvement over the original, thus serving as a type of ratchet to improve the organism and keep it matched to its changing environment. The entropy cost of the second law is paid as the energy required to produce those individuals that did not survive. The net result is that life opportunistically saves, builds upon, and improves whatever will function. At first glance, this may appear to conflict with the second law of thermodynamics, but the apparent conflict is not real. Therefore, no divinely precoded plan or mystical "vital force" is needed. Life and evolution are natural phenomena.

REFERENCE

Morrison, P. 1978. In *On aesthetics in science*, ed. J. Wechsler, 69. Cambridge, MA: MIT Press.

Excerpted from William Thwaites and Frank Awbrey, Biological evolution and the second law. *Creation/Evolution* 2, no. 4 (1981): 5–7. Used with permission.

• •

RADIOMETRIC DATING: READINGS SUPPORTING CREATIONIST CLAIMS

Radiometric Dating

In attempting to determine the real age of the earth, it should always be remembered, of course, that recorded history began only several thousand years ago. Not even uranium dating is capable of experimental verification, since no one would actually watch uranium decaying for millions of years to see what happens.

In order to obtain a prehistoric date, therefore, it is necessary to use some kind of physical process that operates slowly enough to measure and steadily enough to produce significant changes. If certain assumptions are made about it, then it can yield a date which could be called the *apparent age*. Whether or not the apparent age is the *true age* depends completely on the validity of the assumptions. Since there is no way in which the assumption can be tested, there is no *sure* way (except by divine revelation) of knowing the true age of any geologic formation. The processes that are most likely to yield dates, which approximate the true dates, are those for which the assumptions are least likely to be in error. . . .

As far as the age of geological formations and of the earth itself are concerned, only radioactive decay processes are considered useful today by evolutionists. There are a number of these, but the most important ones are: (1) the various uranium-thorium-lead methods; (2) the rubidium-strontium method, and (3) the potassium-argon method. In each of these systems, the parent (e.g., uranium) is gradually changed to the daughter (e.g., lead) component of the system, and the relative proportions of the two are considered to be an index of the time since initial formation of the system.

For these or other methods of geochronometry, one should note carefully that the following assumptions must be made:

1. *The system must have been a closed system.* That is, it cannot have been altered by factors extraneous to the dating process; nothing inside the system could have been removed, and nothing outside the system added to it.
2. *The system must initially have contained none of its daughter component.* If any of the daughter component were present initially, the initial amount must be corrected in order to get a meaningful calculation.
3. *The process rate must always have been the same.*

Similarly, if the process rate has ever changed since the system was established, then this change must be known and corrected for if the age calculation is to be of any significance.

Other assumptions may be involved for particular methods, but the three listed above are always involved and are critically important. In view of this fact, the highly speculative nature of all methods of geochronometry becomes apparent when one realizes that *not one* of the above assumptions is valid! None are provable, or testable, or even reasonable.

Excerpted from Henry M. Morris, ed. *Scientific creationism* (San Diego: Creation-Life Publishers, 1974). Used with permission.

What about Carbon Dating?

Other Radiometric Dating Methods

There are various other radiometric dating methods used today to give ages of millions or billions of years for rocks. These techniques, unlike carbon dating, mostly use the relative concentrations of parent and daughter products in radioactive decay chains. For example, potassium-40 decays to argon-40; uranium-238 decays to lead-206 via other elements like radium; uranium-235 decays to lead-207; rubidium-87 decays to strontium-87; etc. These techniques are applied to igneous rocks, and are normally seen as giving the time since solidification.

The isotope concentrations can be measured very accurately, but isotope concentrations are not dates. To derive ages from such measurements, unprovable assumptions have to be made such as:

1. The starting conditions are known (for example, that there was no daughter isotope present at the start, or that we know how much was there).
2. Decay rates have always been constant.
3. Systems were closed or isolated so that no parent or daughter isotopes were lost or added....

"Bad" Dates

When a "date" differs from that expected, researchers readily invent excuses for rejecting the result. The common application of such posterior reasoning shows that radiometric dating has serious problems. Woodmorappe (1999) cites hundreds of examples of excuses used to explain "bad" dates.

For example, researchers applied posterior reasoning to the dating of *Australopithecus ramidus* fossils (Wolde Gabriel et al. 1994). Most samples of basalt closest to the fossil-bearing strata give dates of about 23 Ma (mega annum, million years) by the argon-argon method. The authors decided that was "too old," according to their beliefs about the place of the fossils in the evolutionary grand scheme of things. So they looked at some basalt further removed from the fossils and selected 17 of 26 samples to get an acceptable maximum age of 4.4 Ma. The other nine samples again gave much older dates but the authors decided they must be contaminated and discarded them. That is how radiometric dating works. It is very much driven by the existing long-age world view that pervades academia today.

A similar story surrounds the dating of the primate skull known as KNM-ER 1470. This started with an initial 212 to 230 Ma, which, *according to the fossils*, was considered way off the mark (humans "weren't around then"). Various other attempts were made to date the volcanic rocks in the area. Over the years an age of 2.9 Ma was settled upon because of the agreement between several different published studies (although the studies involved selection of "good" from "bad" results, just like *Australopithecus ramidus*, above).

However, preconceived notions about human evolution could not cope with a skull like 1470 being "that old." A study of pig fossils in Africa readily convinced most anthropologists that the 1470 skull was much younger. After this was widely accepted, further studies of the rocks brought the radiometric age down to about 1.9 Ma—again several studies "confirmed" *this* date. Such is the dating game.

Are we suggesting that evolutionists are conspiring to massage the data to get what they want? No, not generally. It is simply that all observations must fit the prevailing paradigm. The paradigm, or belief system, of molecules-to-man evolution over eons of time, is so strongly entrenched it is not questioned—it is a "fact." So every observation must fit this paradigm. Unconsciously, the researchers, who are supposedly "objective scientists" in the eyes of the public, select the observations to fit the basic belief system.

We must remember that the past is not open to the normal processes of experimental science, that is, repeatable experiments in the present. A scientist cannot do experiments on events that happened in the past. Scientists do not measure the age of rocks, they measure isotope concentrations, and these can be measured extremely accurately. However, the "age" is calculated using assumptions about the past that cannot be proven.

We should remember God's admonition to Job, "Where were you when I laid the foundations of the earth?" (Job 38:4).

Those involved with unrecorded history gather information in the present and construct stories about the past. The level of proof demanded for such stories seems to be much less than for studies in the empirical sciences, such as physics, chemistry, molecular biology, physiology, etc.

Williams, an expert in the environmental fate of radioactive elements, identified 17 flaws in the isotope dating reported in just three widely respected seminal papers that supposedly established the age of the earth at 4.6 billion years (Williams 1992). John Woodmorappe has produced an incisive critique of these dating methods (Woodmorappe 1999). He exposes hundreds of myths that have grown up around the techniques. He shows that the few "good" dates left after the "bad" dates are filtered out could easily be explained as fortunate coincidences.

What Date Would You Like?

The forms issued by radioisotope laboratories for submission with samples to be dated commonly ask how old the sample is expected to be. Why? If the techniques were absolutely objective and reliable, such information would not be necessary. Presumably, the laboratories know that anomalous dates are common, so they need some check on whether they have obtained a "good" date.

Testing Radiometric Dating Methods

If the long-age dating techniques were really objective means of finding the ages of rocks, they should work in situations where we know the age. Furthermore, different techniques should consistently agree with one another.

Methods Should Work Reliably on Things of Known Age

There are many examples where the dating methods give "dates" that are wrong for rocks of known age. One example is K-Ar "dating" of five historical andesite lava flows from Mount Nguaruhoe in New Zealand. Although one lava flow occurred in 1949, three in 1954, and one in 1975, the "dates" range from less than 0.27 to 3.5 Ma (Snelling 1998).

Again, using hindsight, it is argued that "excess" argon from the magma (molten rock) was retained in the rock when it solidified. The secular scientific literature lists many examples of excess argon causing dates of millions of years in rocks of known historical age (Snelling 1998). This excess appears to have come from the upper mantle, below the earth's crust. This is consistent with a young world—the argon has had too little time to escape (Snelling 1998). If excess argon can cause exaggerated dates for rocks of known age, then why should we trust the method for rocks of unknown age?

Other techniques, such as the use of isochrons, make different assumptions about starting conditions, but there is a growing recognition that such "foolproof" techniques can also give "bad" dates. So data are again selected according to what the researcher already believes about the age of the rock.

Geologist Dr. Steve Austin sampled basalt from the base of the Grand Canyon strata and from the lava that spilled over the edge of the canyon. By evolutionary reckoning, the latter should be a billion years younger than the basalt from the bottom. Standard laboratories analyzed the isotopes. The rubidium-strontium isochron technique suggested that the recent lava flow was 270 Ma older than the basalts beneath the Grand Canyon—an impossibility.

Different Dating Techniques Should Consistently Agree

If the dating methods are an objective and reliable means of determining ages, they should agree. If a chemist were measuring the sugar content of blood, all valid methods for the determination would give the same answer (within the limits of experimental error). However, with radiometric dating, the different techniques often give quite different results.

In the study of the Grand Canyon rocks by Austin, different techniques gave different results (Austin 1994: 84–85). Again, all sorts of reasons can be suggested for the "bad" dates, but this is again posterior reasoning. Techniques that give results that can be dismissed just because they don't agree with what we already believe cannot be considered objective.

Method	Age
Six potassium-argon model ages	10,000 years to 117 Ma
Five rubidium-strontium ages	1,270–1,390 Ma
Rubidium-strontium isochron	1,340 Ma
Lead-lead isochron	2,600 Ma

REFERENCES (more available in original article)

Austin, S. A., ed. 1994. *Grand Canyon: Monument to catastrophe.* Santee, CA: Institute for Creation Research.

Snelling, A. A. 1998. The cause of anomalous potassium-argon "ages" for recent andesite flows at Mt. Nguaruhoe, New Zealand, and the implications for potassium-argon "dating." In *Proceedings of the 4th International Creation Conference,* 503–525.

Williams, A. R. 1992. Long-age isotope dating short on credibility. *CEN Technical Journal* 6 (1): 2–5.

Wolde Gabriel, G., et al. 1994. Ecological and temporal placement of early Pliocene hominids at Aramis, Ethiopia. *Nature* 371: 330–333.

Woodmorappe, J. 1999. *The Mythology of Modern Dating Methods.* San Diego: Institute for Creation Research.

Excerpted from Don Batten, Ken Ham, Jonathan D. Sarfati, and Carl Wieland, eds. *The revised and expanded answers book: The 20 most-asked questions about creation, evolution and the book of Genesis answered!*, rev. ed. (Green Forest, AR: Master Books 2000), 120–123. Used with permission.

● ●

RADIOMETRIC DATING: READING OPPOSING CREATIONIST CLAIMS

Common Misconceptions Regarding Radiometric Dating Methods

1. Radiometric dating is based on index fossils whose dates were assigned long before radioactivity was discovered.

This is not at all true, though it is implied by some young-Earth literature. Radiometric dating is based on the half-lives of the radioactive isotopes. These half-lives have been measured over the last 40–90 years. They are not calibrated by fossils.

2. No one has measured the decay rates directly; we only know them from inference.

Decay rates have been directly measured over the last 40–100 years. In some cases a batch of the pure parent material is weighed and then set aside for a long time and then the resulting daughter material is weighed. In many cases it is easier to detect radioactive decays by the energy burst that each decay gives off. For this a batch of the pure parent material is carefully weighed and then put in front of a Geiger counter or gamma-ray detector. These instruments count the number of decays over a long time.

3. If the half-lives are billions of years, it is impossible to determine them from measuring over just a few years or decades.

The example given in the section titled "The Radiometric Clocks" shows that an accurate determination of the half-life is easily achieved by direct counting of decays over a decade or shorter. This is because (a) all decay curves have exactly the same shape [figure omitted], differing only in the half-life, and (b) trillions of decays can be counted in one year even using only a fraction of a gram of material with a half-life of a billion years. Additionally, lavas of historically known ages have been correctly dated even using methods with long half-lives.

4. The decay rates are poorly known, so the dates are inaccurate.

Most of the decay rates used for dating rocks are known to within two percent. Uncertainties are only slightly higher for rhenium (5 %), lutetium (3 %), and beryllium (3 %), discussed in connection with Table 1 [table omitted]. Such small uncertainties are no reason to dismiss radiometric dating. Whether a rock is 100 million years or 102 million years old does not make a great deal of difference....

6. Decay rates can be affected by the physical surroundings.

This is not true in the context of dating rocks. Radioactive atoms used for dating have been subjected to extremes of heat, cold, pressure, vacuum, acceleration, and strong chemical reactions far beyond anything experienced by rocks, without any significant change. The only exceptions, which are not relevant to dating rocks, are discussed under the section "Doubters Still Try."...

10. To date a rock one must know the original amount of the parent element. But there is no way to measure how much parent element was originally there.

It is very easy to calculate the original parent abundance, but that information is not needed to date the rock. All of the dating schemes work from knowing the present abundances of the parent and daughter isotopes. The original abundance N_0, of the parent, is simply $N_0 = N_e{}^{kt}$, where N is the present abundance, t is time, and k is a constant related to the half-life.

11. There is little or no way to tell how much of the decay product, that is, the daughter isotope, was originally in the rock, leading to anomalously old ages.

A good part of this article is devoted to explaining how one can tell how much of a given element or isotope was originally present. Usually it involves using more than one sample from a given rock. It is done by comparing the ratios of parent and daughter isotopes relative to a stable isotope for samples with different relative amounts of the parent isotope. For example, in the rubidium-strontium method one compares rubidium-87/strontium-86 to strontium-87/strontium-86 for different minerals. From this one can determine how much of the daughter isotope would be present if there had been no parent isotope. This is the same as the initial amount (it would not change if there were no parent isotope to decay). Figures 4 and 5 [omitted], and the accompanying explanation, tell how this is done most of the time. While this is not absolutely 100% foolproof, comparison of several dating methods will always show whether the given date is reliable.

12. There are only a few different dating methods.

This article has listed and discussed a number of different radiometric dating methods and has also briefly described a number of non-radiometric dating methods. There are actually many more methods out there. Well over forty different radiometric dating methods are in use, and a number of non-radiogenic methods not even mentioned here....

14. A young-Earth research group reported that they sent a rock erupted in 1980 from Mount Saint Helens volcano to a dating lab and got back a potassium-argon age of several million years. This shows we should not trust radiometric dating.

There are indeed ways to "trick" radiometric dating if a single dating method is improperly used on a sample. Anyone can move the hands on a clock and get the wrong time. Likewise, people actively looking for incorrect radiometric dates can in fact get them. Geologists have known for over forty years that the potassium-argon method cannot be used on rocks only twenty to thirty years old. Publicizing this incorrect age as a completely new finding was inappropriate. . . . Be assured that multiple dating methods used together on igneous rocks are almost always correct unless the sample is too difficult to date due to factors such as metamorphism or a large fraction of xenoliths. . . .

18. We know the Earth is much younger because of non-radiogenic indicators such as the sedimentation rate of the oceans.

There are a number of parameters which, if extrapolated from the present without taking into account the changes in the Earth over time, would seem to suggest a somewhat younger Earth. These arguments can sound good on a very simple level, but do not hold water when all the factors are considered. Some examples of these categories are the decaying magnetic field (not mentioning the widespread evidence for magnetic reversals), the saltiness of the oceans (not counting sedimentation!), the sedimentation rate of the oceans (not counting Earthquakes and crustal movement, that is, plate tectonics), the relative paucity of meteorites on the Earth's surface (not counting weathering or plate tectonics), the thickness of dust on the moon (without taking into account brecciation over time), the Earth-Moon separation rate (not counting changes in tides and internal forces), etc. While these arguments do not stand up when the complete picture is considered, the case for a very old creation of the Earth fits well in all areas considered. . . .

20. Different dating techniques usually give conflicting results.

This is not true at all. The fact that dating techniques most often agree with each other is why scientists tend to trust them in the first place. Nearly every college and university library in the country has periodicals such as *Science*, *Nature*, and specific geology journals that give the results of dating studies. The public is usually welcome to (and should!) browse in these libraries. So the results are not hidden; people can go look at the results for themselves. Over a thousand research papers are published a year on radiometric dating, essentially all in agreement. [Figures and tables omitted.]

Excerpted from Roger C. Weins, Radiometric dating: A Christian perspective. ASA Resources, 2002. Retrieved July 26, 2003, from http://www.asa3.org/ASA/resources/Wiens.html#page%2023. Used with permission.

• •

THE ANTHROPIC PRINCIPLE: READING SUPPORTING CREATIONIST CLAIMS

The Harmony of the Spheres

Physicists recognize four fundamental forces. These largely determine the way in which one bit of matter or radiation can interact with another. In effect, these four forces determine the main characteristics of the universe (Trimble 1977). They are the gravitational force, the electromagnetic force, the strong or nuclear force, and the weak force.

An extraordinary feature of these four fundamental forces is that their strength varies enormously over many orders of magnitude. In the table below they are given in international standard units (Boslough 1985).

The forces of nature

Gravitational force	$5.9 \cdot 10^{-39}$
Nuclear or strong force	15
Electromagnetic force	$3.05 \cdot 10^{-12}$
Weak force	$7.03 \cdot 10^{-3}$

The fact that the gravitational force is fantastically weaker than the strong nuclear force by an unimaginable thirty-eight orders of magnitude is critical to the whole cosmic scheme and particularly to the existence of stable stars and planetary systems (Boslough 1985). If, for example, the gravitational force was a trillion times stronger, then the universe would be far smaller and its life history far shorter. An average star would have a mass a trillion times less than the sun and a life span of about one year—far too short a time for complex life to develop and flourish. On the other hand, if gravity had been less powerful, no stars of galaxies would ever have formed. As Hawking points out, the growth of the universe—so close to the border of collapse and external expansion that man has not been able to measure it—has been at just the proper rate to allow galaxies and stars to form (Boslough 1985).

The other relationships are no less critical. If the strong force had been just slightly weaker, the only element that would be stable would be hydrogen. No other atoms could exist. If it had been slightly stronger in relation to electromagnetism, then an atomic nucleus consisting of only two protons would be a stable feature of the universe—which would mean there would be no hydrogen, and if any stars or galaxies evolved, they would be very different from the way they are (Gribben and Rees 1989).

Clearly, if these various forces and constants did not have precisely the values they do, there would be no stars, no supernovae, no planets, no atoms, no life. . . .

In short, the laws of physics are supremely fit for life and the cosmos gives every appearance of having been specifically and optimally tailored to that end: to ensure the generation of stable stars and planetary systems; to ensure that these will be far enough apart to avoid gravitational interactions which would destabilize planetary orbits; to ensure that a nuclear furnace is generated in the interior of stars in which hydrogen will be converted into the heavier elements essential for life; to ensure that a proportion of stars will undergo supernovae explosions to release the key elements into interstellar space; to ensure that galaxies last several times longer than the lifetime of an average star, for only then will there be time for the atoms scattered by an earlier generation of supernovae within any one galaxy to be gathered into second generation solar systems; to ensure that the distribution and frequency of supernovae will not be so frequent that planetary surfaces would be repeatedly bathed in lethal radiation but not so infrequent that there would be no heavier atoms manufactured and gathered on the surface of newly formed planets; to ensure in the cosmos's vastness and in the trillions of its suns and their accompanying planetary systems a stage immense enough and a time long enough to make certain that the great evolutionary drama of life's becoming will inevitably be manifest sometime, somewhere on an earthlike planet.

And so we are led toward life and our own existence via a vast and ever-lengthening chain of apparently biocentric adaptations in the design of the cosmos in which each adaptation seems adjusted with almost infinite precision toward the goal of life.

REFERENCES

Boslough, J. 1985. *Stephen Hawking's universe*. New York: Quill.
Gribben, J. R., and M. J. Rees. 1989. *Cosmic coincidences*, chap. 10. New York: Bantam Books.
Trimble, V. 1977. Cosmology: Man's place in the universe. *American Scientist* 65: 76–86.

Excerpted from Michael Denton, *Nature's destiny: How the laws of biology reveal purpose in the universe* (New York: Free Press, 1998), 12–14. Used with permission.

• •

THE ANTHROPIC PRINCIPLE: READING OPPOSING CREATIONIST CLAIMS

Anthropic Design: Does the Universe Show Evidence of Purpose?

The fine-tuning argument is based on the fact that earthly life is very sensitive to the values of several fundamental physical constants. Making the tiniest change in any of these, and life as we know it would not exist. The delicate connections between physical constants and life are called the anthropic coincidences (Carter 1974; Barrow and Tipler 1986). The name is a misnomer. Human life is not singled out in any special way. At most, the coincidences show that the production of carbon and the other elements that make earthly life possible required a sensitive balance of physical parameters. . . .

The interpretation of the anthropic coincidences in terms of purposeful design should be recognized as yet another variant of the ancient argument from design that has appeared in many different forms over the ages. The anthropic design argument asks: how can the universe possibly have obtained the unique set of physical constants it has, so exquisitely fine-tuned for life as they are, except by purposeful design—design with life and perhaps humanity in mind?

This argument, however, has at least one fatal flaw. It makes the wholly unwarranted assumption that only one type of life is possible—the particular form of carbon-based life we have here on earth. Even if this is an unlikely result of chance, some form of life could still be a likely result. It is like arguing that a particular card hand is so improbable that it must have been foreordained.

Based on recent studies in the sciences of complexity and "Artificial Life" computer simulations, sufficient complexity and long life appear to be primary conditions for a universe to contain some form of reproducing, evolving structures. This can happen with a wide range of physical parameters, as has been demonstrated (Stenger 1995). The fine-tuners have no basis in current knowledge for assuming that life is impossible except for a very narrow, improbable range of parameters. . . . The inflationary big bang offers a plausible, natural scenario for the uncaused origin and evolution of the universe, including the formation of order and structure—without the violation of any laws of physics. These laws themselves are now understood far more deeply than before, and we are beginning to grasp how they too could have come about naturally....

So how did our universe happen to be so "fine-tuned" as to produce wonderful, self-important carbon structures...? If we have no reason to assume ours is the only life form, we also have no reason to assume that ours is the only universe. Many universes can exist, with all possible combinations of physical laws and constants. In that case, we just happen to be in the particular one that was suited for the evolution of our form of life. When cosmologists refer to the anthropic principle, this is all they usually mean. Since we live in this universe, we can assume it possesses qualities suitable for our existence. Humans evolved eyes sensitive to the region of electromagnetic spectrum from red to violet because the atmosphere is transparent in that range. Yet some would have us think that the causal action was the opposite, that the atmosphere of the earth was designed to be transparent from red to violet because human eyes are sensitive in that range. Stronger versions of the anthropic principle, which assert that the universe is somehow actually required to produce intelligent "information processing systems" (Barrow and Tipler 1986), are not taken seriously by most scientists or philosophers....

The existence of many universes is consistent with all we know about physics and cosmology (Smith 1990; Smolin 1992, 1997; Linde 1994; Tegmark 1997). Some theologians and scientists dismiss the notion as a gross violation of Occam's razor (see, for example., Swinburne 1990). It is not. No new hypothesis is needed to consider multiple universes. In fact, it takes an added hypothesis to rule them out—a super law of nature that says only one universe can exist. But we know of no such law, so we would violate Occam's razor to insist on only one universe.

REFERENCES

Barrow, John D., and Frank J. Tipler. 1986. *The anthropic cosmological principle*. Oxford: Oxford University Press.

Carter, Brandon. 1974. Large number coincidences and the anthropic principle in cosmology. In *Confrontation of cosmological theory with astronomical data*, ed. M. S. Longair, 291–298. Dordrecht: Reidel.

Linde, Andre. 1994. The self-reproducing inflationary universe. *Scientific American* (November), 48–55.

Morris, H. M., ed. 1974. *Scientific creationism*. San Diego: Creation-Life Publishers.

Smith, Quentin. 1990. A natural explanation of the existence and laws of our universe. *Australasian Journal of Philosophy* 68: 22–43.

Smolin, Lee. 1992. Did the universe evolve? *Classical and Quantum Gravity* 9: 173–191.

Smolin, Lee. 1997. *The life of the cosmos*. Oxford: Oxford University Press.

Stenger, Victor J. 1995. *The unconscious quantum: Metaphysics in modern physics and cosmology*. Amherst, NY: Prometheus Books.

Swinburne, Richard. 1990. Argument from the fine-tuning of the universe. In *Physical cosmology and philosophy*, J. Leslie, editor, New York: Collier-Macmillan (154–173).

Tegmark, Max. 1998. Is "the theory of everything" merely the ultimate ensemble theory? *Annals of Physics* 270: 1–51.

Excerpted from Victor Stenger, Anthropic design: Does the universe show evidence of purpose? *Skeptical Inquirer* 23, no. 4 (1999): 40–43.

CHAPTER 9

•••••••••••••

Patterns and Processes
of Biological Evolution

INTRODUCTION

Biological evolution, the inference of common ancestry, is uniformly rejected by young-Earth, old-Earth, and most intelligent design creationists. In this chapter, we will look at how creationists and evolutionary biologists look at biological evolution's patterns (the relationships of living things through time) and processes (the forces or factors that are thought to bring about evolution's cumulative changes). Three themes pertinent to the pattern of evolutionary biology are commonly encountered in antievolutionist literature: the presence of gaps in the fossil record, the Cambrian Explosion, and the question of design. Processes of evolution may conceptually be separated from the patterns of evolution, but as you will read, writers often discuss both of them together.

The key question of process to antievolutionists is whether natural selection and other microevolutionary processes can account for what they call macroevolution. This is discussed later in the chapter under the heading "Micro/Macro." Both creation science and intelligent design proponents also specifically challenge natural selection as being able to do anything more impressive than redistribute genes in a gene pool, with the assumption that the gene pool belongs to fixed, created kinds. Readings on this subject are also included in this chapter.

•••••••••••••••••••••••••••

GAPS IN THE FOSSIL RECORD

Introduction

Evolutionary biologists and antievolutionists are united in one respect: both agree that there are gaps in the fossil record. The record of life as seen in stone does not present a smooth, intergrading continuum from earliest times until the present, nor is there a continuum of variation of form between all living things. Darwin accounted for the fact that most living species do not grade into one another (although some do) because extinction removes intermediate forms. But Darwin himself expected that as the fossil record became better known (it was scarcely investigated in 1859), fossils showing links between groups—transitional fossils—would be discovered.

Creationists claim that there is a systematic lack of transitional fossils; their view is that no transitional groups would occur if the basic kinds had been specially created. Evolutionary biologists point to myriad intermediates—generally unknown or unaccepted by creationists—and are unperturbed by the presence of gaps when they occur.

Part of the difference between the two positions is conceptual and definitional: creationists and evolutionists define and understand evolution differently. Creationists view evolution as being progressive and ladderlike, as in the great chain of being discussed in chapter 4. This and their incomplete understanding of natural selection predict a slow and gradual change of one species into another, resulting in a graded succession of living things. (Hence the familiar, "If man evolved from monkeys, why are there still monkeys?" as if all of the monkey kind evolved into the human kind). If this was indeed how evolution transpired, the fossil record of such a succession probably also would be finely graded, showing myriads of intermediates.

But this is not how evolutionary biologists view the process of evolution. When a new species emerges, it is most likely the result of a population or segment of a species budding off and becoming reproductively isolated: rarely would it be expected that an entire species would evolve into another entire species, leaving no members of the parent species around. Rather, evolutionary biologists stress the branching and splitting of lineages through time—common ancestry is the hallmark of evolution, and therefore the tree of life resembles a bush more than a ladder. Similarly, creationists focus on natural selection almost to the exclusion of other evolutionary mechanisms, whereas evolutionary biologists recognize that there are many factors (e.g., isolating mechanisms that produce a new population that is genetically different from its parent). Evolutionary biologists also don't demand that the rates of change need be gradual: change can occur rapidly. For many reasons, then, evolutionary biologists do not expect that the fossil record will show a smooth and continuous trail of intermediates linking all of life. The casualness with which evolutionary biologists accept gaps also reflects their recognition that the fossil record represents only a fraction of a fraction of all the species that have ever lived; gaps are to be expected.

● ●

GAPS: READING SUPPORTING CREATIONIST CLAIMS

Of Pandas and People

I had intended to excerpt a passage from the intelligent design textbook *Of Pandas and People* (1993), but the authors denied permission. In the section of *Pandas* titled "Fossil Stasis and Gaps Within the Phyla" (100–101), Percival W. Davis and Dean H. Kenyon begin with the claim that much of the fossil record is characterized by a lack of transitions: there is no continuous transitional series between, for example, the earliest horse and the contemporary horse, or between reptiles and mammals. Davis and Kenyon quote the paleontologists David Raup, Stephen Jay Gould, and Steven M. Stanley, and the morphologist Harold C. Bold, all writing in the 1960s or 1970s, for support. Turning to a specific example—the series from reptiles to mammals by way of a group of fossils called therapsids—they address a discussion by James Hopson, writing in *The American Biology Teacher*.

Considering a series of eight therapsids and comparing them to the early mammal *Morganucodon*, Hopson itemizes five ways in which they are increasingly mammalian: (1) in the connection of the limbs, (2) in the mobility of the head, (3) in the fusing of the palate, (4) in the musculature of the jaw, and (5) in the migration of certain bones from the jaw to the middle ear. In response, Davis and Kenyon note parenthetically that soft tissues, such as those in the circulatory and reproductive systems, are not recorded in the fossils. Moreover, they argue, Hopson's series is not a lineage—a single path of genealogical descent—but merely a structural or morphological series. Quoting the biologist Douglas Futuyma, they note that it is impossible to tell which of the numerous therapsid species represented in the fossil record were in fact the ancestors of mammals, and consequently pose two questions. First, if only one therapsid lineage is ancestral to mammals, but several therapsid lineages have the same mammal-like features as the actual ancestral lineage, how powerful are those mammal-like features as evidence of ancestry? Second, if several therapsid lineages independently evolved into mammals, how plausible is it that they independently converged on the distinctive mammalian ear? To the suggestion that the mammalian ear might have been contained, unexpressed, in the genome of the earliest therapsid, they counter that natural selection acts only on expressed traits, and conclude that the evidence is in favor of the existence of "a common blueprint not developed by descent" (1993:101).

Readers are encouraged to read the source for themselves. It is Percival W. Davis and Dean H. Kenyon, *Of Pandas and People*, 2nd ed. (Dallas, TX: Haughton, 1993).

● ●

GAPS: READINGS OPPOSING CREATIONIST CLAIMS

Are There Transitional Forms in the Fossil Record?

. . . Want to start a barroom fight? Ask another patron if he can produce proof of his unbroken patrilineal ancestry for the last four hundred years. Failing your challenge, the legitimacy of his birth is to be brought into question. At this insinuation, tables are overturned, convivial beverages spilled, and bottles fly. No fair, claims the gentle reader. This goes beyond illogic to impoliteness, because you are not only placing on the other patron an unreasonable burden of proof, you are questioning his integrity if he fails. But isn't that what creationists do when they claim that our picture of evolution in the fossil record must be fraudulent because we have so many gaps between forms?

. . . In the search for fossil forms that are ancestral to others, it is commonly assumed that such forms were the *actual individuals* from which living or later forms were descended. This definition is impossible to establish, unlikely on statistical grounds, and unnecessarily restrictive in concept.

... Anthropologists distinguish ... between *lineal* (direct) ancestors and *collateral* (side-branch) ancestors, and it is useful to borrow this concept to discuss real and apparent gaps in the fossil record. Collateral ancestors can still tell us much about the features, habits, and other characteristics of ancestors whose records may be lost but who would still be similar in most respects to those whose records we do have. Your grandfather is your lineal ancestor, whereas your great-uncle is a collateral ancestor; but were their lives and times necessarily much different? This is as true in paleontology as in anthropology. The most basal known member of a taxon (the one who retains the most "primitive" characteristics, and lacks the most derived ones) does not have to be the direct ancestor of the more derived ones; we can accept it as a collateral ancestor, and learn from it a great deal about the features of the actual (though hypothetical) unknown direct ancestor. However, we need to consider the most effective methods for approaching this kind of analysis.

... Phylogenetic [cladistic] analysis, again, provides a solution. In the phylogenetic system, emphasis is placed not on discovering ancestral taxa, but on inferring ancestral (or general) and derived features. Shared derived features (synapomorphies) are the currency of phylogenetic reconstruction. If a synapomorphy is found in two or more related organisms, it is inferred to have been present in their common ancestor. (It could, of course, be independently evolved in each, and this question can be approached by adding more characters and taxa into the analysis.) So, rather than looking for fossils of lineal ancestors, we are now looking for synapomorphies that link collateral ancestors.

... Among living terrestrial vertebrates there is perhaps no clade as distinctive and easily recognizable as the mammals. A variety of anatomical, physiological, osteological, and behavioral characteristics sets mammals apart from other groups of tetrapods.... The diagnosis given by Linnaeus when he coined the term *Mammalia* can still be used to differentiate between living mammals and other tetrapods. However, when we begin to take into account many of the early fossil relatives of mammals, things become much more confusing; it becomes harder to draw a clear distinction between what is a mammal and what is not. This problem stems in part from the fact that the various characteristics that so clearly delineate extant mammals did not all evolve at the same time. Instead, they evolved in a stepwise fashion, with some character states appearing before or after others.

... [Fossil relatives of mammals] are frequently referred to in the popular and scientific literature as "mammal-like reptiles," but this term is misleading and does not reflect our understanding of the relationships between mammals and reptiles. Mammals and the "mammal-like reptiles" are all members of the clade Synapsida and are characterized by having a single opening on the side of the skull through which jaw musculature passes. The clade Mammalia is hierarchically nested within Synapsida, and any synapsid that does not have the synapomorphies that diagnose mammals can be called a nonmammalian synapsid. Early nonmammalian synapsids are only somewhat similar to early reptiles, and not at all like extant ones. For example, their lower jaws are made up of a number of bones, like those of reptiles, instead of the single bone found in modern mammals. But these similarities were inherited by both lineages from their common amniote ancestor, so because they are shared primitive character states ... they are not useful for grouping some synapsids within Reptilia and others within Mammalia. The lineal and collateral ancestors of mammals were never reptiles (reptiles are a separate lineage of amniotes), and the description of nonmammalian synapsids as "mammal-like reptiles" is a holdover of "ladder thinking."

... Unfortunately, critics of evolution such as Denton and Johnson never bother to understand or clarify this distinction, because it is more to their purpose to suggest that we are vainly chasing non-existent transitions between "reptiles" and "mammals" than to show their audiences that the groups of mammals and their relative nestle nicely within a hierarchy of successively more inclusive phylogenetic groups.

Excerpted from Kevin Padian and Kenneth D. Angielczyk, Are there transitional forms in the fossil record? In *The evolution-creationism controversy II: Perspectives on science, religion and geological education*, ed. P. H. Kelley, J. R. Bryan, and T. A. Hansen (Fayetteville, AR: Paleontological Society, 1999), 49–68. Used with permission.

Common Descent, Transitional Forms, and the Fossil Record

Limits of the Fossil Record

... Soft-bodied or thin-shelled organisms have little or no chance of preservation, and the majority of species in living marine communities are soft-bodied. Consider that there are living today about 14 phyla of "worms" comprising nearly half of all animal phyla, yet only a few (e.g., annelids and priapulids) have even a rudimentary fossil record.

... Even those organisms with preservable hard parts are unlikely to be preserved under "normal" conditions. Studies of the fate of clamshells in shallow coastal waters reveal that shells are rapidly destroyed by scavenging, boring, chemical dissolution, and breakage. Occasional burial during major storm events is one process that favors the incorporation of shells into the sedimentary record, and their ultimate preservation as fossils. Getting terrestrial vertebrate material into the fossil record is even more difficult. The terrestrial environment is a very destructive one: with decomposition and scavenging together with physical and chemical destruction by weathering.

The limitations of the vertebrate fossil record can be easily illustrated. The famous fossil *Archaeopteryx*, occurring in a rock unit renowned for its fossil preservation, is represented by only seven known specimens, of which only two are essentially complete. Considering how many individuals of this genus probably lived and died over the thousands or millions of years of its existence, these few known specimens give some feeling for how few individuals are actually preserved as fossils and subsequently discovered.... Complete skeletons are exceptionally rare. For many fossil taxa, particularly small mammals, the only fossils are teeth and jaw fragments. If so many fossil vertebrate species are represented by single specimens, the number of completely unknown species must be greater still!

... In addition to these preservational biases, the erosion, deformation, and metamorphism of originally fossiliferous sedimentary rocks has eliminated significant portions of the fossil record over geologic time. Furthermore, much of the fossil-bearing sedimentary record is hidden in the subsurface, or located in poorly accessible or little studied geographic areas. For these reasons, only a small portion of those once living species actually preserved in the fossil record have been discovered and described by science.

Climbing Down the Tree of Life

. . . A long-standing misperception of the fossil record of evolution is that fossil species form single lines of descent with unidirectional trends. Such a simple linear view of evolution is called orthogenesis ("straight origin"), and has been rejected by paleontologists as a model of evolutionary change. The reality is much more complex than that, with numerous branching lines of descent and multiple anatomical trends. The fossil record reveals that the history of life can be understood as a densely branching bush with many short branches (short-lived lineages). The well-known fossil horse series, for example, does not represent a single, continuous, evolving lineage. Rather, it records more or less isolated twigs of an adapting and diversifying limb of the tree of life. While incomplete, this record provides important insights into the patterns of morphological divergence and the modes of evolutionary change.

Curiously, some critics of evolution view the record of fossil horses from *Hyracotherium* ("Eohippus"), the earliest known representative of this group, to the modern *Equus* as trivial. However, that is only because the intermediate forms are known. Without them, the anatomical gap would be very great. *Hyracotherium* was a very small (some species only 18 inches long) and generalized herbivore (probably a browser). In addition to the well-known difference in toe number (4 toes in front, 3 in back), *Hyracotherium* had a narrow, elongate skull with a relatively small brain and eyes placed well forward in the skull. It possessed small canine teeth, simple tricuspid premolars, and low-crowned simple molars. Over geologic time and within several lines of descent, the skull became much deeper, the eyes moved back, and the brain became larger. The incisors were widened, premolars took the form of molars, and both premolars and molars became very high-crowned with a highly complex folding of the enamel.

The significance of the fossil record of horses becomes clearer when it is compared with that of the other members of the odd-toed ungulates (hoofed mammals). The fossil record of the extinct brontotheres is quite good, and the earliest representatives of this group are very similar to *Hyracotherium*. Likewise, the earliest members of the tapirs and rhinos were also very much like the earliest horses. All these very distinct groups of terrestrial vertebrates can be traced back through a sequence of forms to a group of very similar small, generalized ungulates in the early Eocene. The fossil record thus supports the derivation of horses, rhinos, tapirs, and brontotheres from a common ancestor resembling *Hyracotherium*. Furthermore, moving farther back in time to the late Paleocene, the earliest representatives of the odd-toed ungulates, even-toed ungulates (deer, antelope, cattle, pigs, sheep, camels, etc.), and the proboscideans (elephants and their relatives) were also very similar to each other.

Similar patterns are seen when looking at the fossil record of the carnivores. One group of particular interest is the pinnipeds (seals, sea lions, and walruses). These aquatic carnivores have been found to be closely related to the bears, and transitional forms are known from the early and middle Miocene. More broadly, the living groups of carnivores are divided into two main branches, the Feliformia (cats, hyenas, civets, and mongooses) and the Caniformia (dogs, raccoons, bears, pinnipeds, and weasels). The earliest representatives of these two carnivore branches are very similar to each other, and likely derived from a primitive Eocene group called the miacids. Of the early carnivores, an eminent vertebrate paleontologist has stated: "Were we living at the beginning of the Oligocene, we should probably consider all these small carnivores as members of a single family." This statement also illustrates the point that the erection of a higher taxon is done in retrospect, after sufficient divergence has occurred to give particular traits significance.

... The complex of transitional fossil forms has created significant problems for the definition of the class Mammalia. For most workers, the establishment of a dentary-squamosal jaw articulation is considered one of the primary defining characters for mammals. The transition in jaw articulation associated with the origin of mammals is particularly illustrative of the appearance of a "class-level" morphologic character. In nonmammalian vertebrates, the lower jaw contains several bones, and a small bone at the back of the jaw (the articular) articulates with a bone of the skull (the quadrate). In mammals, the lower jaw consists of only a single bone, the dentary, and it articulates with the squamosal bone of the skull. Within the cynodont lineage, the dentary bone becomes progressively larger and the other bones are reduced to nubs at the back. In one group of advanced cynodonts, the dentary bone has been brought nearly into contact with the squamosal. The earliest known mammals, the morganucodonts, retain the vestigial lower jaw bones of the earlier cynodonts. These small bones still formed a reduced, but functional, jaw joint adjacent to the new dentary-squamosal mammalian articulation. These animals possessed simultaneously both "reptilian" and mammalian jaw articulations! The "reptilian" jaw elements were subsequently detached completely from the jaw to become the bones of the mammalian middle ear. Better intermediate character states could hardly be imagined!

As with most transitions between higher taxonomic categories, there is more than one line of descent that possesses intermediate morphologies. Again, this is consistent with both the expectations of evolutionary theory and the nature of the fossil record. The prediction would be for a bush of many lineages, most of which would be dead ends. [References and illustrations omitted; see original article.]

Excerpted from Keith B. Miller, Common descent, transitional forms, and the fossil record, in *Perspectives on an evolving creation*, edited by K. B. Miller (Grand Rapids, MI: Eerdmans, 2003), 162–168. Used with permission.

• •

THE CAMBRIAN EXPLOSION

Introduction

The Cambrian Explosion began about 535 million years ago when basic features of body plans of invertebrates first appear in the fossil record—shells as found in mollusks, jointed limbs as found in arthropods, exoskeletons, and so on. Most Precambrian animal fossils looked quite different from living invertebrates, although some relatives of modern forms occur during that span of time. Creationists believe that the rapidity with which the Cambrian fauna appear rules out the possibility of natural selection producing these varieties; to them, God created the different kinds separately. The Cambrian problem is largely the product of creationists misunderstanding the evolutionary biologist position, especially their insistence that evolution is always slow and gradual.

Paleontologists consider the Cambrian Explosion to be an interesting scientific puzzle, but by no means a problem for evolution or even for evolution by natural selection. Most paleontologists conclude that invertebrate body plans have a history extending well before the Precambrian-Cambrian boundary—though fossil evidence for this is scarce. But as will be noted here, the fossil record is not the only source of information on relationships among invertebrate groups.

• •

THE CAMBRIAN EXPLOSION: CREATIONIST CLAIMS

Attack and Counterattack: The Fossil Record

...There are two huge gaps in the fossil record that are so immense and indisputable that any further discussion of the fossil record becomes superfluous. These are the gap between microscopic, single-celled organisms and the complex, multicellular invertebrates, and the vast gap between these invertebrates and fish. There are now many reports in the scientific literature claiming the discovery of fossil bacteria and algae in rocks supposedly as old as 3.8 billion years. Paleontologists generally consider that the validity of these claims is beyond dispute. In rocks of the so-called Cambrian period, which evolutionists believe began to form about 600 million years ago, and which supposedly formed during about 80 million years, are found the fossils of a vast array of very complicated invertebrates—sponges, snails, clams, brachiopods, jellyfish, trilobites, worms, sea urchins, sea cucumbers, sea lilies, etc. Unnumbered billions of these fossils are known to exist. Supposedly, these complex invertebrates had evolved from a single-celled organism.

The rocks that generally underlie the Cambrian rocks are simply called Precambrian rocks. Some are thousands of feet thick, and many are undisturbed—perfectly suitable for the preservation of fossils. If it is possible to find fossils of microscopic, single-celled, soft-bodied bacteria and algae, it should certainly be possible to find fossils of the transitional forms between those organisms and the complex invertebrates. Many billions times billions of the intermediates would have lived and died during the vast stretch of time required for the evolution of such a diversity of complex organisms. The world's museums should be bursting at the seams with enormous collections of the fossils of transitional forms. As a matter of fact, not a single such fossil has ever been found! Right from the start, jellyfish have been jellyfish, trilobites have been trilobites, sponges have been sponges, and snails have been snails. Furthermore, not a single fossil has been found linking, say, clams and snails, sponges and jellyfish, or trilobites and crabs, yet all of the Cambrian animals supposedly have been derived from common ancestors.

For a time, evolutionists believed that the Ediacaran Fauna, originally discovered in Australia but now known to be worldwide in distribution, contained creatures that, even though already very complex in nature, might be ancestral to many of the Cambrian animals. Some of the Ediacaran creatures were placed in the same categories as the Cambrian jellyfish, worms, and corals. According to Adolph Seilacher, a German paleontologist, the Ediacaran creatures are, however, basically different from all of the Cambrian animals, and so could not possibly have been ancestral to them. It is believed that all of the Ediacaran creatures became extinct without leaving any evolutionary offspring (Gould 1984). Thus, the Cambrian "explosion," as it is commonly called, remains an unsolved mystery for evolutionists.

. . . Eldredge's main argument is that evolution does not necessarily proceed slowly and gradually, but that some episodes in evolution may, geologically speaking, proceed very rapidly (Eldredge 1982). Thus, just before the advent of the Cambrian, for some reason or other, there was an evolutionary burst—a great variety of complex multicellular organisms, many with hard parts, suddenly evolved. This evolution occurred so rapidly (perhaps in a mere fifteen to twenty million years, more or less) there just wasn't enough time for the intermediate creatures to leave a detectable fossil record.

This notion of explosive evolution is really not a new idea at all, as it has been employed in the past to explain the absence of transitional forms (Simpson 1949). This notion will not stand up under scrutiny, however. First, what is the only evidence for these postulated rapid bursts of evolution? The absence of transitional forms! Thus, evolutionists, like Eldredge, Simpson, and others, are attempting to snatch away from creation scientists what these scientists consider to be one of the best evidences for creation, that is, the absence of transitional forms, and use it as support for an evolutionary scenario!

. . . Later in the book by Eldredge quoted above, Eldredge suggests the most incredible notion of all to explain away the vast Cambrian explosion. He states:

> We don't see much evidence of intermediates in the Early Cambrian because the intermediates had to have been soft-bodied, and thus extremely unlikely to become fossilized. (Eldredge 1982: 130)

It is difficult to believe that Eldredge or any other scientist could have made such a statement. Whatever they were, the evolutionary predecessors of the Cambrian animals had to be complex. A single-celled organism could not possibly have suddenly evolved into a great variety of complex invertebrates without passing through a long series of intermediates of increasing complexity. Surely, if paleontologists are able to find numerous fossils of microscopic, single-celled, soft-bodied bacteria and algae, as Eldredge does not doubt they have, then they could easily find fossils of all the stages intermediate between these microscopic organisms and the complex invertebrates of the Cambrian. Furthermore, in addition to the many reported findings of fossil bacteria and algae, there must be many hundreds of finds of soft-bodied, multicellular creatures, such as worms and jellyfish, in the scientific literature. The creatures of the Ediacaran Fauna, which have been reported from five continents, are soft-bodied.

REFERENCES

Eldredge, Niles. 1982. *The monkey business.* New York: Washington Square Press.
Gould, S. J. 1984. The Ediacaran experiment: Insights into mass extinction theory. *Natural History* 93: 14.
Simpson, G. G. 1949. *The meaning of evolution.* New Haven, CT: Yale University Press.

Excerpted from Duane T. Gish, *Creation scientists answer their critics* (El Cajon, CA: Institute for Creation Research, 1993), 115–119. Used with permission.

Biological Evidence Supports the Theory of Intelligent Design

In recent years the fossil record has also provided new support for the design hypothesis. Fossil studies reveal a "biological big bang" near the beginning of the Cambrian period 530 million years ago. At that time roughly forty separate major groups of organisms or "phyla" (including most all the basic body plans of modern animals) emerged suddenly without evident precursors. Although neo-Darwinian theory requires vast periods of time for the step-by-step development of new biological organs and body plans, fossil finds have repeatedly confirmed a pattern of explosive appearance and prolonged stability in living forms. As I recently argued in an extensive scientific review article published in the peer-reviewed *Proceedings of the Biological Society of Washington*, the emergence of the biological information needed to build these new organisms points strongly to intelligent design. As the information theorist Henry Quastler once observed, "information habitually arises from conscious activity." Thus, I argue that the large infusion of biological information that arises in the Cambrian fossil record points strongly to intelligent design. Moreover, the fossil record also shows a "top-down" hierarchical pattern of appearance in which major structural themes or body plans emerge before minor variations on those themes. Not only does this pattern directly contradict the "bottom-up" pattern predicted by neo-Darwinism, but as I have argued with paleontologist Marcus Ross, philosopher of biology Paul Nelson and University of San Francisco marine paleobiologist Paul Chien, in a 20,000-word scientific review article, "the pattern in the fossil record strongly resembles the pattern evident in the history of human technological design." Thus, we argue that this pattern suggests actual (i.e., intelligent) design as the best explanation for evidence in the fossil record. (Reference omitted.)

REFERENCE

Meyer, S. C., M. Ross, et al. 2003. The Cambrian Explosion: Biology's Big Bang. In *Darwinism, design, and public education,* ed. Stephen C. Meyer and John Angus Campbell, 323–402. East Lansing, MI: Michigan State University Press.

Excerpted from revised report of Stephen C. Meyer, Ph.D., May 19, 2005, in the case of *Kitzmiller v. Dover Area School District*, Case No. 04-CV-2688 (19).

● ●

THE CAMBRIAN EXPLOSION: READINGS OPPOSING CREATIONIST CLAIMS

The Cambrian Explosion

... The gist of Wells's argument [in *Icons of Evolution*, 2000] is that the Cambrian Explosion happened too fast to allow large-scale morphological evolution to occur by natural selection ("Darwinism"), and that the Cambrian Explosion shows "top-down" origination of taxa ("major" "phyla" level differences appear early in the fossil record rather than develop gradually), which he claims is the opposite of what evolution predicts. He asserts that phylogenetic trees predict a different pattern for evolution than what we see in the Cambrian Explosion. These arguments are spurious and show his lack of understanding of basic aspects of both paleontology and evolution.

Wells mistakenly presents the Cambrian Explosion as if it were a single event. The Cambrian Explosion is, rather, the preservation of a series of faunas that occurs over a 15–20 million year period starting around 535 million years ago (MA). A fauna is a group of organisms that live together and interact as an ecosystem; in paleontology, "fauna" refers to a group of organisms that are fossilized together because they lived together. The first fauna that shows extensive body plan diversity is the Sirius Passet fauna of Greenland, which is dated at around 535 MA. The organisms preserved become more diverse by around 530 MA, as the Chenjiang fauna of China illustrates.

... The diversification continues through the Burgess shale fauna of Canada at around 520 MA, when the Cambrian faunas are at their peak. Wells makes an even more important paleontological error when he does not explain that the "explosion" of the middle Cambrian is preceded by the less diverse "small shelly" metazoan faunas, which appear at the beginning of the Cambrian (545 MA). These faunas are dated to the early Cambrian, not the Precambrian as stated by Wells. This enables Wells to omit the steady rise in fossil diversity between the beginning of the Cambrian and the Cambrian Explosion.

In his attempt to make the Cambrian Explosion seem instantaneous, Wells also grossly mischaracterizes the Precambrian fossil record. In order to argue that there was not enough time for the necessary evolution to occur, Wells implies that there are no fossils in the Precambrian record that suggest the coming diversity or provide evidence of more primitive multicellular animals than those seen in the Cambrian Explosion.... Wells ... asserts that there is no evidence for metazoan life until "just before" the Cambrian Explosion, thereby denying the necessary time for evolution to occur. Yet Wells is evasive about what counts as "just before" the Cambrian. Cnidarian and possible arthropod embryos are present 30 million years "just before" the Cambrian. There is also a mollusc, *Kimberella*, from the White Sea of Russia dated approximately 555 million years ago, or 10 million years "just before" the Cambrian. This primitive animal has an uncalcified "shell," a muscular foot and a radula inferred from "mat-scratching" feeding patterns surrounding fossilized individuals. These features enable us to recognize it as a primitive relative of molluscs, even though it lacks a calcified shell. There are also Precambrian sponges as well as numerous trace fossils indicating burrowing by wormlike metazoans beneath the surface of the ocean's floor. Trace fossils demonstrate the presence of at least one ancestral lineage of bilateral animals nearly 60 million years "just" before the Cambrian. Sixty million years is approximately the same amount of time that has elapsed since the extinction of non-avian dinosaurs, providing plenty of time for evolution. In treating the Cambrian Explosion as a single event preceded by nothing, Wells misrepresents fact—the Cambrian Explosion is not a single event, nor is it instantaneous and lacking in any precursors. . . . Wells invokes a semantic sleight of hand in resurrecting a "top-down" explanation for the diversity of the Cambrian faunas, implying that phyla appear first in the fossil record, before lower categories. However, his argument is an artifact of taxonomic practice, not real morphology. In traditional taxonomy, the recognition of a species implies a phylum. This is due to the rules of taxonomy, which state that if you find a new organism, you have to assign it to all the necessary taxonomic ranks. Thus when a new organism is found, either it has to be placed into an existing phylum or a new one has to be erected for it. Cambrian organisms are either assigned to existing "phyla" or new ones are erected for them, thereby creating the effect of a "top-down" emergence of taxa.

. . . [T]he "higher" taxonomic groups appear at the Cambrian Explosion . . . because the Cambrian Explosion organisms are often the first to show features that allow us to relate them to living groups. The Cambrian Explosion, for example, is the first time we are able to distinguish a chordate from an arthropod. This does not mean that the chordate or arthropod lineages evolved then, only that they then became recognizable as such.

. . . Similarly, before the Cambrian Explosion, there were lots of "worms," now preserved as trace fossils (i.e., there is evidence of burrowing in the sediments). However, we cannot distinguish the chordate "worms" from the mollusc "worms" from the arthropod "worms" from the worm "worms." Evolution predicts that the ancestor of all these groups was wormlike, but which worm evolved the notochord, and which the jointed appendages? . . . If the animal does not have the typical diagnostic features of a known phyla [sic], then we would be unable to place it and (by the rules of taxonomy) we would probably have to erect a new phylum for it. When paleontologists talk about the "sudden" origin of major animal "body plans," what is "sudden" is not the appearance of animals with a particular body plan, but the appearance of animals that we can recognize as having a particular body plan. Overall, however, the fossil record fits the pattern of evolution: we see evidence for wormlike bodies first, followed by variations on the worm theme. Wells seems to ignore a growing body of literature showing that there are indeed organisms of intermediate morphology present in the Cambrian record and that the classic "phyla" distinctions are becoming blurred by fossil evidence.

Finally, the "top-down" appearance of body plans is, contrary to Wells, compatible with the predictions of evolution. The issue to be considered is the practical one that "large-scale" body-plan change would of course evolve before minor ones. (How can you vary the lengths of the beaks before you have a head?) The difference is that many of the "major changes" in the Cambrian were initially minor ones. Through time they became highly significant and the basis for "body plans." For example, the most primitive living chordate *Amphioxus* is very similar to the Cambrian fossil chordate *Pikaia*. Both are basically worms with a stiff rod (the notochord) in them. The amount of change between a worm and a worm with a stiff rod is relatively small, but the presence of a notochord is a major "body-plan" distinction of a chordate. Further, it is

Figure 9.1
Evolutionary changes. The evolution of the vertebrate body plan can be visualized as a series of stages where simpler forms with fewer features precede later forms with more of the characteristics associated with this group of organisms (i.e., a notochord, head, paired vertebrae, fusiform body, and finally scales.) Examples of these stages have been found in the fossil record. Courtesy of Alan Gishlick.

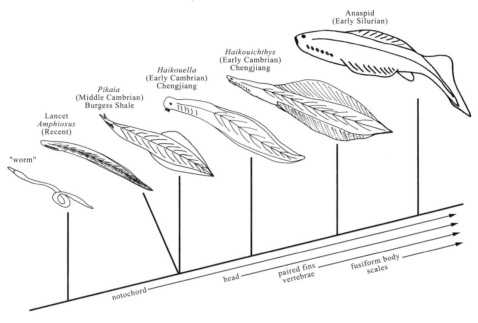

just another small step from a worm with a stiff rod to a worm with a stiff rod and a head (e.g., *Haikouella*) or a worm with a segmented stiff rod (vertebrae), a head, and fin folds (e.g., *Haikouichthys*). Finally add a fusiform body, fin differentiation, and scales: the result is something resembling a "fish" (Figure 9.1). But, as soon as the stiff rod evolved, the animal was suddenly no longer just a worm but a chordate—representative of a whole new phylum! Thus these "major" changes are really minor in the beginning, which is the Precambrian–Cambrian period with which we are concerned.

Excerpted from Alan D. Gishlick, *Icons of evolution? Why much of what Jonathan Wells writes about evolution is wrong* (Icon 2: Darwin's "Tree of Life"). Oakland, CA: National Center for Science Education, 2003. Retrieved October 30, 2008, from http://www.ncseweb. org/creationism/analysis/icon-2-darwins-tree-life. Used with permission.

● ●

DESIGN

Introduction

Living things often show remarkable fit to environments: the grasping foot of a wren allows it to perch on a branch, while the webbed foot of a duck allows it efficiently to propel itself through water. As discussed in chapter 4, before Darwin, the explanation for adaptations was that God had specially created them to allow organisms to survive in their environments: adaptation was seen as a product of Providence. One of the early religious objections to natural selection was that it replaced the necessity of the direct hand of God to explain adaptation. Many antievolutionists continue to believe that if natural selection *could* explain this sort of design, then God is necessarily removed from creation. The argument from design remains an important component of the creationism and evolution controversy.

Creationists regularly (and incorrectly) equate evolution with chance, and because it is absurd to imagine that the complexity of the universe and living things could have come about though random behavior of matter, evolution is rejected as being too improbable. Probability arguments, such as that used by Oller, below, are therefore YEC and ID staples. Both traditional creationists and (especially) intelligent design creationists argue that both the origin of life and anatomical or biochemical/cellular structural complexity are so improbable that natural processes cannot explain them. Note that the structure of these argument focuses upon the alleged shortcomings of evolution, rather than positive evidence for design by God.

● ●

DESIGN: CREATIONIST VIEWS

Not According to Hoyle

In his well-illustrated and impressive new book, *The Intelligent Universe* (London: Michael Joseph, 1983, 256 pp.), Hoyle says:

> As biochemists discover more and more about the awesome complexity of life, it is apparent that its chances of originating by accident are so minute that they can be completely ruled out. Life cannot have arisen by chance. (11–12)

Does this mean that Hoyle has become a creationist? Well, not exactly, and he doesn't expect to either. To forestall any speculation about his apparent "conversion," he says bluntly: "I am not a Christian, nor am I likely to become one as far as I can tell" (251). Still, Hoyle argues that there must have been some "intelligence" behind the emergence of life on Earth. Setting aside the question of what sort of intelligence, he offers an interesting line of argument.

The probability that the simplest life-form could just accidentally arrange itself from particles floating in an ideally prepared primordial soup is very slim. To appreciate just how slim, Hoyle proposes an analogy. He asks how long it would take a blindfolded person to solve a Rubik Cube. Suppose he worked very fast; say, a move a second without resting. According to Hoyle's figuring it would take approximately 67.5 times the estimated age of the universe (allowing the generous figure of 20 billion years since the big bang) for him to reach a solution—about 1.35 trillion years. Judging from the life expectancy of human beings we could say that a solution of the Rubik Cube could not be achieved at all by a blindfolded person. Yet this is just about the same difficulty as the accidental formation of just one of the chains of amino acids necessary to living cells. In the human cell, Hoyle points out, there are about 200,000 such proteins.

The chance of getting all 200,000 by accident is really small. In fact, even if an ideal primordial soup existed, and if it were repeatedly jolted by electrical charges (as in the famous Miller-Urey experiment), the time required for the formation of any one of the requisite 200,000 proteins would be roughly equivalent to 293.5 times the estimated age of the Earth (set at the standard 4.6 billion years).

Yet the odds against the accidental formation of a living organism are considerably worse than the odds against a blindfolded solution of the Rubik Cube—the latter being estimated by Hoyle to be about 50 billion trillion to 1. The trouble is that even a simple protozoan, or a bacterium, requires the prior formation of about 2,000 enzymes, themselves also complex proteins, which are critical to the successful formation of all the other 198,000 or so requisite proteins. The odds in favor of the accidental formation of all 2,000 by accident (never mind the other 198,000), without which no living organism could have come into existence, approaches a truly infinitesimal magnitude. The odds would be similar to those against 2,000 blindfolded persons working Rubik Cubes independently and just accidentally coming to perfect solutions simultaneously—according to Hoyle, roughly 10^{40000} to 1. Or, to give a more graspable notion of the improbability, Hoyle says, it would be roughly comparable to rolling double-sixes 50,000 times in a row with unloaded dice. Looking at it from the point of view of the expected time lapse before reaching a solution, the predicted heat death of our solar system would have occurred early on, and our Milky Way galaxy would have rolled itself up like a scroll long before a solution could be hoped for . . . [References omitted.]

Excerpted from John W. Oller Jr., Not according to Hoyle. *Impact* 138 (1984): 1–4. Used with permission.

The Challenge of Irreducible Complexity

Scientists use the term *black box* for a system whose inner workings are unknown. To Charles Darwin and his contemporaries, the living cell was a black box because its fundamental mechanisms were completely obscure. We now know that, far from being formed from a kind of simple, uniform protoplasm (as many nineteenth-century scientists believed), every living cell contains many ultrasophisticated molecular machines.

How can we decide whether Darwinian natural selection can account for the amazing complexity that exists at the molecular level? Darwin himself set the standard when he acknowledged, "If it could be demonstrated that any complex organ existed which could not possibly have been formed by numerous, successive, slight modifications, my theory would absolutely break down."

Some systems seem very difficult to form by such successive modifications—I call them irreducibly complex. An everyday example of an irreducibly complex system is the humble mousetrap. It consists of (1) a flat wooden platform or base; (2) a metal hammer, which crushes the mouse; (3) a spring with extended ends to power the hammer; (4) a catch that releases the spring; and (5) a metal bar that connects to the catch and holds the hammer back. You can't catch a mouse with just a platform, then add a spring and catch a few more mice, then add a holding bar and catch a few more. All the pieces have to be in place before you catch any mice. Natural selection can only choose among systems that are already working so irreducibly complex biological systems pose a powerful challenge to Darwinian theory.

Irreducibly complex systems appear very unlikely to be produced by numerous, successive, slight modifications of prior systems, because any precursor that was missing a crucial part could not function. We frequently observe such systems in cell organelles, in which the removal of one element would cause the whole system to cease functioning. The flagella of bacteria are a good example. They are outboard motors that bacterial cells can use for self-propulsion. They have a long, whiplike propeller that is rotated by a molecular motor. The propeller is attached to the motor by a universal joint. The motor is held in place by proteins that act as a stator. Other proteins act as bushing material to allow the driveshaft to penetrate the bacterial membrane. Dozens of different kinds of proteins are necessary for a working flagellum. In the absence of almost any of them, the flagellum does not work or cannot even be built by the cell.

Another example of irreducible complexity is the system that allows proteins to reach the appropriate subcellular compartments. In the eukaryotic cell there are a number of places where specialized tasks, such as digestion of nutrients and excretion of wastes, take place. Proteins are synthesized outside these compartments and can reach their proper destinations only with the help of "signal" chemicals that turn other reactions on and off at the appropriate times. This constant, regulated traffic flow in the cell comprises another remarkably complex, irreducible system. All parts must function in synchrony or the system breaks down. Still another example is the exquisitely coordinated mechanism that causes blood to clot.

Biochemistry textbooks and journal articles describe the workings of some of the many living molecular machines within our cells, but they offer very little information about how these systems supposedly evolved by natural selection. Many scientists frankly admit their bewilderment about how they may have originated, but refuse to entertain the obvious hypothesis: that perhaps molecular machines appear to look designed because they really are designed.

I am hopeful that the scientific community will eventually admit the possibility of intelligent design, even if that acceptance is discreet and muted. My reason for optimism is the advance of science itself, which almost every day uncovers new intricacies in nature, fresh reasons for recognizing the design inherent in life and the universe.

Excerpted from Michael Behe, The challenge of irreducible complexity. *Natural History* (April 2002), 74; retrieved May 10, 2008, from http://www.naturalhistorymag.com/darwinanddesign.html. Copyright © Natural History Magazine, Inc. 2002.

● ●

DESIGN: SELECTIONS OPPOSING CREATIONIST VIEWS

Probability

[Creationists claim that the] proteins necessary for life are very complex. The odds of even one simple protein molecule forming by chance are 1 in 10^{113}, and thousands of different proteins are needed to form life.

Response:

1. The calculation of odds assumes that the protein molecule formed by chance. However, biochemistry is not chance, making the calculated odds meaningless. Biochemistry produces complex products, and the products themselves interact in complex ways. For example, complex organic molecules are observed to form in the conditions that exist in space, and it is possible that they played a role in the formation of the first life (Spotts 2001).
2. The calculation of odds assumes that the protein molecule must take one certain form. However, there are innumerable possible proteins that promote biological activity. Any calculation of odds must take into account all possible molecules (not just proteins) that might function to promote life.
3. The calculation of odds assumes the creation of life in its present form. The first life would have been very much simpler.
4. The calculation of odds ignores the fact that innumerable trials would have been occurring simultaneously.

REFERENCE

Spotts, Peter N. 2001. Raw materials for life may predate Earth's formation. *Christian Science Monitor*, January 30, 2001, http://search.csmonitor.com/durable/2001/01/30/fp2s2-csm.shtml

Excerpted from Mark Isaak, *The counter-creationism handbook* (Westport, CT: Greenwood Press, 2005), 43–44.

Miracles and Molecules

. . . Complex biochemical systems, then, bear the molecular stamp of their evolutionary origins. Often, these systems can be found, in one or another organism, in a primitive, less complex state—a state that functions adequately, even if not as efficiently as the more complex state that evolved in other lineages. The eye of a mammal is wondrously, perhaps "irreducibly," complex, but an eye without a lens, capable at least of distinguishing light from dark, is better than no eye at all. Likewise, a lamprey's hemoglobin, even if less efficient than that of a jawed vertebrate, suffices to keep lampreys alive. Yet it is doubtful that a mammal could survive with a lampreylike hemoglobin, for the physiological functions that have evolved in mammals, such as maintaining high body temperature, demand oxygen at a rate that can be supplied only by more efficient, tetrameric [four-strand] hemoglobin. Likewise, it is unlikely that a mammalian fetus could survive without its special hemoglobin. What was once merely an advantage has become a necessity. As Orr emphasizes, irreducible complexity is acquired—it evolves.

Among vertebrates, only a subset—the jawed vertebrates that first evolved about 430 million years ago—have tetrameric hemoglobin, and of these only a subset—the mammals whose ancestors became differentiated from other reptiles about 320 million years ago—have fetal hemoglobin. These facts permit two possible explanations. One—Behe's explanation—is that the common ancestor of all vertebrates, or of all life, was equipped with all the molecular machinery any of its descendants would ever use, and that most of the machinery was lost in most lineages. This hypothesis is not only ludicrous, but also, as Orr points out, makes predictions that are contradicted by evidence. The alternative hypothesis is that new molecular complexities came into existence in various lineages of organisms at different points in time.

If this is true, and if we were to follow Behe in denying a natural, evolutionary origin of each such instance, then each origin of a divergent, duplicate hemoglobin requires us to postulate a special intervention by the omnipotent designer. Bear in mind that the several new hemoglobins I have described are only a few of the many, slightly different hemoglobins that, like those of the salmon, contribute to the complex, fine-tuned adaptation of diverse organisms to their environments. And these are but a tiny fraction of the "irreducibly complex" molecular adaptations to be found among vertebrates, insects, plants, and other forms of life. Behe, then, must be forced to see the designer's handiwork everywhere. Life must present him with countless instances of supernatural intervention—of miracles.

When scientists invoke miracles, they cease to practice science. Were a geologist to cite plate tectonics, a chemist hydrogen bonds, or a physicist gravity as an instance of the miraculous, he or she would be laughed out of the profession. Moreover, they would not be doing their job, which is to seek answers by posing and testing explanatory hypotheses. Faced with the unknown, as all scientists are, the scientist who invokes a miracle in effect says "this is unknowable" and admits defeat. It is only through confidence that the unknown is knowable that physical scientists have achieved explanation, and that biologists have advanced understanding of heredity, development, and evolution to heights scarcely hoped for just a few decades ago. Yet Behe, claiming a miracle in every molecule, would urge us to admit the defeat of reason, to despair of understanding, to rest content in ignorance. Even as biology daily grows in knowledge and insight, Behe counsels us to just give up.

Excerpted from Douglas J. Futuyma, Miracles and molecules. *Boston Review* 22, no. 1 (1997): 29–30, Retrieved October 30, 2008 from http://www.bostonreview.net/BR22.1/futuyma.html. Used with permission.

• •

MICROEVOLUTION/MACROEVOLUTION

Introduction

The classic microevolutionary processes are natural selection, mutation, migration, and genetic drift, though some scientists would also include isolating mechanisms and other factors involved in speciation. These are genetically based mechanisms that affect gene pools of species, and that may result in change (adaptation) or stasis. Microevolutionary processes operate at the level of the species or population. When speciation occurs, genetic material is no longer exchanged between groups of individuals, and each group begins an independent evolutionary history; this branching of species over and over through time results in the familiar tree of life. Evolutionary biologists use the term *macroevolution* to refer to the topics relevant to understanding the distribution of patterns that emerge as species and lineages branch through time. Some of these are the rate of evolutionary change (rapid or slow), the pace of evolutionary change (gradual or jerky), adaptive radiation, morphological trends in lineages (e.g., whether body size gets smaller or larger), extinction or branching of a lineage, concepts (not covered in this book) such as species sorting, and the emergence of major new morphological features (such as segmentation, shells, the fusion or loss of bones). Scientists sometimes colloquially refer to macroevolution as "evolution above the species level," but this phrase does not do justice to the complexity of topics included within the concept.

Micro- and macroevolution are thus different levels of analysis of the same phenomenon: evolution. Macroevolution cannot solely be reduced to microevolution because it encompasses so many other phenomena: adaptive radiation, for example, cannot be reduced only to natural selection, though natural selection helps bring it about. Similarly, macroeconomics cannot be reduced to microeconomics: the food distribution system of the United States, involving import and export of goods from around the world, varied transportation systems and their regulation, multinational corporate decisions about supply and demand, investment and market control—and of course political decisions in Washington and in foreign nations—cannot be modeled by expanding the local organic farmers' market!

But the farmers' market is based on money exchanged for food products consumed by people, which is also an important part of the national food distribution system; there are commonalities between micro- and macroeconomics. Similarly, the currency of microevolution is genes, and because evolution is a genealogical relationship of species, genes are highly relevant to understanding macroevolution—though the consideration of genes will not explain all macroevolutionary topics any more than the amount of money involved can explain macroeconomic systems.

Creationists' view of microevolution is similar to that of evolutionary biologists, but the two groups understand macroevolution very differently. Creationists accept microevolutionary processes affecting genetic variation of populations, and most also accept speciation, or the branching of a lineage into reproductively isolated groups. But creationists take literally the evolutionary biologists' definition of macroevolution as evolution above the species level, and infer that major groups of living things such as phyla and classes—the upper taxonomic levels characterized by body plan differences—have a qualitatively different history than lower levels such as populations and species. They view the distinguishing features of phyla and classes as appearing suddenly, denying that such structures as segments, appendages, exoskeletons, and the like could evolve through microevolutionary processes. Their definition of macroevolution thus overlaps only slightly with that of evolutionary biologists because they concentrate only on the emergence of new body plans or major features that distinguish major kinds of living things. Effectively, macroevolution to creationists equates to the inference of common ancestry, which they reject. Their view is that because God created living things as separate kinds, major groups and the features distinguishing them could not have come about through natural processes, microevolutionary or otherwise. Their position is "micro yes, macro no."

There is a robust argument among evolutionary biologists over how new body plans or major new morphological features arose. No one disputes the importance of natural selection: it affects the genetic variation in populations, which may be the basis for a new species (in conjunction with isolating mechanisms). All parties likewise recognize the possibility or even likelihood of other biological mechanisms affecting morphological features that distinguish major groups of organisms. The issue debated in evolutionary biology is how and how much natural selection and other microevolutionary processes are supplemented by other mechanisms (such as regulatory genes operating early in embryological development). This is rather different from the treatment of this topics among creationists, as you will read.

● ●

MICRO/MACRO: READINGS SUPPORTING CREATIONIST CLAIMS

What Is the Difference between Macroevolution and Microevolution?

. . . There is much misinformation about these two words, and yet, understanding them is perhaps the crucial prerequisite for understanding the creation/evolution issue.

Macroevolution refers to major evolutionary changes over time, the origin of new types of organisms from previously existing, but different, ancestral types. Examples of this would be fish descending from an invertebrate animal, or whales descending from a land mammal. The evolutionary concept demands these bizarre changes. Microevolution refers to varieties within a given type. Change happens within a group, but the descendant is clearly of the same type as the ancestor. This might better be called variation, or adaptation, but the changes are "horizontal" in effect, not "vertical." Such changes might be accomplished by "natural selection," in which a trait within the present variety is selected as the best for a given set of conditions, or accomplished by "artificial selection," such as when dog breeders produce a new breed of dog.

The small or microevolutionary changes occur by recombining existing genetic material within the group. As Gregor Mendel observed with his breeding studies on peas in the mid-1800s, there are natural limits to genetic change. A population of organisms can vary only so much. What causes macroevolutionary change?

Genetic mutations produce new genetic material, but do these lead to macroevolution? No truly useful mutations have ever been observed. The one most cited is the disease sickle-cell anemia, which provides an enhanced resistance to malaria. How could the occasionally deadly disease of SSA ever produce big-scale change?

Evolutionists assume that the small, horizontal microevolutionary changes (which are observed) lead to large, vertical macroevolutionary changes (which are never observed). This philosophical leap of faith lies at the core of evolution thinking. A review of any biology textbook will include a discussion of microevolutionary changes. This list will include the variety of beak shape among the finches of the Galápagos Islands, Darwin's favorite example. Always mentioned is the peppered moth in England, a population of moths whose dominant color shifted during the Industrial Revolution, when soot covered the trees. Insect populations become resistant to DDT, and germs become resistant to antibiotics. While in each case, observed change was limited to microevolution, the inference is that these minor changes can be extrapolated over many generations to macroevolution.

In 1980 about 150 of the world's leading evolutionary theorists gathered at the University of Chicago for a conference entitled "Macroevolution." Their task: "to consider the mechanisms that underlie the origin of species" (Lewin, *Science,* vol. 210, pp. 883–887). "The central question of the Chicago conference was whether

the mechanisms underlying microevolution can be extrapolated to explain the phenomena of macroevolution . . . the answer can be given as a clear, No." Thus the
scientific observations support the creation tenet that each basic type is separate and
distinct from all others, and that while variation is inevitable, macroevolution does
not and did not happen.

Excerpted from John D. Morris, What is the difference between macroevolution and microevolution? *Back to Genesis* October (1996): 94b. Used with permission.

A Real Education in Evolution

. . . *2. Learn to use terms precisely and consistently. Evolution* is a term of many meanings, and the meanings have a way of changing without notice. Dog breeding and
finch-beak variations are frequently cited as typical examples of evolution. So is the
fact that all the differing races of humans descend from a single parent, or even that
Americans today are larger on average than they were a century ago (due to better
nutrition). If relatively minor variations like that were all evolution were about, there
would be no controversy, and even the strictest biblical fundamentalists would be
evolutionists.

Of course evolution is about a lot more than in-species variation. The important
issue is whether the dog breeding and finch-beak examples fairly illustrate the process that created animals in the first place. Using the single term *evolution* to cover
both the controversial and the uncontroversial aspects of evolution is a recipe for
misunderstanding.

At a minimum students must learn to distinguish between microevolution (cyclical
variation within the type, as in the finch-beak example) and macroevolution (the
vaguely described process that supposedly creates innovations such as new complex
organs or new body parts). Don't be impressed by claims that in a few borderline cases
microevolution may have produced, or almost produced, new "species." The definition
of "species" is flexible and sometimes means no more than "isolated breeding group."
By such a definition a fruit fly that breeds in August rather than June may be considered
a new species, although it remains a fruit fly. The question is how we get insects and
other basic groups in the first place. Darwinists typically (but not always) claim that
macroevolution is just microevolution continued over a very long time. The claim is
very controversial, and students should learn why.

3. Keep your eye on the mechanism of evolution; it's the all-important thing. . . . Darwin's
mechanism was natural selection. Today, despite many efforts to find an alternative,
there still isn't really a competitor to the two-part Darwinian mechanism of random
variation (mutation) and natural selection. Darwinists argue with each other about
the relative importance of chance and selection, but some combination of these two
elements is just about the only game in town.

Remember that the mechanism has to be able to design and build very complex
structures like wings and eyes and brains. Remember also that it has to have done
this reliably again and again. Despite offhand references in the literature to possible
alternatives, Darwinian natural selection remains the only serious candidate for a
mechanism that might be able to do the job.

That, by the way explains why many Darwinists are reluctant to make a clear distinction between microevolution and macroevolution. They have evidence for a mechanism for minor variations, as illustrated by the finch-beak example, but have no distinct mechanism for the really creative kind of evolution, the kind that builds new body plans and new complex organs. Either macroevolution is just microevolution continued over a longer time, or it's a mysterious process with no known mechanism. A process like that isn't all that different from a miraculous or God-guided process, and it certainly wouldn't support those expansive philosophical statements about evolution being purposeless and undirected.

Excerpted from Phillip E. Johnson, *Defeating Darwinism by opening minds* (Downers Grove, IL: InterVarsity Press, 1997) 57–59. Used with permission.

● ●

MICRO/MACRO: READINGS OPPOSING CREATIONIST CLAIMS

Creationist Arguments

. . . 4. Creationists, however, deny that mutation, recombination, and natural selection can form new, complex features.

. . . It is not true, however, that mutations are almost universally harmful. Whether mutations that alter the metabolic abilities of bacteria, confer insecticide resistance on a fly, or change the height and growth form of a plant are harmful or beneficial depends on the environment. Evolutionary theory does not postulate that "mutations must be primarily beneficial"—only that some are. Put a culture of bacteria, fungi, or flies into a novel environment and within a few generations it will have evolved improved adaptation, even if, as is easily done with these organisms, you begin with a population of genetically identical individuals, and even if the majority of the mutations in that population are negative.

. . . The crux of the creationist objection, though, lies in the emphasis on "real novelties." Creationists have responded to the fact that biologists have observed genetic change in organisms by inventing the idea that each "kind" was created with a great variety of genes. However, "Modern molecular biology, with its penetrating insight into the remarkable genetic code, has further confirmed that normal variations operate only within the range specified by the DNA for the particular type of organism, so that no truly novel characteristics, producing higher degrees of order or complexity, can appear" (Morris, 1974: 51).

Modern molecular biology has confirmed no such thing. It has confirmed that mutations can affect a small or a large part of a gene or of a chromosome; that new genetic information can come into existence by the duplication of preexisting genes and by exchanges of nucleotides to form entirely new gene sequences; that mutations can alter the organism's biochemistry a great deal or not at all. Together with the study of development, molecular genetics has shown that even slight genetic changes can provide enzymes with new biochemical functions; can alter the size, shape, and growth rate of every feature of an organism's body; and can produce changes much like those that distinguish different, related species. The "range specified by the DNA for the particular type of organism" is a creationist fiction for which none of molecular biology offers support.

The "higher degrees of order or complexity" that creationists believe cannot evolve are actually impossible to define. Begin with a reptile, for example, and imagine one of the lower jaw bones becoming larger and the other smaller, so that they finally are disconnected. Is this an increase in complexity? It is one of the chief defining features of the class Mammalia. The single-cusp tooth of the reptiles develops small accessory cusps. Is this a higher degree of complexity? The different variations on the multicusp theme are the basis of much of the adaptive radiation of the mammals into different ways of life, and genetic variations in tooth form are common within many species of mammals. Imagine slight variations in the position of the eyes, from the side of the head toward the front. Such variations in shape and orientation are characteristic of almost every feature of organisms, although this particular one is a major adaptive feature of the primates. Is it really more complex than similar ones in "lesser" species? The "higher degrees of order and complexity" that so impress the creationists are, in a sense, illusory. The "complexity" of a horse or a dandelion is actually just a collection of individual features. Each of these can (and usually did) evolve independently, and each is not very drastic remodeling of the features of the ancestor. And the material for remodeling is evident in the variation within species.

The creationists continue to argue that variation cannot transcend the limits of the "kind"—a Biblical term that has no meaning in modern taxonomy. Yet they have no idea how to define or recognize a "kind." Are lizards and snakes different "kinds" because iguanas are so different from cobras, or the same "kind" because there are so many intermediate snakelike lizards and lizardlike snakes? This vagueness is convenient for the creationist argument, of course, because whenever a biologist or paleontologist finds an intermediate between two "kinds," the creationist can claim that they are the same "kind" after all. The argument that genetic changes cannot bring about new, more complex "kinds" of organisms rests on the belief that organisms fall into discrete, higher and lower "kinds." But they do not.

REFERENCE

Morris, Henry M., ed. 1974. *Scientific creationism*. San Diego: Creation-Life Publishers.

Excerpted from Douglas J. Futuyma, *Science on trial: The case for evolution* (Sunderland, MA: Sinauer Associates, 1995). 185–187. Used with permission.

A Fin Is a Limb Is a Wing

Today biologists are beginning to understand the origins of life's complexity—the exquisite optical mechanism of the eye, the masterly engineering of the arm, the architecture of a flower or a feather, the choreography that allows trillions of cells to cooperate in a single organism.

... Studying how complex structures came to be is one of the most exciting frontiers in evolutionary biology, with clues coming at remarkable speed.

Some have emerged from spectacular fossils that reveal the precursors of complex organs such as limbs or feathers. Others come from laboratories, where scientists are studying the genes that turn featureless embryos into mature organisms. By comparing the genes that build bodies in different species, they've found evidence that structures as seemingly different as the eyes of a fly and a human being actually have a shared heritage.

Scientists still have a long way to go in understanding the evolution of complexity, which isn't surprising since many of life's devices evolved hundreds of millions of years ago. Nevertheless, new discoveries are revealing the steps by which complex structures developed from simple beginnings. Through it all, scientists keep rediscovering a few key rules. One is that a complex structure can evolve through a series of simpler intermediates. Another is that nature is thrifty, modifying old genes for new uses and even reusing the same genes in new ways, to build something more elaborate.

. . . A limb, a feather, or a flower is a marvel, but not a miracle.

Catching the Light

. . . Evolution, with all its blunders, made the eye; Darwin himself had no doubt about that. But how?

A full answer has to account for not just our own eye, but all the eyes in the animal kingdom. Not long ago, the evidence suggested that the eyes in different kinds of animals—insects, cats, and octopuses, for example—must have evolved independently, much as wings evolved independently in birds and bats. After all, the differences between, say, a human eye and a fly's are profound. Unlike the human eye with its single lens and retina, the fly's is made up of thousands of tiny columns, each capturing a tiny fraction of the insect's field of vision. And while we vertebrates capture light with cells known as ciliary photoreceptors (for their hairlike projections, called cilia), insects and other invertebrates use rhabdomeric photoreceptors, cells with distinctive folds.

In recent years, however, these differences became less stark as scientists examined the genes that build photoreceptors. Insects and humans use the same genes to tell cells in their embryos to turn into photoreceptors. And both kinds of photoreceptors snag light with molecules known as opsins.

These links suggested that photoreceptors in flies, humans, and most other animals all evolved from a single type of cell that eventually split into two new cell types. If so, some animals might carry both types of photoreceptors. And in 2004, scientists showed that rag worms, aquatic relatives of earthworms, have rhabdomeric photoreceptors in their eyes and ciliary photoreceptors hidden in their tiny brain, where they appear to sense light to set the rag worm's internal clock.

With such discoveries, a new picture of eye evolution is emerging. The common ancestor of most animals had a basic tool kit of genes for building organs that could detect light. These earliest eyes were probably much like those found today in little gelatinous sea creatures like salps: just pits lined with photoreceptor cells, adequate to sense light and tell its direction. Yet they were the handiwork of the same genes that build our own eyes, and they relied on the same light-sensing opsins.

Evolution then used those basic genes to fashion more sophisticated eyes, which eventually acquired a lens for turning light into an image. The lens too did not appear out of nothing. Lenses are made of transparent proteins called crystallins, which can bend light "like protein glass," as one scientist says. And crystallins, it turns out, existed well before evolution put them to work in the eye. They were just doing other jobs.

Scientists have discovered one crystallin, for example, in the central nervous system of sea squirts. Instead of making a lens, it is part of a gravity-sensing organ. A mutation may have caused cells in the early vertebrate eye to make the crystallin as well. There it turned out to do something new and extraordinarily useful: bring the world into focus.

Early Blooming

Like many other Victorian gentlemen, Charles Darwin was fond of plants. He packed his hothouses with sundews, cowslips, and Venus flytraps. He had exotic orchids shipped from the tropics. And yet, as he wrote to a friend in 1879, flowers were for him "an abominable mystery."

Darwin was referring to the sudden, unheralded emergence of flowers in the fossil record. Making the mystery all the more abominable was the exquisite complexity of flowers. Typical flowers have whorls of petals and petal-shaped sepals surrounding the plant's male and female sex organs. Many also produce brilliant pigments and sweet nectars to lure insects, which ferry pollen from flower to flower.

Today the mystery of flowers is less abominable, although big questions still remain. The first flowers must have evolved after the ancestors of flowering plants split from their closest living relatives, the gymnosperms—including pines and other conifers, cycads, and ginkgoes—which produce seeds but not flowers.

Some of the most important clues to this transition come from the genes active each time a plant blossoms. It turns out that before a flower takes shape, sets of genes mark out an invisible map at the tip of the stem—the same kind of map found on animal embryos.

The genes divide the tip into concentric rings. "It's like a stack of doughnuts on top of the stem," says Vivian Irish of Yale. Guided by the genes, cells in each ring develop into different flower parts—sepals in the outer ring, for example, and sex organs in the innermost rings.

As is so often the case with complexity, the genes that build flowers are older than the flowers themselves. Gymnosperms turn out to carry flower-building genes even though they don't make flowers. Scientists have yet to determine what those genes do in gymnosperms, but their presence indicates that these genes probably existed in the common ancestor of gymnosperms and flowering plants.

In the flowering plant lineage those genes were borrowed to map out the structure of the flower. The first flowers were simple. But over time, the genes were duplicated accidentally, freeing one copy to take on a new role in flower development. Flowers grew more complex, and some of their parts gained new functions, such as luring insects with bright colors and fragrance.

This flexibility may help explain the success of flowering plants. Some 250,000 known species of flowering plants exist today. Gymnosperms, their flowerless relatives, are stuck at just over 800.

Complexity in Miniature

Some of life's most marvelous structures are its smallest: the minute clockwork of molecules that make cells tick. *E. coli*, a bacterium found in the gut, swims with a tiny spinning tail made up of several dozen different proteins, all working together. Doubters of evolution are fond of pointing out that the flagellum, as this tail is called, needs every one of its parts to function. They argue that it could not have evolved bit by bit; it must have been created in its present form.

But by comparing the flagellar proteins to those in other bacterial structures, Mark Pallen of the University of Birmingham in England and his colleagues have found clues to how this intricate mechanism was assembled from simpler parts. For example, *E. coli* builds its flagellum with a kind of pump that squirts out proteins. The pump is nearly identical, protein for protein, to another pump found on many disease-causing bacteria, which use it not for building a tail but for priming a molecular syringe that injects toxins into host cells. The similarity is, in Pallen's words, "an echo of history, because they have a common ancestor."

Scientists have discovered enough of these echoes to envision how *E. coli*'s flagellum could have evolved. Pallen proposes that its pieces—all of which have counterparts in today's microbes—came together step-by-step over millions of years. It all started with a pump-and-syringe assembly like those found on pathogens. In time, the syringe acquired a long needle, then a flexible hook at its base. Eventually it was linked to a power source: another kind of pump found in the cell membranes of many bacteria. Once the structure had a motor that could make it spin, the needle turned into a propeller, and microbes had new mobility.

Whether or not that's the full story, there is plenty of other evidence that natural selection has been at work on the flagellum. Biologists have identified scores of different kinds of flagella in various strains of bacteria. Some are thick and some are thin; some are mounted on the end of the cell and some on the side; some are powered by sodium ions and some by hydrogen ions. It's just the kind of variation that natural selection is expected to produce as it tailors a structure to the needs of different organisms.

Darwin also argued that complex features can decay over time. Ostriches are descended from flying birds, for example, but their wings became useless as they evolved into full-time runners. It turns out that microbial tails can become vestigial as well. Although *E. coli* is believed to make only one kind of tail, it also carries the remnants of genes for a second type. "You expect to see the baggage of history," says Pallen.

Evolution, ruthless and practical, is equally capable of building the most wonderful structures and tossing them aside when they're no longer needed.

Excerpted from Carl Zimmer, A fin is a limb is a wing, *National Geographic* 210 (2006): 110–135. Used with permission.

•••••••••••••••••••••••••••

NATURAL SELECTION AS IMPOTENT

Introduction

First proposed by Darwin in 1859, natural selection is viewed by evolutionary biologists as the most powerful engine of evolutionary change, and the best explanation for adaptation. There are many other factors that affect evolution that have been discovered since Darwin's time; for example, we know a great deal more about the sources of genetic variation on which natural selection operates. We also know more about the process of speciation, surely a critically important component of evolution that rarely is directly related to natural selection. Nonetheless, natural selection is the best explanation for the adaptations that fit organisms to their environments, resulting in shifts of morphology and biochemistry that when tracked, show us the branching pattern of species through time that we recognize as evolution.

Because of the primacy of natural selection in evolutionary biology theory, creationists have attacked it as a means of attacking evolution, confusing the big idea of evolution with the mechanisms of evolution (see chapter 2). In their view, natural selection can only produce small-scale adaptation of species to environments (for example, shifting finch beak size in conditions of drought); it is not "powerful" enough to produce differences between kinds, such as the difference between a segmented worm and an arthropod. Natural selection in the creationist perspective is capable only of moving genes around within a population or species; it is too weak a mechanism to do the important job ascribed to it by evolutionary biologists. The impotence of natural selection is related to the claimed qualitative distinction between macroevolution and microevolution discussed in the previous section, but in this section of the chapter, we will focus more narrowly on views of natural selection itself.

Michael Behe's concept of irreducible complexity as an argument for design, as discussed in chapter 6, depends on the relative impotence of natural selection to produce new information and complex structures.

• •

NATURAL SELECTION AS IMPOTENT: CREATIONIST VIEW

1.2.3.1.2. The lack of rigorous, detailed Darwinian explanations for the appearance of design in biology

Proponents of Darwin's theories such as Richard Dawkins are convinced that natural selection can count for the strong appearance of design in biology. However, like proponents of complexity theory, proponents of intelligent design theory are skeptical of the Darwinian claim, and deny that random mutation and natural selection have been shown to account for some complex aspects of life.

Some scientists who are not proponents of intelligent design freely admit that Darwinian theory has so far been unable to give rigorous, detailed explanations for the complex biochemical machinery discovered in the cell by modern science. For example, in *The Way of the Cell*, published by Oxford University Press in 2001, while considering the claims of intelligent design proponents, Colorado State University emeritus microbiologist Franklin M. Harold wrote that:

> [W]e must concede that there are presently no detailed Darwinian accounts of the evolution of any biochemical system, only a variety of wishful speculations.

When reviewing my book, *Darwin's Black Box: the Biochemical Challenge to Evolution*, which argued that Darwinian explanations have not yet been given for complex bio-chemical systems, for the science journal *Nature*, University of Chicago evolutionary biologist Jerry Coyne wrote:

> There is no doubt that the pathways described by Behe are dauntingly complex, and their evolution will be hard to unravel. . . . We may forever be unable to envision the first proto-pathways. (Coyne 1996)

The point is that some scientists who are not at all sympathetic to ID nonetheless admit the Darwinian theory has not given detailed, testable explanations for the Dar-winian evolution by random mutation and natural selection of complex biochemical systems in the cell. *Thus the path is open to alternative explanations.*

3. What are the gaps and problems with the Darwinian theory of evolution?

3.1 The problem of the origin of new, complex biological features

It is my scientific opinion that the primary problem with Darwin's theory of evo-lution is the lack of detailed, testable, rigorous explanations for the origin of new, complex, biological features, as explained above in section 1.2.3.1.2. This problem was recognized in the 19th century, shortly after Darwin published *The Origin of Species*, by biologists such as St. George Mivart ("What is to be brought forward [against Darwin's theory] may be summed up as follows: That "Natural Selection" is incompetent to account for the incipient stages of useful structures . . ."), and continues to be a prob-lem today. Although vague stories and speculations are sometimes offered, rarely are such stories testable in a way that could falsify the claim that the complex feature was produced in a Darwinian fashion. For example, as stated above in section 1.2.3.1.2, in the case of the molecular machinery found in cells Franklin M. Harold wrote that: "[W]e must concede that there are presently no detailed Darwinian accounts of the evolution of any biochemical system, only a variety of wishful speculations." And Jerry Coyne wrote "there is no doubt that the pathways described by Behe are dauntingly complex, and their evolution will be hard to unravel. . . . We may forever be unable to envision the first proto-pathways." *It is extremely difficult or impossible to test—or even meaningfully critique—"wishful speculations" or envisioned proto-pathways.*

It should be strongly emphasized that under this broad category of difficulties lies much of the structure and development of life, including: the existence of the genetic code; transcription of DNA; translation of mRNA; the structure and function of the ribosome; the structure of the cytoskeleton; nucleosome structure; the development of new protein-protein interaction; the existence of the proteosome; the existence of the endoplasmic reticulum; the existence of motility organelles such as the bacterial flagellum and the eukaryotic cilium; the development of the pathways for the con-struction of the cilium and flagellum; the existence of the defensive apparatus such as the immune system and blood clotting system; and much else. The existence of such unresolved difficulties for Darwinian theory at the molecular level of life makes it reasonable to wonder if a Darwinian framework is the right way to approach such questions. It also makes it reasonable to wonder if Darwinian processes explain major new features of life at higher levels, such as the level of organs and organisms.

REFERENCES

Coyne, Jerry. 1996. God in the details. *Nature* 383: 227–228.
Harold, F. M. 2001. *The way of the cell.* Oxford: Oxford University Press.
Mivart, S. 1871. *On the genesis of species*, 2nd ed. London: Macmillan.

Excerpted from expert report of Michael Behe, Ph.D., March 24, 2005, in the case of *Kitzmiller v. Dover Area School District*, Case No. 04-CV-2688 (13–17).

• •

NATURAL SELECTION AS IMPOTENT: OPPOSING CREATIONIST VIEW

Evolution: God as Genetic Engineer

In *Darwin's Black Box: The Biochemical Challenge to Evolution* (1), Behe had forwarded the notion that certain biochemical systems were "irreducibly complex," could not have evolved stepwise by Darwinian mechanisms, and thus were intelligently designed. Since that earlier book, Behe has played a key role in the intelligent design (ID) movement, including a star turn as a defense witness in the 2005 Dover school board case. Despite his testimony—or, I should say, partly because of what he said (2)—ID was ruled to be a religious concept and its teaching in public schools unconstitutional.

. . . Behe also explores some examples of Darwinian evolution at the molecular level, including an extensive treatment of the evolutionary "trench warfare" fought between humans and malarial parasites over the millennia—all in the context of what Darwinian evolution "can do." So what's the problem?

The problem is what Behe asserts Darwinian evolution can't do: produce more "complex" changes than those that have enabled humans to battle malaria or allowed malarial parasites to evade the drugs we throw at them. Behe's main argument rests on the assertion that two or more simultaneous mutations are required for increases in biochemical complexity and that such changes are, except in rare circumstances, beyond the limit of evolution. He concludes that "most mutations that built the great structures of life must have been nonrandom." In short, God is a genetic engineer, somehow designing changes in DNA to make biochemical machines and higher taxa.

But to arrive at this conclusion, Behe relies on invalid assertions about how genes and proteins evolve and how proteins interact, and he completely ignores a huge amount of experimental data that directly contradicts his faulty premises. Unfortunately, these errors are of a technical nature and will be difficult for lay readers, and even some scientists (those unfamiliar with molecular biology and evolutionary genetics), to detect. Some people will be hoodwinked. My goal here is to point out the critical flaws in Behe's key arguments and to guide readers toward some references that illustrate why what he alleges to be beyond the limits of Darwinian evolution falls well within its demonstrated powers.

Behe's chief error is minimizing the power of natural selection to act cumulatively as traits or molecules evolve stepwise from one state to another via intermediates. Behe states correctly that in most species two adaptive mutations occurring instantaneously at two specific sites in one gene are very unlikely and that functional changes in proteins often involve two or more sites. But it is a non sequitur to leap to the conclusion, as Behe does, that such multiple-amino acid replacements therefore can't happen. Multiple replacements can accumulate when each single amino acid replacement affects performance, however slightly, because selection can act on each replacement individually and the changes can be made sequentially.

Behe begrudgingly allows that only "rarely, several mutations can sequentially add to each other to improve an organism's chances of survival." Rarely? This, of course, is the everyday stuff of evolution. Examples of cumulative selection changing multiple sites in evolving proteins include tetrodotoxin resistance in snakes (3), the tuning of color vision in animals (4), cefotaxime antibiotic resistance in bacteria (5), and pyrimethamine resistance in malarial parasites (6)—a notable omission given Behe's extensive discussion of malarial drug resistance.

Behe seems to lack any appreciation of the quantitative dimensions of molecular and trait evolution. He appears to think of the functional features of proteins in qualitative terms, as if binding or catalysis were all or nothing rather than a broad spectrum of affinities or rates. Therefore, he does not grasp the fundamental reality of a mutational path that proteins follow in evolving new properties.

This lack of quantitative thinking underlies a second, fatal blunder resulting from the mistaken assumptions Behe makes about protein interactions. The author has long been concerned about protein complexes and how they could or, rather, could not evolve. He argues that the generation of a single new protein-protein binding site is extremely improbable and that complexes of just three different proteins "are beyond the edge of evolution." But Behe bases his arguments on unfounded requirements for protein interactions. He insists, based on consideration of just one type of protein structure (the combining sites of antibodies), that five or six positions must change at once in order to make a good fit between proteins—and, therefore, good fits are impossible to evolve. An immense body of experimental data directly refutes this claim. There are dozens of well-studied families of cellular proteins (kinases, phosphatases, proteases, adaptor proteins, sumoylation enzymes, etc.) that recognize short linear peptide motifs in which only two or three amino acid residues are critical for functional activity [reviewed in (7–9)]. Thousands of such reversible interactions establish the protein networks that govern cellular physiology.

Very simple calculations indicate how easily such motifs evolve at random. If one assumes an average length of 400 amino acids for proteins and equal abundance of all amino acids, any given two-amino acid motif is likely to occur at random in every protein in a cell. (There are 399 dipeptide motifs in a 400-amino acid protein and $20 \times 20 = 400$ possible dipeptide motifs.) Any specific three-amino acid motif will occur once at random in every 20 proteins and any four-amino acid motif will occur once in every 400 proteins. That means that, without any new mutations or natural selection, many sequences that are identical or close matches to many interaction motifs already exist. New motifs can arise readily at random, and any weak interaction can easily evolve, via random mutation and natural selection, to become a strong interaction (9). Furthermore, any pair of interacting proteins can readily recruit a third protein, and so forth, to form larger complexes. Indeed, it has been demonstrated that new protein interactions (10) and protein networks (11) can evolve fairly rapidly and are thus well within the limits of evolution.

Is it possible that Behe does not know this body of data? Or does he just choose to ignore it? Behe has quite a record of declaring what is impossible and of disregarding the scientific literature, and he has clearly not learned any lessons from some earlier gaffes. He has again gone "public" with assertions without the benefit (or wisdom) of first testing their strength before qualified experts.

For instance, Behe once wrote, "if random evolution is true, there must have been a large number of transitional forms between the Mesonychid [a whale ancestor] and the ancient whale. Where are they?" (12). He assumed such forms would not or could not be found, but three transitional species were identified by paleontologists within a year of that statement. In *Darwin's Black Box*, he posited that genes for modern complex biochemical systems, such as blood clotting, might have been "designed billions of years ago and have been passed down to the present . . . but not 'turned on'." This is known to be genetically impossible because genes that aren't used will degenerate, but there it was in print. And Behe's argument against the evolution of flagella and the immune system have been dismantled in detail (13, 14) and new evidence continues to emerge (15), yet the same old assertions for design reappear here as if they were uncontested.

The continuing futile attacks by evolution's opponents reminds me of another legendary confrontation, that between Arthur and the Black Knight in the movie *Monty Python and the Holy Grail*. The Black Knight, like evolution's challengers, continues to fight even as each of his limbs is hacked off, one by one. The "no transitional fossils" argument and the "designed genes" model have been cut clean off, the courts have debunked the "ID is science" claim, and the nonsense here about the edge of evolution is quickly sliced to pieces by well-established biochemistry. The knights of ID may profess these blows are "but a scratch" or "just a flesh wound," but the argument for design has no scientific leg to stand on.

REFERENCES

1. M. J. Behe, *Darwin's black box: The biochemical challenge to evolution*. New York: Free Press, 1996.
2. *Kitzmiller et al. v. Dover Area School District et al.*, Memorandum Opinion, December 20, 2005; http://www.pamd.uscourts.gov/kitzmiller/decision.htm.
3. S. L. Geffeney et al., *Nature* 434, 759 (2005).
4. S. B. Carroll, *The making of the fittest: DNA and the ultimate forensic record of evolution*. New York: Norton, 2006.
5. D. M. Weinreich, N. F. Delaney, M. A. DePristo, and D. L. Hartl, *Science* 312, 111 (2006).
6. W. Sirawaraporn et al., *Proceedings of the National Academy of Science USA* 94, 1124 (1997).
7. V. Neduva et al., *PLoS Biology* 3, e405 (2005).
8. R. P. Bhattacharyya, A. Reményi, B. J. Yeh, and W. A. Lim, *Annual Review of Biochemistry* 75, 655 (2006).
9. V. Neduva and R. B. Russell, *FEBS Letters* 579, 3342 (2005).
10. Y. V. Budovskaya, J. S. Stephan, S. J. Deminoff, and P. K. Herman, *Proceedings of the National Academy of Science USA* 102, 13933 (2005).
11. P. Beltrao and L. Serrano, *PLoS Computational Biology* 3, e25 (2007).
12. M. J. Behe, in *Darwinism, science or philosophy?*, ed. J. Buell and V. Hearn, 60–71. Richardson, TX: Foundation for Thought and Ethics, 1994.

13. A. Bottaro, M. A. Inlay, and N. J. Matzke, *Nature Immunology* 7, 433 (2006).
14. M. J. Pallen and N. J. Matzke, *Nature Reviews Microbiology* 4, 784 (2006).
15. R. Liu and H. Ochman, *Proceedings of the National Academy of Science USA* 104, 7126 (2007).

Excerpted from Sean B. Carroll, Evolution: God as genetic engineer. *Science* 316, no. 5830 (2007): 1427–1428. Reprinted with permission from AAAS.

CHAPTER 10

• • • • • • • • • • • • • •

Legal Issues

INTRODUCTION

To understand the legal issues involving the creationism and evolution controversy, one must first understand the three components of the First Amendment of the United States Constitution. The Religion Clause states, "Congress shall make no law respecting an establishment of religion, or prohibiting the free exercise thereof." The Free Speech Clause adds, "or abridging the freedom of speech, or of the press." The third clause proclaims "the right of the people peaceably to assemble, and to petition the government for a redress of grievances." All of the legal decisions generated by the creationism and evolution controversy have been decided based on interpretations of the Religion and Free Speech clauses of the First Amendment.

The Religion Clause has two elements, referred to as the Establishment and the Free Exercise clauses. Taken together, these mean that public institutions have to be religiously neutral. Public schools, for example, can neither advance nor inhibit religion. Neutrality has meant different things to different people: to William Jennings Bryan in the 1920s neutrality meant not teaching evolution. To creation science proponents during the 1970s and 1980s (see chapter 5), neutrality meant that evolution should be "balanced" with the teaching of creationism. To opponents of creationism, neutrality means teaching only scientific views—keeping sectarian religion out of the classroom.

The 1971 Supreme Court case *Lemon v. Kurtzman* applies a three-part test to laws and regulations to determine if they violate the Establishment Clause. The *Lemon* test requires that a bill or practice must have a secular rather than a religious purpose; it must not have an effect that either promotes or inhibits religion, and it must not create undue entanglement between government and religion. Failure on any of the three prongs of *Lemon* means the bill is unconstitutional. All of the creationism cases decided after 1971 have referred to the *Lemon* test.

Tension exists between the Establishment and the Free Exercise clauses. An early (if unsuccessful) legal strategy of antievolutionists was to claim that the teaching of evolution violated a child's free exercise of religion because teaching evolution supposedly was an attack on religious belief (Larson 2003: 131). On the other hand, presenting creationism in science class—encouraged by antievolutionists as a means to counter the allegedly antireligious effect of teaching evolution—violates the Establishment Clause. The Free Speech Clause has also been invoked in support of a teacher's right to teach evolution by both John Scopes (see chapter 5) and by Susan Epperson (see chapter 5 and below). More recently, creationist teachers have argued a free speech and academic freedom right to add creationism to the curriculum (see *Webster*, below), but courts have held that teachers at the K–12 level must follow the curriculum set by the district. Precollege teachers have far less academic freedom than do university professors; hence the free speech argument has not fared well in court. "Academic Freedom Bills" purporting to give teachers "freedom" to introduce "all views" (i.e., the "evidence against evolution" suggested by Justice Scalia in his dissent to *Edwards*, discussed in chapter 7) have made their appearance in the early part of the twenty-first century.

When it comes to religion in schools, courts since the 1980s have favored the Establishment Clause over the other clauses of the First Amendment. Because attendance in public schools is mandatory, and because parents have the right to guide their children's religious views, courts have firmly restricted teachers from proselytizing students. For example, teachers cannot lead prayers or be active participants in religious after-school clubs. Because creationism is an inherently religious concept, advocating it—whether in biblical, creation science, or intelligent design form—is considered proselytization, and is therefore unconstitutional.

At the same time, the teaching of creationism with evolution is a popular idea in the United States, and many citizens believe that a judge's job is to reflect popular sentiment. Yet the rule of law requires that previous decisions ("precedents") be seriously considered before a decision is rendered. Among the readings in this introductory section is an excerpt from an article by Judge John E. Jones, the federal district court judge in the *Kitzmiller v. Dover* trial, who reflects on what is meant by "judicial independence" and the importance of precedent.

In most of the trials concerning the creationism and evolution controversy, science itself has played a role: whether creationism can be a legitimate science depends on what the definition of science is. If it can be considered science, is it a valid science? A question arises, then, who defines science, or decides whether a claim is scientific? To what degree are sitting judges, the majority of whom do not have scientific training, competent to make these decisions? But science enters into the legal arena in more contexts than merely the creationism/evolution controversy: many aspects of law (e.g., liability, patent law, environmental law, forensics) require judges to make decisions about the nature or content of science. A Supreme Court decision, *Daubert v. Dow Chemical*, established some general rules for the judiciary in dealing with science issues in the courtroom. This first group of readings, then, also reflect on *Daubert*, in consideration of the larger question of how legal decisions are made.

REFERENCE

Larson, Edward J. 2003. *Trial and error: The American controversy over creation and evolution*, 3rd ed. New York: Oxford University Press.

Science and the Law

In 1993, the U.S. Supreme Court issued a landmark decision in the case of *Daubert v. Merrell Dow Pharmaceuticals*. *Daubert* articulated standards judges should use (falsifiability, error rate, peer review, and general acceptance) to determine the admissibility of expert testimony in court. It affirmed that judges had a responsibility to be gatekeepers, keeping evidence that did not meet these standards out of the courtroom. For example, applying the *Daubert* guidelines, judges have excluded handwriting analysis as evidence in a number of cases (Adams 2003).

One of the issues raised by the *Daubert* decision was whether judges could fulfill their new gatekeeping function. Did they know enough about science and the scientific method to be able to apply the *Daubert* guidelines? A few years ago, a team of researchers attempted to find out (Dobbin et al. 2002). To assess how well judges understood the four standards prescribed in *Daubert*, the researchers surveyed 400 state trial court judges in all 50 states. A majority of the judges clearly understood peer review and general acceptance, but only a fraction clearly understood falsifiability and error rate [figure omitted]. The survey results suggest that "many judges may not be fully prepared to deal with the amount, diversity and complexity of the science presented in their courtrooms" and that "many judges did not recognize their lack of understanding" (Gatowski et al. 2001)....

Furthermore, a group of judges recently asked renowned physics professor Robert L. Park for guidance on how to recognize questionable scientific claims. The author of a landmark book on the subject, Park came up with "seven warning signs" that a scientific claim is probably bogus (Park 2002):

> The discoverer pitches the claim directly to the media (thus bypassing the peer review process by denying other scientists the opportunity to determine the validity of the claim).
> The discoverer claims that a powerful establishment is trying to suppress his or her work. (The mainstream science community may be deemed part of a larger conspiracy that includes industry and government.)
> The scientific effect involved is always at the very limit of detection.
> The evidence for a discovery is anecdotal.
> The discoverer says a belief is credible because it has endured for centuries.
> The discoverer has worked in isolation.
> The discoverer must propose new laws of nature to explain an observation.

REFERENCES

Adams, C. 2003. Is handwriting analysis legit science? *The Straight Dope*, April 18.
Dobbin, S. A., S. I. Gatowski, J. T. Richardson, G. P. Ginsburg, M. L. M. Ginsburg, and V. Dahir. 2002. Applying Daubert: How well do judges understand science and scientific method? *Judicature* 85 (5): 244–247.

Gatowski, S. I., S. A. Dobbin, J. T. Richardson, G. P. Ginsburg, M. L. Merlino, and V. Dahir. 2001. Asking the gatekeepers: A national survey of judges on judging expert evidence in a post-Daubert world. *Journal of Law and Human Behavior* 25 (5): 433–458.

Park, R. L. 2003. The seven warning signs of bogus science. *Chronicle of Higher Education*, January 31, p. B20.

Excerpted from National Science Board, Science and the law, in *Science and engineering indicators 2004* (Arlington, VA, National Science Foundation, 2004), 7–18.

Our Constitution's Intelligent Design

In the context of the *Kitzmiller* case, there have been several seminal cases handed down by the Supreme Court that establish applicable legal precedent relating to Establishment Clause jurisprudence. They are of course binding on me as a federal trial judge. The cases of *Lemon v. Kurtzman* and *County of Allegheny versus A.C.L.U.* [citations omitted] developed respectively what has come to be known as the "Lemon" and "Endorsement" tests for judges to apply to the particular governmental policies at issue. The Lemon test instructs the court to examine the stated purpose of the subject policy to determine whether it is in fact secular or religious. If it is found to be secular, the court moves to an analysis of the effect of the policy to evaluate whether a stated secular purpose is a sham. The third prong, which deals with excessive entanglement, was not applicable to the *Kitzmiller* case. The subsequently developed endorsement test essentially collapses the Lemon analysis into one unified exercise. In what I described in my opinion as a "belt and suspenders" approach, I applied both tests in deciding the case.

Although these tests have been the subject of a great deal of scholarly debate, not to mention dissent within the Supreme Court itself, they represented good law and controlling precedent at the time *Kitzmiller* was decided There was little if any ink spilled or broadcast time devoted to explaining the role of precedent by the critics of my decision, but, to the extent there was any mention of the case law, the clear implication was that I could (and should) have disregarded these well-established tests and in effect have crafted my own as they related to the facts of the case. This situation was not unique.

Earlier in 2005 my colleague on the federal bench, Judge James Whittemore, decided the very controversial Terri Schiavo right-to-die case. . . .

The Schiavo case unloosed a reaction that emanated not just from the public and press but also from some different quarters than in my case. In this regard it was also instructive. By the time of its conclusion, no fewer than 19 separate Florida state judges, as well as the federal courts, had exhaustively reviewed every possible claim and avenue pertaining to an admittedly emotional and controversial issue. Although all reached the same conclusion, in the aftermath of Schiavo's death, members of Congress, including most notably former House Majority Leader Tom DeLay, decried the various judicial determinations. . . . Coming as it did from the highest echelons of Congress, the public statement once again serve to stoke the public that in the main has no basic understanding of how courts operate at any level in the United States.

These examples illustrate that many people believe that judges should rule according to the popular will as it exists at any given moment. Put another way, and in the context of the *Kitzmiller* decision, polls have shown that, notwithstanding the current state of the law, many Americans believe that it is acceptable—and in fact ought to be legal—to teach creationism alongside Darwin's theory of evolution in our public schools. The result is that the public believes that judges should get with the program and implement public sentiment in their decisions. With the issue once again viewed against the backdrop of my case, letters to the editor in many newspapers wondered why the federal trial judge was thwarting what the people truly wanted. This, of course, directly conflicts with the true role of the judiciary and disregards the Founding Fathers' and constitutional framers' intention that the federal courts serve as a check and bulwark against blind implementation of the popular will when it conflicts with our laws.

In deciding the *Kitzmiller* case consistent with existing precedent, I was certainly not an activist judge. Instead, I was applying binding constitutional precedent to the facts before me, which required me to find that the policy violated the Establishment Clause because it clearly had a religious purpose. For example, the record established that as early as two years before the enactment of the policy, the Dover school board discussed teaching creationism as an alternative to the theory of evolution. In 2004, as the policy was being developed, various school board members during meetings made blatant religious statements while stating their support for the introduction of intelligent design into the science curriculum. A local church solicited contributions to purchase copies of the intelligent design textbook, *Of Pandas and People*, which were then donated to the school's library for students to consult in aid of their understanding of intelligent design. The textbook use the word *creationism* in its first edition: after case law clearly held at the teaching of creationism violated the Establishment Clause, later editions simply replaced that word with *intelligent design*. Testimony from several leading members of the school board revealed that they could not in any meaningful way distinguish intelligent design from creationism. Finally, overwhelming expert testimony demonstrated that intelligent design does not represent reputable science and is predominantly a religious argument.

As I have frequently stated since rendering my decision, had I disregarded the rather clear facts regarding the intentions of the Dover school board members in passing the policy, or if I had ignored the well-established tests contained in Supreme Court precedent, then I would have engaged in true judicial activism. This would be equally true if I had injected my personal beliefs into the process in order to reach a holding in line with what I perceive would please either segment of the public, or those I might have considered my political benefactors. Of course I did none of those things. The point that must not be lost is that the majority of the public *thought* that I could reach a decision by this process instead of following the rule of law.

Excerpted from John E. Jones, III. Our Constitution's intelligent design. *Litigation* 33, no. 3 (2007): 3–6, 57 (from 5–6, 57). Used with permission.

The remainder of this chapter is divided into three parts, reflecting the history of the creationism and evolution controversy. As discussed in chapters 5, 6, and 7, antievolutionists first attempted to ban evolution, then to teach creation science to balance it; and now we are in the current, neocreationist period, during which a variety of efforts to denigrate evolution are being made.

● ●

BANNING EVOLUTION

William Jennings Bryan on Evolution

I object to the theory for several reasons. First, it is a dangerous theory. If a man link himself in generations with the monkey, it then becomes an important question whether he is going towards him or coming from him—and I have seen them going in both directions. I don't know of any argument that can be used to prove that man is an improved monkey that may not be used just as well to prove that the monkey is a degenerate man, and the latter theory is more plausible than the former... ("The Prince of Peace," 41).

Go back as far as we may, we cannot escape from the creative act, and it is just as easy for me to believe that God created man as he is as to believe that, millions of years ago, He created a germ of life and endowed it with power to develop into all that we see to-day. I object to the Darwinian theory, until more conclusive proof is produced, because I fear we shall lose the consciousness of God's presence in our daily life, if we must accept the theory that through all the ages no spiritual force has touched the life of man or shaped the destiny of nations. But there is another objection. The Darwinian theory represents man as reaching his present perfection by the operation of the law of hate—the merciless law by which the strong crowd out and kill off the weak.... I prefer to believe that love rather than hatred is the law of development ("The Prince of Peace," 42).

Our first objection to Darwinism is that it is not true. I may add here, so I will not have to refer to it again, that I am answering theistic evolution as well as atheistic evolution; I do not make any difference between them for the theistic evolutionist and the atheistic evolutionist walk along hand in hand until they reach the beginning of life. They are nearer together than either of them is to the Christian... ("Is the Bible True?" 82).

Let us look at some of their guesses.... Do you know how the eye came?... They guess that an animal that did not have any eyes, away back yonder had a piece of pigment or freckle on the skin—it just happened—and when the sun's rays were traveling over the animal's body and came to that piece of pigment or freckle, they converged there more than elsewhere and that made it warmer there than elsewhere, and that irritated the skin there instead of elsewhere, and that brought a nerve there instead of somewhere else, and the nerve developed into an eye. And then another freckle, and another eye; in the right place and at the right time. Can you beat it? ("Is the Bible True?" 82–83).

Excerpted from R. M. Cornelius, ed., *Selected orations of William Jennings Bryan: The Cross of Gold centennial edition* (Dayton, TN: William Jennings Bryan College, 1996). Used with permission.

Epperson v. Arkansas (1968)

The Arkansas law makes it unlawful for a teacher in any state-supported school or university "to teach the theory or doctrine that mankind ascended or descended from a lower order of animals," or "to adopt or use in any such institution a textbook that teaches" this theory. Violation is a misdemeanor and subjects the violator to dismissal from his position.

[T]he law must be stricken because of its conflict with the constitutional prohibition of state laws respecting an establishment of religion or prohibiting the free exercise thereof. The overriding fact is that Arkansas' law selects from the body of knowledge a particular segment which it proscribes for the sole reason that it is deemed to conflict with a particular religious doctrine; that is, with a particular interpretation of the Book of Genesis by a particular religious group. . . .

Government in our democracy, state and national, must be neutral in matters of religious theory, doctrine, and practice. It may not be hostile to any religion or to the advocacy of no-religion; and it may not aid, foster, or promote one religion or religious theory against another or even against the militant opposite. The First Amendment mandates governmental neutrality between religion and religion, and between religion and nonreligion.

While study of religions and of the Bible from a literary and historic viewpoint, presented objectively as part of a secular program of education, need not collide with the First Amendment's prohibition, the State may not adopt programs or practices in its public schools or colleges which "aid or oppose" any religion. This prohibition is absolute. It forbids alike the preference of a religious doctrine or the prohibition of theory which is deemed antagonistic to a particular dogma. As Mr. Justice Clark stated in *Joseph Burstyn, Inc. v. Wilson*, "the state has no legitimate interest in protecting any or all religions from views distasteful to them. . . ."

Arkansas' law cannot be defended as an act of religious neutrality. Arkansas did not seek to excise from the curricula of its schools and universities all discussion of the origin of man. The law's effort was confined to an attempt to blot out a particular theory because of its supposed conflict with the Biblical account, literally read. Plainly, the law is contrary to the mandate of the First, and in violation of the Fourteenth, Amendment to the Constitution. [Internal citations deleted.]

Selection excerpted from *Epperson v. Arkansas* 393 U.S. 97 (1968).

• •

EQUAL TIME FOR CREATION SCIENCE

In 1981, Arkansas was the first state to pass a law dictating equal time for creation science and evolution. The federal district court struck it down in *McLean v. Arkansas*. As discussed in chapter 5, the *McLean* decision was not appealed, so its conclusions did not become precedent beyond its district. However, its reasoning was highly influential in several cases, including a later Supreme Court case, *Edwards v. Aguillard*, that struck down equal-time laws nationwide. An excerpt from *Edwards* follows this one from *McLean*.

McLean v. Arkansas Board of Education (1982)

On March 19, 1981, the Governor of Arkansas signed into law Act 590 of 1981, entitled "Balanced Treatment for Creation-Science and Evolution-Science Act.". . . Its essential mandate is stated in its first sentence: "Public schools within this State shall give balanced treatment to creation-science and to evolution-science." On May 27, 1981, this suit was filed challenging the constitutional validity of Act 590 on three distinct grounds.

First, it is contended that Act 590 constitutes an establishment of religion prohibited by the First Amendment to the Constitution, which is made applicable to the states by the Fourteenth Amendment. Second, the plaintiffs argue the Act violates a right to academic freedom which they say is guaranteed to students and teachers by the Free Speech Clause of the First Amendment. Third, plaintiffs allege the Act is impermissibly vague and thereby violates the Due Process Clause of the Fourteenth Amendment.

The individual plaintiffs include the resident Arkansas Bishops of the United Methodist, Episcopal, Roman Catholic and African Methodist Episcopal Churches, the principal official of the Presbyterian Churches in Arkansas, other United Methodist, Southern Baptist and Presbyterian clergy, as well as several persons who sue as parents and next friends of minor children attending Arkansas public schools. One plaintiff is a high school biology teacher. All are also Arkansas taxpayers. Among the organizational plaintiffs are the American Jewish Congress, the Union of American Hebrew Congregations, the American Jewish Committee, the Arkansas Education Association, the National Association of Biology Teachers and the national Coalition for Public Education and Religious Liberty, all of which sue on behalf of members living in Arkansas.

The defendants include the Arkansas Board of Education and its members, the Director of the Department of Education, and the State Textbooks and Instructional Materials Selecting Committee. . . .

The unusual circumstances surrounding the passage of Act 590, as well as the substantive law of the First Amendment, warrant an inquiry into the stated legislative purposes. The author of the Act has publicly proclaimed the sectarian purpose of the proposal. The Arkansas residents who sought legislative sponsorship of the bill did so for a purely sectarian purpose. These circumstances alone may not be particularly persuasive, but when considered with the publicly announced motives of the legislative sponsor made contemporaneously with the legislative process; the lack of any legislative investigation, debate or consultation with any educators or scientists; the unprecedented intrusion in school curriculum; and official history of the State of Arkansas on the subject, it is obvious that the statement of purpose has little, if any, support in fact. The State failed to produce any evidence which would warrant an inference or conclusion that at any point in the process anyone considered the legitimate educational value of the Act. It was simply and purely an effort to introduce the Biblical version of creation into the public school curricula. The only inference which can be drawn from these circumstances is that the Act was passed with the specific purpose by the General Assembly of advancing religion. The Act therefore fails the first prong of the three-pronged test, that of secular legislative purpose, as articulated in *Lemon v. Kurtzman.* . . .

The evidence establishes that the definition of "creation science" contained in 4(a) has as its unmentioned reference the first 11 chapters of the Book of Genesis. Among the many creation epics in human history, the account of sudden creation from nothing, or *creation ex nihilo*, and subsequent destruction of the world by flood is unique to Genesis. The concepts of 4(a) are the literal Fundamentalists' view of Genesis. Section 4(a) is unquestionably a statement of religion, with the exception of 4(a)(2), which is a negative thrust aimed at what the creationists understand to be the theory of evolution.

Both the concepts and wording of Section 4(a) convey an inescapable religiosity. Section 4(a)(1) describes "sudden creation of the universe, energy and life from nothing." Every theologian who testified, including defense witnesses, expressed the opinion that the statement referred to a supernatural creation which was performed by God.

The facts that creation-science is inspired by the Book of Genesis and that Section 4(a) is consistent with a literal interpretation of Genesis leave no doubt that a major effect of the Act is the advancement of particular religious beliefs. The legal impact of this conclusion will be discussed further at the conclusion of the Court's evaluation of the scientific merit of creation-science.

The approach to teaching "creation-science" and "evolution-science" found in Act 590 is identical to the two-model approach espoused by the Institute for Creation Research and is taken almost verbatim from ICR writings. It is an extension of Fundamentalists' view that one must either accept the literal interpretation of Genesis or else believe in the godless system of evolution.

The two-model approach of the creationists is simply a contrived dualism which has no scientific factual basis or legitimate educational purpose. It assumes only two explanations for the origins of life and existence of man, plants, and animals: it was either the work of a creator or it was not. Application of these two models, according to creationists, and the defendants, dictates that all scientific evidence which fails to support the theory of evolution is necessarily scientific evidence in support of creationism and is, therefore, creation-science "evidence" in support of Section 4(a). . . .

The methodology employed by creationists is another factor which is indicative that their work is not science. A scientific theory must be tentative and always subject to revision or abandonment in light of facts that are inconsistent with, or falsify, the theory. A theory that is by its own terms dogmatic, absolutist, and never subject to revision is not a scientific theory.

The creationists' methods do not take data, weigh it against the opposing scientific data, and thereafter reach the conclusions stated in Section 4(a). Instead, they take the literal wording of the Book of Genesis and attempt to find scientific support for it. . . .

Implementation of Act 590 will have serious and untoward consequences for students, particularly those planning to attend college. Evolution is the cornerstone of modern biology, and many courses in public schools contain subject matter relating to such varied topics as the age of the earth, geology, and relationships among living things. Any student who is deprived of instruction as to the prevailing scientific thought on these topics will be denied a significant part of science education. Such a deprivation through the high school level would undoubtedly have an impact upon the quality of education in the State's colleges and universities, especially including the pre-professional and professional programs in the health sciences. . . .

The defendants presented Dr. Larry Parker, a specialist in devising curricula for public schools. He testified that the public school's curriculum should reflect the subjects the public wants in schools. The witness said that polls indicated a significant majority of the American public thought creation-science should be taught if evolution was taught. . . .

The application and content of First Amendment principles are not determined by public opinion polls or by a majority vote. Whether the proponents of Act 590 constitute the majority or the minority is quite irrelevant under a constitutional system of government. No group, no matter how large or small, may use the organs of government, of which the public schools are the most conspicuous and influential, to foist its religious beliefs on others. [Internal citations omitted.]

Excerpted from *McLean v. Arkansas Board of Education*, 529 F. Supp. 1255 (E.D. La. 1982).

Edwards v. Aguillard (1987)

Louisiana's "Creationism Act" forbids the teaching of the theory of evolution in public elementary and secondary schools unless accompanied by instruction in the theory of "creation science." The Act does not require the teaching of either theory unless the other is taught. It defines the theories as "the scientific evidences for [creation or evolution] and inferences from those scientific evidences. . . ."

Held:

1. The Act is facially invalid as violative of the Establishment Clause of the First Amendment, because it lacks a clear secular purpose.

(a) The Act does not further its stated secular purpose of "protecting academic freedom." It does not enhance the freedom of teachers to teach what they choose and fails to further the goal of "teaching all of the evidence." Forbidding the teaching of evolution when creation science is not also taught undermines the provision of a comprehensive scientific education. Moreover, requiring the teaching of creation science with evolution does not give schoolteachers a flexibility that they did not already possess to supplant the present science curriculum with the presentation of theories, besides evolution, about the origin of life. Furthermore, the contention that the Act furthers a "basic concept of fairness" by requiring the teaching of all of the evidence on the subject is without merit. Indeed, the Act evinces a discriminatory preference for the teaching of creation science and against the teaching of evolution by requiring that curriculum guides be developed and resource services supplied for teaching creationism but not for teaching evolution, by limiting membership on the resource services panel to "creation scientists," and by forbidding school boards to discriminate against anyone who "chooses to be a creation-scientist" or to teach creation science, while failing to protect those who choose to teach other theories or who refuse to teach creation science. A law intended to maximize the comprehensiveness and effectiveness of science instruction would encourage the teaching of all scientific theories about human origins. Instead, this Act has the distinctly different purpose of discrediting evolution by counter-balancing its teaching at every turn with the teaching of creationism.

(b) The Act impermissibly endorses religion by advancing the religious belief that a supernatural being created humankind. The legislative history demonstrates that the term "creation science," as contemplated by the state legislature, embraces this religious teaching. The Act's primary purpose was to change the public school science curriculum to provide persuasive advantage to a particular religious doctrine that rejects the factual basis of evolution in its entirety. Thus, the Act is designed either to promote the theory of creation science that embodies a particular religious tenet or to prohibit the teaching of a scientific theory disfavored by certain religious sects. In either case, the Act violates the First Amendment. . . .

The Court has been particularly vigilant in monitoring compliance with the Establishment Clause in elementary and secondary schools. Families entrust public schools with the education of their children, but condition their trust on the understanding that the classroom will not purposely be used to advance religious views that may conflict with the private beliefs of the student and his or her family. Students in such institutions are impressionable and their attendance is involuntary. The State exerts great authority and coercive power through mandatory attendance requirements, and because of the students' emulation of teachers as role models and the children's susceptibility to peer pressure. Furthermore, "[t]he public school is at once the symbol of our democracy and the most pervasive means for promoting our common destiny. In no activity of the State is it more vital to keep out divisive forces than in its schools. . . ."

Consequently, the Court has been required often to invalidate statutes which advance religion in public elementary and secondary schools. Therefore, in employing the three-pronged *Lemon* test, we must do so mindful of the particular concerns that arise in the context of public elementary and secondary schools. . . .

We do not imply that a legislature could never require that scientific critiques of prevailing scientific theories be taught. Indeed, the Court acknowledged in *Stone* that its decision forbidding the posting of the Ten Commandments did not mean that no use could ever be made of the Ten Commandments, or that the Ten Commandments played an exclusively religious role in the history of Western Civilization. In a similar way, teaching a variety of scientific theories about the origins of humankind to schoolchildren might be validly done with the clear secular intent of enhancing the effectiveness of science instruction. But because the primary purpose of the Creationism Act is to endorse a particular religious doctrine, the Act furthers religion in violation of the Establishment Clause. . . .

The Louisiana Creationism Act advances a religious doctrine by requiring either the banishment of the theory of evolution from public school classrooms or the presentation of a religious viewpoint that rejects evolution in its entirety. The Act violates the Establishment Clause of the First Amendment because it seeks to employ the symbolic and financial support of government to achieve a religious purpose. The judgment of the Court of Appeals therefore is Affirmed. [Internal citations omitted.]

Excerpted from *Edwards v. Aguillard*, 482 U.S. 578 (1987).

Recall from chapters 6, 7, and 8 that present-day antievolutionists—aware of the preceding court decisions, and especially the purpose prong of *Lemon*—try to avoid reference to creationism or religion. Instead, evolution is declared a *scientifically* controversial topic and legislation or policies are proposed that would require teachers to give students all the evidence, an approach that appeals to long-standing American cultural standards of fairness. These various neocreationist approaches comprise the remainder of this chapter. As discussed in chapter 6, these approaches include scientific alternatives to evolution, an example of which is intelligent design, and various attempts to denigrate evolution through disclaimers and policies that call for teaching students alleged evidence against evolution, or strengths and weaknesses of evolution, or for a critical analysis of evolution.

• •

THE SANTORUM AMENDMENT AND ITS REPERCUSSIONS

An example of this more subtle approach is an amendment proposed by Senator Rick Santorum to the 2002 No Child Left Behind (NCLB) education bill. Although couched as a critical-thinking statement, only evolution is singled out from all potentially controversial scientific theories. Even though the amendment failed and appears only in altered form in the conference committee report, language from the amendment shows up in antievolution legislation around the country, as indicated below. The Santorum amendment incorrectly has been treated by creationists as having the force of law. The NCLB act does not mandate any aspect of curriculum in science or any other field.

The Original Amendment

It is the sense of the Senate that

1. good science education should prepare students to distinguish the data or testable theories of science from philosophical or religious claims that are made in the name of science; and
2. where biological evolution is taught, the curriculum should help students to understand why the subject generates so much continuing controversy, and should prepare the students to be informed participants in public discussions regarding the subject.

Final version: Item 78, Appearing in Conference Committee Report

The conferees recognize that a quality science education should prepare students to distinguish the data and testable theories of science from religious or philosophical claims that are made in the name of science. Where topics are taught that may generate controversy (such as biological evolution), the curriculum should help students to understand the full range of scientific views that exist, why such topics may generate controversy, and how scientific discoveries can profoundly affect society.

LEGISLATION INSPIRED BY THE SANTORUM LANGUAGE

Georgia HB 1563 (2002)

...In recognition of the fact that a quality science education should prepare students to distinguish the data and testable theories of science from philosophical claims that are made in the name of science, the State Board of Education is authorized to promulgate rules and regulations and develop a curriculum for topics that may generate controversy, such as biological evolution, to help students understand the full range of scientific views that exist, why such topics may generate controversy, and how scientific discoveries can profoundly affect society. [Died in committee.]

Ohio HB 481 (2002)

(A) Encourage the presentation of scientific evidence regarding the origins of life and its diversity objectively and without religious, naturalistic, or philosophic bias or assumption;

(B) Require that whenever explanations regarding the origins of life are presented, appropriate explanation and disclosure shall be provided regarding the historical nature of origins science and the use of any material assumption which may have provided a basis for the explanation being presented;

(C) Encourage the development of curriculum that will help students think critically, understand the full range of scientific views that exist regarding the origins of life, and understand why origins science may generate controversy. [Died in committee.]

● ●

SCIENTIFIC ALTERNATIVES TO EVOLUTION: INTELLIGENT DESIGN

Introduction

Since the Supreme Court's 1987 decision declaring that teaching creation science was unconstitutional, the most popular scientific alternative to evolution has been intelligent design. Reminiscent of the efforts for equal time for creation science of the late 1970s and 1980s, attempts to mandate the teaching of ID have occurred at both the state level (in legislation) and local-school-board level. Often language echoing the Santorum amendment is included, and of course, it is not uncommon that intelligent design is combined with some form of denigration of evolution in the policy. A school board in Dover, Pennsylvania, required the teaching of ID in its curriculum and attempted to have the ID textbook *Of Pandas and People* adopted as a supplemental textbook. After a lengthy lawsuit (*Kitzmiller v. Dover*), a federal district court declared the teaching of ID unconstitutional (see chapter 7). Readings presenting pro-ID and anti-ID views are presented subsequently.

• •

READINGS/POLICIES SUPPORTING INTELLIGENT DESIGN

Michigan HB 4382 (2001), HB 4946 (2003)

In the science standards for middle school and high school, all references to "evolution" and "natural selection" shall be modified to indicate that these are unproven theories by adding the phrase "describe how life may be the result of the purposeful, intelligent design of a Creator." (Died in committee in 2001; reintroduced in 2003, died.)

Blount County (TN) School Board Policy on Teaching Intelligent Design

As Members of the Blount County School Board WE BELIEVE:

As a local public school board we have the authority and responsibility to approve curricula.

The teachers in our school system should enjoy academic freedom to present appropriate materials to their students pertaining to the subject matter being covered.

Students should not be deprived of current pertinent scientific information. The omission or denial of such information may unfairly deprive students of the opportunity to examine the full range of scientific theories about biological origins.

Presently our high school biology textbooks only present the theory of Darwinian evolution and omits teaching a variety of scientific theories about origins. By restricting content subject matter without academic freedom, biology education becomes indoctrination.

Further, we, as members of the Blount County School Board confirm:

1. Teaching a variety of scientific theories about origins may be done with the clear secular intent of enhancing the effectiveness of science instruction.

2. Design theory, in particular, constitutes an inference from biological data, and is not an inference or conclusion from religious authority.

3. It is constitutionally lawful for teachers and school boards to expose students to scientific problems with current Darwinian theory as well as to other scientific alternatives with respect to theories about biological origins.

4. It is unconstitutional under the Free Speech Clause of the First Amendment to exclude these ideas from a public forum simply because of the content of these ideas.

5. Biological origins is an open forum for free speech and as such cannot be censored based solely on the content of the speech.

6. With respect to biological origins and scientific alternatives our local school board has the freedom to exercise discretion in the selection of curriculum materials.

7. Therefore, we hereby encourage our biology teachers to teach the controversy with respect to biological origins. The theory of intelligent design may be taught as part of the current controversy. We also encourage the inclusion of intelligent origins in the State approved textbooks.

Excerpted from minutes of the Blount County Board of Education, January 13, 2005, Retrieved on October 28, 2008 from http://www.blountk12.org/BdofEd/bd_minutes.htm.

Dover, Pennsylvania (2004) Curriculum Guide

Students will be made aware of gaps/problems and Darwin's theory and of other theories of evolution including, but not limited to Intelligent Design. Reference: *Of Pandas and People*.

Excerpted from Dover Area School District Biology I Planned Instruction/Curriculum Guide, board approved October 18, 2004.

• •

READING OPPOSING INTELLIGENT DESIGN

Kitzmiller v. Dover

In addition to Dr. Padian, Dr. Miller also testified that *Pandas* presents discredited science. Dr. Miller testified that *Pandas'* treatment of biochemical similarities between organisms is "inaccurate and downright false" and explained how *Pandas* misrepresents basic molecular biology concepts to advance design theory through a series of demonstrative slides. Consider, for example, that he testified as to how *Pandas* misinforms readers on the standard evolutionary relationships between different types of animals, a distortion which Professor Behe, a "critical reviewer" of *Pandas* who wrote a section within the book, affirmed. In addition, Dr. Miller refuted *Pandas'* claim that evolution cannot account for new genetic information and pointed to more than three dozen peer-reviewed scientific publications showing the origin of new genetic information by evolutionary processes. In summary, Dr. Miller testified that *Pandas* misrepresents molecular biology and genetic principles, as well as the current state of scientific knowledge in those areas in order to teach readers that common descent and natural selection are not scientifically sound.

Accordingly, the one textbook to which the Dover ID Policy directs students contains outdated concepts and badly flawed science, as recognized by even the defense experts in this case.

... The evidence presented in this case demonstrates that ID is not supported by any peer-reviewed research, data or publications. Both Drs. Padian and Forrest testified that recent literature reviews of scientific and medical-electronic databases disclosed no studies supporting a biological concept of ID. On cross-examination, Professor Behe admitted that: "There are no peer reviewed articles by anyone advocating for intelligent design supported by pertinent experiments or calculations which provide detailed rigorous accounts of how intelligent design of any biological system occurred." Additionally, Professor Behe conceded that there are no peer-reviewed papers supporting his claims that complex molecular systems, like the bacterial flagellum, the blood-clotting cascade, and the immune system, were intelligently designed. In that regard, there are no peer-reviewed articles supporting Professor Behe's argument that certain complex molecular structures are "irreducibly complex." In addition to failing to produce papers in peer-reviewed journals, ID also features no scientific research or testing.

After this searching and careful review of ID as espoused by its proponents, as elaborated upon in submissions to the Court, and as scrutinized over a six week trial, we find that ID is not science and cannot be adjudged a valid, accepted scientific theory as it has failed to publish in peer-reviewed journals, engage in research and testing, and gain acceptance in the scientific community. ID, as noted, is grounded in theology, not science. Accepting for the sake of argument its proponents', as well as Defendants' argument that to introduce ID to students will encourage critical thinking, it still has utterly no place in a science curriculum. Moreover, ID's backers have sought to avoid the scientific scrutiny which we have now determined that it cannot withstand by advocating that the controversy, but not ID itself, should be taught in science class. This tactic is at best disingenuous, and at worst a canard. The goal of the IDM [Intelligent Design Movement] is not to encourage critical thought, but to foment a revolution which would supplant evolutionary theory with ID.

To conclude and reiterate, we express no opinion on the ultimate veracity of ID as a supernatural explanation. However, we commend to the attention of those who are inclined to superficially consider ID to be a true "scientific" alternative to evolution without a true understanding of the concept the foregoing detailed analysis. It is our view that a reasonable, objective observer would, after reviewing both the voluminous record in this case, and our narrative, reach the inescapable conclusion that ID is an interesting theological argument, but that it is not science. [Internal citations omitted.]

Excerpted from *Kitzmiller v. Dover Area School District*, 400 F.Supp.2d 707 (M.D. Pa. 2005) (from 744–746).

• •

EVIDENCE AGAINST EVOLUTION (CRITICAL ANALYSIS OF EVOLUTION)

Introduction

The strategy of evidence against evolution or critical analysis of evolution has deep roots in creationism. As discussed in chapters 5, 6, and 7, creation science and intelligent design both promote a two-model approach (in the terminology of creation science). Evidence against evolution is considered evidence for creationism; hence most of the literature presents alleged weaknesses of evolution rather than positive evidence for special creation. So in the face of failure to ban evolution, or balance it with either creation science or intelligent design, the fallback position of the antievolutionists has been to propose that the teaching of evolution be balanced with the teaching of evidence against evolution. Suggestions as to the content of such instruction can be found in Jonathan Wells's book *Icons of Evolution* (2000) and in a more recent book published by Fellows of the Discovery Institute, *Explore Evolution* (Meyer et al. 2007). These include staples of creationism such as gaps in the fossil record, attacks on the concept of homology as support for common ancestry, attacks on developmental biology (embryology) as support for common ancestry, and attacks on natural selection's ability to produce complexity.

Teaching evidence against evolution, a critical analysis of evolution, or the strengths and weaknesses of evolution, as such policies have been proposed, have an inherent attractiveness given American traditions of democracy. *Teach the controversy* is a phrase popularized by the Discovery Institute. The controversy is not the creationism/evolution controversy but a supposed controversy *within science* over whether evolution has occurred.

The first reading here refers to a "Sticker." The reading is from an amicus ("friend of the court") brief submitted on behalf of the defendant school district in Cobb County, Georgia, which had been sued over the legality of inserting a disclaimer sticker (see Figure 7.2, p.155) in high school biology textbooks.

● ●

CRITICAL ANALYSIS OF EVOLUTION: READING SUPPORTING CREATIONIST CLAIMS

Amicus Brief in Cobb County, Georgia

D. The Sticker's placement is reasonable in light of the growing scientific controversy over Darwin's theory

1. A growing number of scientists now question key aspects of the theories of chemical and biological evolution on scientific grounds.

A number of scientists who accept key aspects of Darwin's theory of evolution disagree about the mechanism by which it may have occurred. The scientists have raised the possibility that none of the explanations now offered for the mechanism of evolution will prove accurate. The Sticker is a reminder of the need for further research, serving an appropriate educational purpose.

Peer-reviewed science literature discusses many questions and criticisms of aspects of Darwin's theory, including: whether natural selection acting on random mutations and variations sufficiently explains new genetic information, new organs, the new complex body plans, and responses to environmental stimuli; how the pattern in the fossil record does not conform to neo-Darwinian expectations of life's history; and how many homologous structures come from nonhomologous genes while nonhomologous structures come from similar genes. The literature also discusses problems with chemical evolutionary scenarios for the origin of life.

Some scientists who support Darwinian evolution have noted that many textbook presentations of evidence for the theory are inadequate or erroneous. The First Amendment can hardly be understood to preclude the school board from encouraging students to study these issues or learn about the scientific controversy.

Finally, the students have a right to be informed that a significant number of scientists have published works calling into question all or several key aspects of Darwinian theory. For instance, more than 300 biologists, biochemists, and other doctoral scientists have signed a statement expressing their skepticism of the central tenet of Darwin's theory of evolution on scientific grounds, which reads: "We are skeptical of claims for the ability of random mutation and natural selection to account for the complexity of life. Careful examination of the evidence for Darwinian theory should be encouraged." These dissenting scientists include professors and researchers at a [sic] academic institutions such as Yale, Princeton, MIT, the Smithsonian, University of California Berkeley, and the University of Texas. This call for examination of scientific evidence for Darwin's theory (apart from consideration of any alternative scientific theory, such as intelligent design, or from any religious concepts such as creationism), supports the Board's decision. . . .

E. Darwin's theory of evolution can be subjected to scientific criticism, apart from any consideration of alternative scientific theories

In *Edwards*, the Supreme Court implicitly recognized that the ability to teach scientific critiques of scientific theories does not require consideration of alternative theories of origins. The Sticker, on its face, calls for student awareness of scientific critiques of evolution; it does not address alternative theories.

As noted, a critical analysis of evolution can be analogized to a judge's evaluation of whether, under *Daubert*, scientific evidence is admissible. The test of admissibility is not whether there is a variety of alternative theories with regard to a factual assertion. Instead, the proponent of scientific evidence need only establish a sufficient basis for relying upon the scientific opinion or theory being offered. Likewise, the student taught to critically analyze the scientific evidence supporting the theory can simply inquire as to the sufficiency of the evidence supporting that particular theory, apart from consideration of alternative theories. [Legal and other references omitted; the full brief can be found on NCSE's Web site, http://www.ncseweb.org/creatism/legal]

Excerpted from Amicus Brief of Parents for Truth in Education, in the U.S. District Court, Northern District of Georgia Atlanta division, *in re Jeffrey Michael Selman, plaintiff, v. Cobb County School District, et al., defendants*. David K. DeWolf, Marjorie Rogers, and William R. Johnson, November 5, 2004.

● ●

CRITICAL ANALYSIS OF EVOLUTION: READING OPPOSING CREATIONIST CLAIMS

Analyzing "Critical Analysis": The Fallback Antievolutionist Strategy

The beginning of the ID movement's shift towards "critical analysis" can be traced to two factors. The first is *Icons of Evolution* (2000), a blustering antievolution polemic written by Discovery Institute fellow Jonathan Wells, which succeeded in intimidating many educators by attacking biology textbooks for allegedly crude mistakes or worse. The second is the so-called Santorum Amendment, a nonbinding "sense of the Senate" provision that Pennsylvania Senator Rick Santorum attempted to insert into the No Child Left Behind Act in June 2001. Santorum's amendment (written, in fact, by ID leader Phillip Johnson) did not mention creationism or ID, but it did single out evolution as controversial science. The amendment was eventually stripped from the statute language. Instead, a watered down version appeared in the joint explanatory statement of the House-Senate conference committee. This statement had no force as law, but it was nevertheless trumpeted by antievolutionists as a signal victory for their cause.

The Santorum language was cited in subsequent battles over "critical analysis" policies, and it paved the way for the ID movement's two major state-level victories: "critical analysis" language in the science standards of Ohio (2002) and Kansas (2005). Creationists and ID advocates have also pushed unsuccessfully for critical analysis language in the science standards in Minnesota, New Mexico, Pennsylvania, South Carolina, West Virginia, Arizona, and Georgia. Of the proposed antievolution legislation tracked by the National Center for Science Education in 2005, many of the bills employed "critical analysis" or similar euphemisms for the deprecation of evolutionary science. Furthermore, many local school boards have been the scenes of battles over similar policies—prominent examples include Roseville, CA; Darby, MT; Grantsburg, WI; and Cobb County, GA.

Critical analysis of "critical analysis"

Shouldn't we, though, encourage truly critical thought even about currently dominant scientific theories? The answer is, of course, "yes." But that is not what these policies are doing. They are, instead, uncritically promoting creationism-driven pseudoscience, in full accord with the traditional creationist goal: getting public school science classes to teach as science the theological doctrine that God intervened to create each "kind" of organism, including humankind. This is effected by singling out evolution *alone* for special criticism, ignoring all other major scientific theories, including those about which there really is a current argument (the relation between quantum mechanics and gravity, for example). To complete the job, long-ago-debunked creationist criticisms of evolution are presented as though they are real, current science.

The problem for science education is that "critical analysis" policies promote creationist pseudoscience, or just bad science known to be fallacious. Students will simply be misinformed if these claims are taught as if they are accepted science. Supporters of first-rate science education need to be aware of "critical analysis" claims and some of the problems with these claims, because they are often disguised in science-style phrases and promoted as cutting-edge science. To illustrate, we offer here some actual, rather than sham, critical analysis of the "critical analysis" items presented in the 2005 Kansas science standards.

Inconsistencies in phylogenetic trees

The new creationism offers a standard list of challenges to the evidence that modern organisms share common ancestry. The Kansas Science Education Standards that were adopted on November 8, 2005, list several. The first deals with disagreements between phylogenetic trees (lineages, or patterns of descent) constructed from DNA or protein similarity analyses:

> The view that living things in all the major kingdoms are modified descendants of a common ancestor (described in the pattern of a branching tree) has been challenged in recent years by:

i. Discrepancies in the molecular evidence (e.g., differences in relatedness inferred from sequence studies of different proteins) previously thought to support that view.

The claim is that phylogenetic trees based on different datasets conflict so badly as to call common ancestry into question. The usual creationist procedure is to dig through the scientific literature to find cases where studies disagree on the *exact* phylogenetic relationships of organisms, and then to trumpet these as inexplicable discrepancies that refute common ancestry. ID creationists universally fail to acknowledge that the similarity of phylogenetic trees can be measured, statistically, and that trees derived from independent datasets typically have extremely *strong* statistical correlations. Such findings, which are very common indeed, support the notion that there are real phylogenetic trees, and that scientists are mapping them. The touted "disagreements," measured quantitatively, are rather like the disagreement between two independent dating methods for the age of the earth, one giving 4.50 billion years, and another giving 4.55 billion years—very similar measurements with a small amount of experimental error. Even phylogenies derived independently from morphological (anatomical) and molecular (chemical) datasets typically show a high degree of correlation. Any ID/creationist claim that phylogenetic trees show "discrepancies" is worthless unless they report proper similarity statistics, and this they have never done. By contrast, a recent striking example of molecular and morphological (fossil) data coming into astonishing agreement is documentation of connection between the ancestors of whales and those of hippos.

Haeckel's embryo drawings

The Kansas Science Standards state,

[Common ancestry is called into doubt by] Studies that show animals follow different rather than identical early stages of embryological development.

This is a key claim from Jonathan Wells's book *Icons of Evolution*. The argument is that evolution is said to be evidenced by embryological similarities as shown in Ernst Haeckel's famous embryo drawings, but that Haeckel "faked" the drawings to make the embryos more similar than they actually are, and that this "fake evidence" for evolution is reproduced in textbooks for school use.

The facts: Haeckel did exaggerate similarities in very early embryos of different species, and his figures, or derivatives of them, have appeared in a few textbooks (3 of the 10 textbooks that Wells examined). But *photographs* of embryos show strong and unquestionable similarities. The embryos of reptiles, birds, and mammals all resemble one another other much more strongly than do the adult forms, exactly as Darwin noted in the *Origin of Species*. The similarities, moreover, are not just superficial. They involve most of the fundamental pathways and structures of embryogenesis. Darwin and Haeckel asked why such different adult forms should all be modifications of what amounts to the same embryological plan—if organisms were specially created, they could just as well each develop directly into the adult forms with no embryological resemblance and no cumbersome remodelings during late embryonic life. Michael Richardson, the specialist who, in an exhaustive critique of Haeckel's work, re-examined all the drawings, observes:

> On a fundamental level, Haeckel was correct: All vertebrates develop a similar body plan (consisting of notochord, body segments, pharyngeal pouches, and so forth). This shared developmental program reflects shared evolutionary history. It also fits with overwhelming recent evidence that development in different animals is controlled by common genetic mechanisms.(Richardson et al. 1998: 983)

The cry of "fake" from Wells and friends is a completely manufactured scandal.

The origin of "information" in DNA

The Kansas Science standards state that

> The sequence of the nucleotide bases within genes is not dictated by any known chemical or physical law. (Kansas Science Standards, p. 73)

This assertion is copied from the 1980s creationists who wrote *Of Pandas and People* and later founded the ID movement. Dean Kenyon, for example, included it in his 1984 affidavit in defense of the creationist Louisiana Balanced Treatment Act, as it wound its way up to the Supreme Court. It formed the basis of books and articles written by two leading ID proponents: Charles Thaxton, academic editor of the *Pandas* project, and Stephen Meyer.

Their argument: The order of the chemical "letters" in DNA is not dictated by any known physical or chemical law; therefore, the "information" in DNA cannot be explained by natural processes; therefore, the "information" in DNA must have a supernatural cause. Stephen Meyer explains in his chapter for the 1994 anthology *The Creation Hypothesis*:

> [S]cientists have attempted to explain how purely natural processes could have given rise to the unlikely and yet functionally specified systems found in biology systems that comprise, among other things, massive amounts of coded genetic information. The origin of such information, whether in the first protocell or at those discrete points in the fossil record that attest to the emergence of structural novelty, remains essentially mysterious on any current naturalistic evolutionary account. (Meyer 1994: 68)

It would be no exaggeration to say that this argument is at the heart of the ID movement. The only problem is that it is scandalously *wrong*. Competent scientists know how new genetic information arises: a variety of well-understood mutational mechanisms copy and modify the DNA letter sequence that makes up a gene. If the new sequence is advantageous to the organism, natural selection spreads the new gene through the population via well-understood processes of population genetics. This shows where new genetic information comes from, and it fully explains, as a bonus, the otherwise puzzling fact that most genes belong to large families and superfamilies of similar composition.

One particularly useful paper published in *Nature Reviews Genetics* in 2003, by Manyuan Long of the University of Chicago, reviews all the mutational processes involved in the origin of new genes, and then lists dozens of examples in which research groups have reconstructed the genes' origins. The paper lists 122 references, virtually all of them published in the last ten years. None has ever been mentioned by the ID movement, let alone rebutted. Dr. Long has devoted his whole career to studying the origin of new genes; his online résumé lists some two dozen recent publications on the topic. [References omitted.]

The other problem with the argument of the Kansas science standards is the obfuscation "any known chemical or physical law." It is deviously phrased to have two meanings: it could simply mean that no laws of chemistry or physics, strictly defined, specify the order of the chemical "letters" in DNA. In that limited sense, the statement is (approximately, not completely) correct, but pointless. The shape of the Grand Canyon is also not *strictly* specified by any chemical or physical process—so what? The shape of the Grand Canyon is in fact specified by complex but reasonably well-understood interactions of erosion, rock structure, weather patterns, plate tectonics, and the like. But the Kansas science standards statement is *meant* to imply that *no natural explanation exists* for genetic information. That is a radical and false claim. Like the Grand Canyon, no simple physical "law" determines the DNA sequence, but the complex interacting processes that do explain it, described above, are well known. The game being played here is that the radical claim will be taught to students, but when the standard comes up for criticism, the limited, mostly true claim will be used to defend the phrase (pp. 40–43).

Conclusion

Although supporters of strong science education should not expect that the courts will always ride to the rescue, there are reasons to be hopeful in the case of "critical analysis of evolution" standards. First, although "critical analysis" consists not of critical analysis in the ordinary sense, but of a haphazard collection of objections to evolution, it emerges on closer inspection that *every single* "critical analysis" argument can be traced directly to established, long discredited creation science and intelligent design claims. This is important because courts have repeatedly emphasized the relevance of the historical lineage of governmental policies. A court that is made aware of this history should and may well conclude that "critical analysis of evolution" is substantially the same as creation science and intelligent design, and see it as one more in an unending series of attempts to privilege the same sectarian view—creationism—with a new and deceptive label.

Second, derived as they are from creationist literature, "critical analysis" arguments share all the problems of their creationist ancestors. The objections to evolution are not serious scientific arguments; they are superficially investigated and poorly reasoned talking points. Like all creationist arguments, they are aimed at uninformed audiences: they sound good in op-eds, media soundbites, and sermons, but they disintegrate upon detailed examination in court or in serious and extended public discussion.

REFERENCES

Meyer, Stephen C. 1994. The methodological equivalence of design and descent: Can there be a scientific 'theory of creation'? In *The creation hypothesis: Scientific evidence for an intelligent designer*, ed. J. P. Moreland, 67–112. Downers Grove, IL: InterVarsity Press.

Richardson, Michael K., James Hanken, Lynne Selwood, Glenda M. Wright, Robert J. Richards, Claude Pieau, and Albert Reynaud. 1998. Letter. *Science* 280 (5366): 983.

Wells, J. 2000. *Icons of evolution: Science or myth?* Washington, D.C.: Regnery.

Excerpted from Nicholas J. Matzke and Paul R. Gross, Analyzing critical analysis: The fallback antievolution strategy, in *Not in our classrooms: Why intelligent design is wrong for our schools*, ed. Eugenie C. Scott and Glenn Branch (Boston: Beacon Press, 2006) 57–82 (from 30–56). Used with permission.

• •

DISCLAIMING EVOLUTION

Introduction

Disclaimers often present evolution as theory, not fact, using *theory* in the popular sense of *guess* or *hunch* rather than in the scientific sense of *explanation*. Attempts to mandate application of a 1994 Alabama State Board of Education disclaimer placed in textbooks have been made in other states and communities (see Figure 10.1). And, from 1974 to 1984, the state of Texas restricted the presentation of evolution in textbooks by requiring that evolution be presented as theory, not fact. In 1984, the Texas state attorney general was requested to rule on the constitutionality of these regulations. Citing the *Lemon* decision, he judged them unconstitutional.

• •

READINGS: PROMOTING DISCLAIMERS

The Texas Textbook Disclaimer, 1974–1984

Textbooks that treat the theory of evolution shall identify it as only one of several explanations of the origins of humankind and avoid limiting young people in their search for meanings of their human existence.

(A) Textbooks presented for adoption which treat the subject of evolution substantively in explaining the historical origins of man shall be edited, if necessary, to clarify that the treatment is theoretical rather than factually verifiable. Furthermore, each textbook must carry a statement on an introductory page that any material on evolution included in the book is clearly presented as theory rather than verified.

(B) Textbooks presented for adoption which do not treat evolution substantively as an instructional topic, but make reference to evolution indirectly or by implication, must be modified, if necessary, to ensure that the reference is clearly to a theory and not to a verified fact. These books will not need to carry a statement on the introductory page.

Figure 10.1
The Alabama Disclaimer.

A MESSAGE FROM THE ALABAMA STATE BOARD OF EDUCATION

This textbook discusses evolution, a controversial theory some scientists present as a scientific explanation for the origin of living things, such as plants, animals and humans.

No one was present when life first appeared on earth. Therefore, any statement about life's origins should be considered as theory, not fact.

The word "evolution" may refer to many types of change. Evolution describes changes that occur within a species. (White moths, for example, may "evolve" into gray moths.) This process is microevolution, which can be observed and described as fact. Evolution may also refer to the change of one living thing to another, such as reptiles into birds. This process, called macroevolution, has never been observed and should be considered a theory. Evolution also refers to the unproven belief that random, undirected forces produced a world of living things.

There are many unanswered questions about the origin of life which are not mentioned in your textbook, including:

- Why did the major groups of animals suddenly appear in the fossil record (known as the "Cambrian Explosion")?

- Why have no new major groups of living things appeared in the fossil record for a long time?

- Why do major groups of plants and animals have no transitional forms in the fossil record?

- How did you and all living things come to possess such a complete and complex set of "Instructions" for building a living body?

Study hard and keep an open mind. Someday, you may contribute to the theories of how living things appeared on earth.

(C) The presentation of the theory of evolution shall be done in a manner which is not detrimental to other theories of origin.

• •

READING OPPOSING DISCLAIMERS

Texas Attorney General's Opinion JM-134, March 12, 1984

The only aspect of "evolution" with which the rule is concerned is that which relates to "the historical origins of man." The rule requires a biology textbook, for example, to carry a disclaimer on its introductory page to the effect that "any material on evolution" included therein is to be regarded as theory rather than as factually verifiable. In the first place, such a disclaimer—which might make sense if applied to all scientific theories—is limited to one aspect—man's origins—of one theory—evolution—of one science—biology. In the context of the controversy between evolutionists and creationists which was before the board at the time of the rule's adoption both in 1974 and 1983, this singling out of one aspect of one theory of one science can be explained only as a response to pressure from creationists.

In the second place, the "theory of evolution," as it is commonly treated in biology texts, is a comprehensive explanation of the development of the various plant and animal species. Only a relatively minor portion is concerned with the "historical origins of man." The latter subject is the primary interest of creationists. Again, the inference is inescapable from the narrowness of the requirement that a concern for religious sensibilities, rather than a dedication to scientific truth, was the real motivation for the rules.

Finally, the rules require that a textbook identify the theory of evolution "as only one of several explanations" of human origins in order to "avoid limiting young people in their search for *meanings* of their human existence." (Emphasis added.) Such language is not conducive to an explanation that the purpose of the rule is to insure that impressionable minds will be able to distinguish between scientific theory and dogma. The "meaning of human existence" is not the stuff of science but rather, the province of philosophy and religion. By its injection into the rules language which is clearly outside the scope of science, the board has revealed the non-secular purpose of its rules.

Clearly, the board made an effort, as it has stated, to "insure neutrality in the treatment of subjects upon which beliefs and viewpoints differ dramatically." In our opinion, however, the board, in its desire not to offend any religious group, has injected religious considerations into an area which must be, at least in the public school context, strictly the province of science. . . .

If the board feels compelled to legislate in this area, it should, in order to avoid the constitutional prohibition, promulgate a rule which is of general application to all scientific inquiry, which does not single out for its requirement of a disclaimer a single theory of one scientific field, and which does not include language suggesting inquiries which lie totally outside the realm of science. The rules submitted, however, when considered in the context of the circumstances of their adoption, fail to evidence a secular purpose, and hence we believe a court would find that they contravene the first and fourteenth amendments to the United States Constitution.

• •

ACADEMIC FREEDOM TO TEACH CREATIONISM/ANTIEVOLUTIONISM

Introduction

In the 1980s and 1990s, after *McLean* and *Edwards*, court battles shifted from state-level regulations to court cases involving individual teachers or school districts seeking to teach some form of creationism or neocreationism. These cases usually involved Establishment Clause concerns, but also free exercise and free speech (academic freedom) issues as well. Continuities with the past include the claim that the teaching of creationism or neocreationism enhances students' rights to learn "all" the material. In the twenty-first century, the academic freedom issue has been revived at the state level, and legislation has been proposed that would exempt teachers from long-standing constraints against introducing creationism into the public school classroom. The Florida bill, excerpted here, is an example. Although creationism is not mentioned by name in the bill, statements made by the bill's author and supporters make it clear that the goal of the legislation was to encourage teachers to bring alternative views like intelligent design into the classroom.

An example of an academic freedom claim made by an individual teacher is the *Webster v. New Lenox* case. In response to student complaints, the New Lenox, Illinois, superintendent of education, Alex Martino, directed the middle school teacher Ray Webster to cease teaching creationism in his classroom. Arguing academic freedom, Webster sued the district for his right to teach creation science. A student, Matthew Dunne, was a coplaintiff, arguing that he had a right to hear about creation science in school. Citing *Edwards*, the U.S. district court ruled against both plaintiffs. The decision was appealed and upheld by the Seventh Circuit Court of Appeals.

Florida SB 2962 "Academic Freedom Act" (2008)

(3) Every public school teacher in the state's K–12 school system shall have the affirmative right and freedom to objectively present scientific information relevant to the full range of scientific views regarding biological and chemical evolution in connection with teaching any prescribed curriculum regarding chemical or biological origins.

(4) A public school teacher in the state's K–12 school system may not be disciplined, denied tenure, terminated, or otherwise discriminated against for objectively presenting scientific information relevant to the full range of scientific views regarding biological or chemical evolution in connection with teaching any prescribed curriculum regarding chemical or biological origins.

(5) Public school students in the state's K–12 school system may be evaluated based upon their understanding of course materials, but may not be penalized in any way because he or she subscribes to a particular position or view regarding biological or

chemical evolution. [Versions of the bill passed by House and Senate, but legislative session ended before both houses reconciled their versions of the bill.]

Webster v. New Lenox (1989)

Raymond Webster is a teacher for New Lenox and as such has certain responsibilities to teach within the framework of curriculum outlined by the District. . . . Webster's rights as a teacher to present certain material within his social studies curriculum is not absolute. . . .

If a teacher in a public school uses religion and teaches religious beliefs or espouses theories clearly based on religious underpinnings, the principles of the separation of church and state are violated as clearly as if a statute ordered the teacher to teach religious theories such as the statutes in *Edwards* did. The school district has the responsibility of ensuring that the Establishment Clause is not violated. Therefore, New Lenox has the responsibility of monitoring the content of its teachers' curricula to ensure that the establishment clause is not violated.

Although Webster denies any improper religious teaching, the question before this court is whether Webster has a first amendment right to teach creation science. As previously discussed, the term "creation science" presupposes the existence of a creator and is impermissible religious advocacy that would violate the first amendment. Webster has not been prohibited from teaching any nonevolutionary theories or from teaching anything regarding the historical relationship between church and state. Martino's letter of October 13, 1987, makes it clear that the religious advocacy of Webster's teaching is prohibited and nothing else. Since no other constraints were placed on Webster's teaching, he has no basis for his complaint and it must fail.

Plaintiff Dunne's claims, if not moot, are without merit. Dunne has not been denied the right to hear about or discuss any information or theory including information as to creation science. He is merely limited to receiving information as to creation science to those locations and settings where dissemination does not violate the first amendment. Dunne's desires to obtain this information in schools are outweighed by defendants' compelling interest in avoiding Establishment Clause violations and in protecting the first amendment rights of other students. Dunne simply fails to state a claim for the violation of any first amendment or other rights and thus his claim must also fail. . . .

The relevant issue here is what Webster was prohibited from teaching. He was prohibited from teaching Creation Science. The U.S. Supreme Court has found that Creation Science is a religiously based theory and that the teaching of this theory in a public school violates the First Amendment. Prohibiting this teaching is thus constitutionally valid.

Since plaintiff Webster has no right to teach Creation Science and plaintiff Dunne has no right to receive information regarding Creation Science in his public school room, both plaintiffs' actions must fail. [Internal citations omitted.]

Excerpted from *Webster v. New Lenox School District Mem. op No. 88 C 2328* (N.D. Ill. 1989).

CHAPTER 11

•••••••••••••

Educational Issues

INTRODUCTION

The antievolution movement has long targeted the teaching of evolution in the public schools, because most young people who encounter evolution will do so during their formal education. Here, then, is where antievolutionists believe they have the best opportunity to ameliorate what they perceive as evolution's negative influence. As discussed in chapters 5 and 6, antievolutionists originally tried to ban evolution from the curriculum, then to "balance" it with the teaching of creationism or creation science. In response to court decisions, antievolutionists currently avoid religious arguments in favor of arguing that evolution is bad science that should be balanced with the presentation of alternative scientific theories such as intelligent design. They also argue that the teaching of evolution should be balanced with evidence against evolution.

In this battle over curricula, four themes recur. One theme is that if a district persists in teaching evolution, students whose parents object should be able to leave the classroom—to opt out of evolution instruction. A second is that teaching creationism (or creation science) and evolution is a matter of fairness or equity—the unadorned fairness pillar of creationism (see Introduction chapter). Because so many people object to their children learning evolution, many believe that if the subject is taught, it at least should be balanced by creationism or creation science. A third theme is to promote the fairness approach as having positive pedagogical value. Here it is claimed that students learning and evaluating all the (alleged) evidence benefit by sharpening their critical thinking skills. Finally, ever since the Scopes era, textbooks' coverage of evolution has been controversial, and the amount and quality of evolution in textbooks has fluctuated over time in response to creationist pressure. These topics are the subject of this chapter.

● ●

OPTING OUT OF EVOLUTION EDUCATION

The OOPSIE Compromise—
A Big Mistake

"Dear NCSE, I have a student whose father wants me to let his son opt out of instruction in evolution this semester. The principal says, 'let him do genetics.' I can't have the kid do Punnett squares for four weeks! What can I do?"

"Dear NCSE, a member of our school board claims that something called the 'Hatch Amendment' requires our teachers to let students opt out of instruction in evolution. Can this possibly be true?"

"Dear NCSE, We don't have a controversial issues policy in my district, so do I have to let a student who says that learning evolution is against her religion opt out of learning about evolution?"

Here at the National Center for Science Education, we receive a steady stream of questions from parents, teachers, science supervisors, and school board members about whether to allow students to opt out of instruction in evolution. Such policies complicate science instruction, of course, and (as we argue below) they have a bad effect on the students who opt out, on their classmates whose studies are disrupted, and especially on their teachers, who cannot fulfill their duty to instruct their charges about biology without emphasizing evolution. Particularly in communities in which creationism is prevalent, allowing students to opt out is often viewed as a satisfactory compromise whereby evolution is taught in general but not inflicted on the unwilling. But is it really satisfactory?

As we use the term, *opt-out policies*—OOPs, for short—are policies that allow students to be withdrawn from activities at school that address topics that they or their parents consider to be offensive or otherwise inappropriate. Not included under the rubric is the practice of allowing students to choose their classes from a choice provided by the school, which might allow them to avoid offensive activities, or the practice of informally steering students who are (or whose parents are) suspected of harboring objections to certain activities to classes where they are minimized or avoided altogether, or the practice of allowing (or even encouraging) students to attend private rather than public schools in order to avoid activities that they or their parents deem offensive.

The sources of OOPs vary. Sometimes OOPs are mandated by state law. In California, for example, the law allows students in the public schools to opt out of sex education and to opt out of animal dissection when they have a moral objection to it ("ick!" isn't enough). Sometimes individual school districts or schools go beyond the demands (if any) of state law, adopting OOPs of their own, which may either be general (policies about "controversial issues" sometimes include OOPs) or refer to specific practices that are of especial concern in their communities. And sometimes, of course, students or their parents may request—or demand—permission to opt out of a particular activity, even in the absence of any formal OOP.

Few OOPs explicitly involve academic topics, but when they do, there is typically a provision that ensures that the students who are opted out will, if possible, have to engage in a substitute activity to acquire the knowledge or ability that the objectionable practice is supposed to impart. In the case of animal dissection, for example, a district may allow the use of detailed plastic models or interactive dissection software. The rationale for such provisions is both obvious and compelling: for basic academic topics, students simply need to learn the material, by hook or by crook.

So when it comes to opt-out policies specifically including evolution—OOPSIEs— the acronym illustrates our view: because of the centrality of evolution to biology, such policies are a bad mistake. As Theodosius Dobzhansky famously wrote thirty-five years ago, "Nothing in biology makes sense except in the light of evolution" (Dobzhansky 1973). Evolution inextricably pervades the biological sciences; it therefore pervades, or at any rate ought to pervade, biology education at the K–12 level. There simply is no alternative to learning about it; there is no substitute activity. A teacher who tries to present biology without mentioning evolution is like a director trying to produce Hamlet without casting the prince. By the same token, a student who is opted out of evolution is likely to regard biology as a tale told by an idiot, full of sound and fury, signifying nothing.

Shakespeare aside, it isn't only students who are opted out of evolution who suffer as a result of OOPSIEs. Accommodating such students is bound to be disruptive to the course as a whole—ironically, the better the treatment of evolution in the course, the worse the disruption. A student opting out of evolution in such a course would have to bob in and out of the classroom several times a month, disappearing, for example, when the structure of the cell is taught (and with it the endosymbiotic origin of mitochondria), and again when taxonomy is taught (and with it phylogenetic systematics), and yet again when genetics is taught (and with it molecular homology), and so on. It is simply unreasonable to expect a teacher to install a revolving door, as it were, to accommodate students who are unwilling to hear the dreaded e-word.

Moreover, OOPSIEs are bad for schools and districts. Students who fail to learn about evolution are not going to perform as well on statewide examinations, which reflects poorly not only on them but also on their schools and districts. Nor are they going to perform as well in their biology classes in colleges and universities, where the faculty expects incoming students to have at least a basic grasp of evolution. Indeed, high school administrators often have to certify that the courses intended to prepare students for college in fact do so; allowing students to opt out of topics that are central to such classes may result in decertification. Schools and districts with OOPSIEs may also find it difficult to attract and retain those science teachers who take their professional responsibilities seriously: given a choice, who would prefer to teach biology at a school where the administrators are unwilling to support the teaching of evolution?

Faced with a proposed or actual OOPSIE, what is a science teacher to do? School and district administrators need to be reminded that science teachers deserve to be treated as professionals, trained in both the content of science and the methods of education; as such, their professional opinions about the necessity of including evolution in the biology curriculum deserve to be heeded. And, of course, their professional groups, such as the National Science Teachers Association, unequivocally endorse "the position that evolution is a major unifying concept in science and should be included in the K–12 science education frameworks and curricula.... If evolution is not taught, students will not achieve the level of scientific literacy they need" (NSTA 2003). Administrators also need to be reminded of the practical repercussions of OOPSIEs: the burdens imposed on teachers, the disruptions caused to the educational process, the damages wreaked on the school's reputation.

Claims that OOPSIE is required by the Constitution, federal law, or state law deserve skepticism. For example, the Protection of Pupil Rights Amendment—sometimes called the "Hatch Amendment" or the "Grassley Amendment"—is occasionally claimed to require school districts to allow students to opt out of various topics, including evolution. In fact, the PPRA is limited in scope, applying only to surveys, analyses, and evaluations funded by the federal Department of Education, and is intended only to protect the privacy of parents and students with regard to such studies; it neither sets limits on the school district's control over its curriculum nor provides any right for students to be opted out from regular classes. (Simpson 1996). But boilerplate citing the PPRA continues to circulate, dismayingly.

In dealing with individual parents who are requesting—or demanding—permission for their children to be opted out of instruction in evolution, not necessarily seeking the institution of a formal OOPSIE, different strategies are appropriate. Such parents are generally going to be conservative Christians who are worried, at bottom, about the prospect that instruction in evolution will challenge or damage their children's faith and that their children will be forced to "believe" in evolution. What helps to alleviate their concern is not a defensive citation of the importance of evolution in the curriculum or the practical repercussions for the school, but a respectful engagement with their worry.

According to teachers on the front lines, it helps to reassure such parents that the aim of education is to impart understanding, not to compel belief. As Ella Ingram and Craig Nelson (2006: 20) put it, "we believe that understanding evolution is more important than accepting evolution, and indeed that, ethically, we should ask students to strive for understanding prior to making decisions regarding acceptance of any theory." Students are asked only to understand evolution and appreciate its evidential basis—not to profess a faith in evolution, much less to renounce their religious views. Indeed, even the major creationist organizations agree that students ought to learn about evolution, although they themselves misunderstand and misrepresent it.

It helps also to explain that there is a range of attitudes about evolution and religion. Although there are those who regard evolution as incompatible with or even threatening to their faith, there are also those who regard it as compatible with or even enriching their faith. Among the latter, cited in the new publication *Science, Evolution, and Creationism* (NAS 2008), are not only religious leaders like the late Pope John Paul II but also prominent scientists like Francis Collins and Kenneth R. Miller, both of whom have written eloquently on evolution and their faith (Collins 2006; Miller 2007). Not all parents will share the religious views of such authors, of course; but they may be impressed enough by their sincere devotion to heed their insistence that to understand modern science, a student needs to learn about evolution.

Ultimately, parents and teachers want the same things: for their children and students to do well in school, learn the subject material, and become educated citizens.

Accomplishing these goals requires that teachers ensure the competence of students in basic subjects such as biology, to which evolution is central. That is why OOPSIEs are not a satisfactory compromise but just a big mistake.

REFERENCES

Collins, Francis S. 2006. *The language of God: A scientist presents evidence for belief*. New York: Free Press.

Dobzhansky, Theodosious. 1973. Nothing in biology makes sense except in the light of evolution. *The American Biology Teacher* 25: 125–129.

Ingram, Ella L., and Craig E. Nelson. 2006. Relationship between achievement and students' acceptance of evolution or creation in an upper-level evolution course. *Journal of Research in Science Teaching* 43 (1): 7–24.

Miller, Kenneth R. 2007. *Finding Darwin's God: A scientist's search for common ground between God and evolution*. New York: Harper Perennial.

National Academy of Sciences. *Science, evolution, and creationism*. 2008. Washington, D.C.: National Academies Press. Retrieved January 11, 2008, from http://www.nap.edu/sec.

National Science Teachers Association. *Position statement: The teaching of evolution, 2003*. Retrieved January 11, 2008, from http://www.nsta.org/about/positions/evolution.aspx.

Simpson, M.D. 1996. What's a teacher to do? *NEA Today* 14 (6): 25. Retrieved January 11, 2008, from http://findarticles.com/p/articles/mi_qa3617/is_199602/ai_n8750005.

Excerpted from E. C. Scott and G. Branch, The OOPSIE compromise—A big mistake. *Evolution: Education and Outreach* 1, no. 2 (2008): 147–149. Used with permission.

• •

"FAIRNESS"

Introduction

Antievolutionists often exhort teachers to present "both views," meaning creationism and evolution. As discussed in chapter 3, there are far more than two views, and no science classroom can possibly devote sufficient time to give justice to even a fraction of the various religious (creationist) explanations of the universe. Creation science proponents, on the other hand, argue that theirs is a legitimate *scientific* view, not merely a religious perspective, and that out of a sense of fairness, both evolution science and creation science should be presented. "Let the children decide" has been the rallying cry of Young Earth-Creationists for decades.

The fairness argument appeals to Americans—indeed, North Americans in general—because of cultural traditions of free speech, individuality, town meetings, democratic traditions, and the like. The argument has been made, however, that science is not a democratic institution, and that cultural standards of equal time do not apply to empirical explanations that have been analyzed and rejected. Opponents of the fairness argument do not question the cultural value of fairness, but rather its application to the science classroom.

● ●

"FAIRNESS": CREATIONIST VIEW

Appendix E: Fairness of a Balanced-Treatment Approach

In view of the fact that evolution and creation are the only two possible concepts of origins, that evolution requires at least as much of a "religious" faith as does creation, and that creation fits all the "scientific" data at least as well as does evolution, it is clear that *both* should be taught in the schools and other public institutions of our country, and that this should be done on an equal-time, equal-emphasis basis, in so far as possible.

This is obviously the only equitable and fair approach to take, the only one con-sistent with traditional American principles of religious freedom, civil rights, freedom of information, scientific objectivity, academic freedom, and constitutionality. That American citizens, when given opportunity to express their opinions, fully support this idea has been proven conclusively in recent carefully conducted, scientifically organized community polls taken in two California school districts.

One of these was a semi-rural district, Del Norte County in northern California. Here, a poll of 1,326 homes revealed 89 percent to favor including creation along with evolution in school curricula (*Acts and Facts*, Vol. 3, April 1974, p. 1). The other was a very cosmopolitan district in the San Jose–San Francisco metropolitan region, Cupertino, the largest elementary school district in the state. In this case, a poll of more than 2,000 homes showed 84 percent to favor including creation (*Acts and Facts*, Vol. 3, August 1974, p. 3). In both cases, the emphasis in the questionnaire was on *scientific* creationism, rather than its religious aspects.

There is little doubt that similar majorities would be obtained in most other school districts across the country, if people were informed on the issue and given opportunity to express their preferences.

Two final points should be stressed. There are really only two scientific models of origins—continuing evolution by natural processes or completed creation by super-natural processes. The latter need not be formulated in terms of Biblical references at all, and is not comparable to the various cosmogonic myths of different tribes and nations, all of which are merely special forms of evolutionism, rather than creationism, rejecting as they do the vital creationist concept of a personal transcendent Creator of all things in the beginning. It is not the Genesis story of creation that should be taught in the schools, of course, but only creationism as a scientific model.

Secondly, the idea of theistic evolution (that is, evolution as God's method of creation) is not in any way a satisfactory compromise between evolution and creation. It is merely an alternate form of evolutionism with no *scientific* distinction from that of naturalistic evolutionism, and is vulnerable to all the scientific, religious and legal objections outlined previously for evolutionism in general.

We conclude, therefore, that *both* creation and evolution should be taught—as scientific models only—in all books and classes where *either* is taught or implied.

Administrators should assume the responsibility of providing adequate training and materials to enable their teachers to accomplish this goal.

Excerpted from Henry M. Morris, Resolution for equitable treatment of both creation and evolution, *Impact* 26 (August 1975): 4. Used with permission.

● ●

"FAIRNESS": OPPOSING CREATIONIST VIEW

It's Only Fair to Teach Creationism

Teachers need to be able to counter the "fairness" argument, which is second only to the idea that evolution is antireligious as a motivator of pro-creationism and antievolution activities. . . . What are some answers when this question arises in a community or classroom?

1. *We determine curricula based on the best scholarship, not because of political pressure.* Both scientists and teachers have rejected "creation science," "intelligent design theory," and other forms of creationism as science. There are many statements from scientific and educational organizations stating that religiously based ideas should not be taught as science; only evolution should be taught as science (these are also on the World Wide Web at www.ncseweb.org/media/voices). Creation science and other forms of creationism should not be taught because they have been evaluated by scientists and educators and found wanting. . . . Just because a pressure group wants to change the curriculum is no reason to abandon scholarly standards.

2. *Science is not a democratic process; nature doesn't work according to our wishes!* It would be a wonderful solution to the problem of universal cheap energy if perpetual motion were a reality, but no matter how many people want this to be so, the laws of physics do not allow it, and we do not teach students perpetual motion as a scientific principle. Even if many individuals in a community favor a special creationist view that the universe came into being in its present form 10,000 years ago, scientific evidence is very much against this, and students should not be taught that special creationism is a viable scientific doctrine.

3. *It is not "fair" to mislead students by presenting scientifically uncontroversial issues as controversial.* Evolution is taught matter-of-factly at any reputable university in this country, including Brigham Young (Mormon), Baylor (Baptist), and Notre Dame (Catholic). It is a controversial subject only to members of the public. It would be "fair" to discuss the creation and evolution controversy as a politically or socially controversial issue, but it shortchanges students to teach them that there is any scientific controversy over the concept that the universe has changed through time, and that living things have descended with modification from earlier ancestors. "Equal time for creationism" policies do just that, handicapping students for future study in college, and decreasing their scientific literacy.

Excerpted from Eugenie C. Scott, "But I don't believe in evolution!" The science teacher's dilemma. *Journal of Religion and Education* 26, no. 2 (1999): 67–75 (from 73–74). Used with permission.

• •

THE PEDAGOGICAL VALUE OF TEACHING ALL THE EVIDENCE: CREATIONIST VIEWS

Summaries of Articles by Meyer and DeWolf

Most North American science standards on the state or provincial level stress the importance of critical thinking. One approach to critical thinking often used by teachers is to have students research and debate a controversial issue. But is the creation and evolution issue an appropriate one for critical thinking exercises? The first two readings take the position that it is.

I had intended to include an editorial written by intelligent design proponent Stephen C. Meyer during a controversy over Ohio's state science education standards (*Cincinnati Enquirer*, March 30, 2002), but permission was denied for the first edition of this book, and a second request for reprinting for this edition went unanswered. Students are encouraged to read the article online at http://www.discovery.org/a/1134.

In this article, Meyer begins by stating a general principle of teaching the controversy: when two groups of experts disagree about a controversial topic relevant to a subject taught in the public schools, teachers should explain the controversy, and the arguments on both sides of it, to their students. Speaking to the Ohio Board of Education while the state was considering new state science standards, Meyer recommended teaching the controversy as a way of resolving disputes about how to teach about evolution and whether or not to teach intelligent design alongside what he calls Darwinism.

In particular, he suggested (1) that teachers ought not to be required to teach intelligent design yet, (2) that teachers ought to teach the scientific controversy about Darwinian evolution, and (3) that teachers ought to be allowed, but not required, to teach intelligent design. He cited five considerations in favor of his recommendation. First, that there is a genuine scientific controversy about Darwinian evolution, as evidenced by a bibliography of scientific articles prepared by the Discovery Institute and submitted to the Ohio Board of Education and by the Discovery Institute's "A Scientific Dissent from Darwinism," a statement on "Darwinism" signed by one hundred scientists. Second, the Supreme Court's decision in *Edwards v. Aguillard* noted that "teaching a variety of scientific theories about origins" was legal. Meyer recommended teaching critiques of "Darwinism" and discussion of "competing" theories. Third, that it is endorsed by the Santorum language in the conference report to the No Child Left Behind Act (see chapter 9). Fourth, that it is overwhelmingly approved of by voters, according to a national Zogby poll. Fifth, that teaching the controversy is educationally sound: it sparks interest among students and motivates them to learn. Here Meyer quoted from Darwin: "A fair result can be obtained only by fully stating and balancing the facts and arguments on both sides of each question." He concluded by saying that the issue in Ohio is not about intelligent design but rather about whether both sides of the scientific controversy over Darwinism will be taught.

I similarly was denied permission for both the first and second editions to reprint a portion of an essay from intelligent design proponent David DeWolf (David K. DeWolf, *Teaching the Origins Controversy: A Guide for the Perplexed*. Discovery Institute Special Report, retrieved October 14, 2008, from http://www.discovery.org/a/48.) Students are encouraged to read the original themselves.

The passage summarized here begins with the same point from the Supreme Court's decision in *Edwards v. Aguillard* emphasized by Meyer: that "teaching a variety of scientific theories . . . might be validly done with the clear secular intent of enhancing the effectiveness of science education." DeWolf suggests that teaching the origins controversy—or, as he says, "making a full, rather than restricted[,] presentation of Darwinism"—would do so. Darwinism, he contends, is typically taught in a way that fails to stimulate students' interest. It ought not to be taught as uncontroversial and as accepted by all reputable scientists, as the National Academy of Sciences suggests. Rather, he suggests, it ought to be presented as a competitor in the scientific arena, vying with other plausible scientific theories for acceptance. Students would thereby be motivated to scrutinize the arguments for and against these theories carefully. Moreover, DeWolf suggests, teaching the origins controversy provides teachers with the opportunity to encourage their students to examine both sides of a controversial issue; even if teachers have strong views on the issue, they can set a good example by forthrightly addressing the view they reject. Learning is facilitated best by active dialogue rather than passive acceptance.

● ●

THE PEDAGOGICAL VALUE OF TEACHING "ALL THE EVIDENCE": OPPOSING CREATIONIST VIEWS

Evolution: What's Wrong with "Teach the Controversy?"

In a piece entitled "Teach the Controversy," Stephen C. Meyer of the Discovery Institute's Center for Science and Culture, the institutional home of the "intelligent design" variety of antievolutionism, writes, "good pedagogy commends this approach. Teaching the controversy about Darwinism as it exists in the scientific community will engage student interest. It will motivate students to learn more about the biological evidence as they see why it matters to a big question." The thought does not originate with the "intelligent design" movement, however. The Institute for Creation Research (ICR), the oldest major antievolutionist organization in the USA recommends that students and teachers be "encouraged to discuss the scientific information that *supports* and *questions* evolution and its underlying assumptions, in order to promote the development of critical thinking skills" (emphasis in original). The intent is not to have students investigate controversies about patterns and processes within evolutionary theory, but to debate whether evolution occurred.

Presenting all sides of a controversial issue appeals to popular values of fairness, openness and equality of opportunity. It thus plays well with the public. But it is important to examine any such appeal carefully, because it is easy to abuse the public's willingness to be swayed by such a call. . . . How is a teacher to decide which controversies are pedagogically valuable?

Criteria for Determining Which Controversies to Teach

We suggest the following five criteria for whether a controversy is appropriate to teach in a public school science class:

The controversy ought to be of interest to students. There is, for example, a raging scientific controversy over whether maximum likelihood or parsimony ought to dominate in phylogenetic interpretation. But we suspect that few students will be fascinated by the controversy, however dear it might be to the readers of *TREE*!

The controversy ought to be primarily scientific, rather than primarily moral, social or religious. The controversy over stem cell research, for example, is not about whether embryos can be manipulated to produce stem cells, but about whether it is morally permissible to do so. Questions about the morality of such research are of course important, but they are not suitable for a science class. Controversies that are primarily religious in nature are especially unsuitable for classes in public schools in the USA due to the Establishment Clause of the First Amendment to the Constitution, which prohibits the government from sponsoring religious advocacy.

The resources for each side of the controversy ought to be comparable in availability. It is difficult to teach the controversy if there is hardly anyone to make the case for one side of it. A teacher who decided to teach the controversy about geocentrism, for example, would find it difficult to locate resources for the geocentric side. (It would, however, be appropriate to teach about the 17th-century controversy as an historical digression.)

The resources for each side of the controversy ought to be comparable in quality. If the arguments for one side of a controversy are generally poor, then students are not likely to profit by studying it. The scientific consensus that AIDS is caused by a virus, for example, is so strong that there is little point to presenting opposing views.

The controversy ought to be understandable by the students. Most of the fascinating controversies over the role of epigenetic factors in development, for example, require a great deal more developmental, morphological and genetic training than a high school student can be expected to master in the time available.

Using these criteria, is the antievolutionists' controversy about evolution one that is worth teaching? We think not. It does satisfy criterion 1: it is probably of interest to students. It also satisfies criterion 3, thanks both to the wide availability of creationist material on the internet and to the advent of "intelligent design," which enjoys a degree of publicity in relatively mainstream venues, although conspicuously not in the peer-reviewed scientific literature. However, the controversy about evolution fails significantly to satisfy the other three criteria.

First, the controversy is not primarily scientific (criterion 2). In spite of their frequent claims to be concerned with the science, for Young-Earth Creationists (such as the ICR) and the "intelligent design" movement alike, the science is essentially a smokescreen for nonscientific concerns. For the ICR, the problem with evolution is its incompatibility with a literal reading of the Bible; for the "intelligent design" movement's guru Phillip Johnson, "This isn't really, and never has been, a debate about science. It's about religion and philosophy." Moreover, as far as students are concerned, the controversy about evolution is essentially religious. Are they going to be able to restrict their concerns solely to the science? Even many college students have difficulty studying religion objectively; at the pre-college level, the problem is worse. And are science teachers willing and able to respond properly to their concerns, without appearing to attack religious beliefs?

... Second, and correspondingly, the scientific quality of the antievolutionist resources is exceedingly poor (criterion 4). The positive claims of Young-Earth Creationism—that the universe and the Earth were created ~10,000 years ago, that the Earth was inundated by Noah's Flood, and that all living things were created by God to reproduce "after their kind," thus setting limits on evolution—are unanimously rejected by the scientific community. Its negative claims (the "evidence against evolution") typically involve either misinterpretation of the scientific literature or arguments from ignorance. The "intelligent design" variety of antievolutionism is strategically noncommittal, limiting its positive program to the claim that it is possible to identify certain natural phenomena as the products of intelligent design; its proponents disagree about the age of the Earth, common ancestry, and a host of other important scientific issues. Its negative claims are already in the repertoire of young-earth creationism, so the same objection applies.

Finally, students are unlikely to be able to understand both sides of the controversy (criterion 5). The evidence for evolution is easy to understand, at least on a basic level. But the antievolutionist critique of evolution ranges freely and opportunistically through the scientific literature, from astronomy to zymurgy, frequently misrepresenting it in the process. Faced with the ICR's tendentious and eclectic list of "questions that could be used to critically examine and evaluate evolutionary theory" or the "Suggested Warning Labels for Biology Textbooks" produced by "intelligent design" proponent Jonathan Wells, even a working research scientist would have a difficult time sorting through the quagmire of misleading and mistaken claims. It is unreasonable to expect teachers, much less their students, to do so. [References omitted.]

Excerpted from Eugenie C. Scott and Glenn Branch. Evolution: What's wrong with "teaching the controversy"? *Trends in Ecology and Evolution* 8, no. 10 (2003): 499–502 (from 499–501). Used with permission.

• •

TEXTBOOK BATTLES

Introduction

For better or for worse, textbooks tend to comprise the de facto curriculum in science classes. When textbooks include evolution, antievolutionists are motivated to modify them either to delete or water down evolution (as during the post-Scopes era) or to balance it with alternatives to evolution (including creationism) or, most recently, to present evolution and the scientific "weaknesses" of evolution. This first selection traces the history of evolution in textbooks.

Effects of the Scopes Trial

The scientific community in the 1920s responded forcefully to the overt attack in the Scopes case. But it failed to follow through. As a result, the teaching of evolution in the high schools—as judged by the content of the average high school biology textbooks—*declined* after the Scopes trial.

. . . The impact of the Scopes trial on high school biology textbooks was enormous. It is easy to identify a text published in the decade following 1925. Merely look up the word *evolution* in the index or the glossary; you almost certainly will not find it.

. . . We do not wish to maintain that the omission of the word *evolution* from the index of a book automatically invalidates the book's treatment of the subject. However, widely used biology textbooks of the 1930s did not treat evolution very well in the text, either. The religious quotations which appear in some of these books, together with the near-disappearance of the theory of evolution and of Darwin's role in establishing it, demonstrate the impact of fundamentalist pressure in general, and the Scopes trial in particular, on the textbook industry.

. . . Publishing high school textbooks is a lucrative business. And the authors and publishers of biology textbooks have to pay attention to their market. Textbook adoption practices vary; some states approve texts for the entire state, while others allow local option. Unfortunately for the market prospects of an evolutionary textbook, most of the states which have at various times practiced statewide textbook adoption are in the south, and no eastern states are included. . . .

Publishers and authors feared that a good treatment of evolution meant the loss of the southern market—a fear which seems to have been justified.

. . . The evolutionists of the 1920s believed they had won a great victory in the Scopes trial. But as far as teaching biology in the high schools was concerned, they had not won; they had lost. Not only did they lose, but they did not even know they had lost. A major reason was that they were unable to understand—sympathetically or otherwise—the strength of the opponents of evolution. It is worth one's while to inquire into what motivates large numbers of people to oppose evolution. Whether one agrees or disagrees with their views, the people and their concerns deserve sympathy and respect. And understanding the opposition to evolution is essential if one is to take any kind of effective action.

. . . Readers may choose their own villain in the story we have told. Like us, some will find the greatest culpability in the scientific community itself, for the large-scale failure to pay attention to the teaching of science in the high schools. Others will blame the textbook authors and publishers for pursuing sales rather than quality. Some will attach blame to the politicians who exploited antievolution sentiment to get into, or remain, in office. Others will blame the conservative Protestant clergy. Some may blame the whole educational system for failing to teach Americans how to evaluate evidence. And many will blame the evolutionists for bringing the matter up in the first place. But whatever the lesson one wishes to draw from the history of biology textbooks since the Scopes trial, we think the story itself is worth knowing. That the textbooks could have downgraded their treatment of evolution with almost nobody noticing is the greatest tragedy of all. [References omitted.]

Excerpted from Judith V. Grabiner and Peter D. Miller, Effects of the Scopes trial, *Science* 185: 4154 (1974): 832–837. Used with permission.

• •

TEXTBOOK BATTLES: CREATIONIST VIEWS OF EVOLUTION IN TEXTBOOKS

Summary of Jonathan Wells's Views about the Peppered Moth Experiments

Jonathan Wells's book *Icons of Evolution* criticizes textbooks' coverage of evolution. He claims that a series of "icons" or common illustrations of evolution or mechanisms of evolution are misleadingly presented in textbooks, even to the extent of presenting fraudulent data. One example is the familiar peppered moth story of natural selection. I requested permission both for the first and current editions of this book to excerpt an article by Wells that summarized his views about the peppered moth. I was denied permission for the first edition, and received no reply to the second request. Readers are encouraged to read the article ("Survival of the Fakest," *The American Spectator*, December 2000/January 2001; http://www.discovery.org/articleFiles/ PDFs/survivalOfTheFakest.pdf) or the relevant chapter in *Icons of Evolution*.

The section of Wells's article titled "Nothing a Little Glue Can't Fix: The Peppered Moths" begins by observing that Darwin's *On the Origin of Species* lacked a concrete example of natural selection in action, and then turns to a discussion of Bernard Kettlewell's research in the 1950s on peppered moths. During the nineteenth century, the British population of these moths changed from being mostly light colored to being mostly dark colored. The cause of the change was thought to be natural selection: due to industrial pollution, the tree trunks on which the peppered moths rest became darker, so light-colored moths were more susceptible to predation by birds. Kettlewell tested this hypothesis by releasing light-colored and dark-colored moths in polluted and unpolluted woods; the moths that were not appropriately camouflaged were indeed more susceptible to predation. Kettlewell's study was widely heralded as a classic example of natural selection and is widely cited in biology textbooks, which typically include photographs of moths resting on tree trunks.

Further research conducted in the 1980s revealed flaws in Kettlewell's study, however: although he released the moths by day onto tree trunks, they normally fly at night and rest under upper tree branches during the day. Consequently, Wells states, many biologists now reject Kettlewell's study, and some even doubt the hypothesis that natural selection caused the change in moth coloration. The textbook photographs were staged; in some cases, the moths were glued to the trees. Although the flaws in Kettlewell's study were discovered in the 1980s, textbooks continue to recount the study uncritically and to reproduce the staged photographs. Wells quotes a textbook author as justifying the continued use of Kettlewell's study by saying, "We want to get across the idea of selective adaptation. Later on, they can look at the work critically." He then observes that Jerry Coyne, a professor of biology at the University of Chicago, learned the truth about the peppered moths only in 1998, showing that the "icons of evolution" (of which the peppered moth example is one) are so deceptive that even experts such as Coyne are apt to be misled.

• •

TEXTBOOK BATTLES: RESPONSES TO CREATIONIST VIEWS

Coyne on Wells and Moths

My only problem with the peppered moth story is that I am not certain whether scientists have identified the precise agent causing the natural selection and evolutionary change. It may well be bird predators, but the experiments leave room for doubt. Creationists such as Jonathan Wells claim that my criticism of these experiments casts strong doubt on Darwinism. But this characterization is false. All of us in the peppered moth debate agree that the moth story is a sound example of evolution produced by natural selection. My call for additional research on the moths has been wrongly characterized by creationists as revealing some fatal flaw in the theory of evolution.

Excerpted from Jerry Coyne, Letter to the editor, *Pratt* (Kansas) *Tribune*, December 6, 2000. Used with permission.

The Peppered Moth: The Proof of Darwinian Evolution

I am sure that most of you know the peppered moth story, but just as a résumé, in brief the story was this: The non-melanic, or *typica* peppered moth is white, liberally speckled with black. In 1848, a black form (*carbonaria*) was recorded in Manchester. By 1895, 98% of the Mancunian peppered moths were black. The *carbonaria* form spread to many other parts of the [UK], reaching high frequencies in [the] industrial centre and regions downwind.

In 1896, the great Victorian lepidopterist, J. W. Tutt, hypothesized that the increase in *carbonaria* was the result of differential bird predation favouring *carbonaria* in polluted regions, but not in unpolluted regions.

. . . In the 1950s, Bernard Kettlewell obtained data from direct predation experiments, and mark-release-recapture experiments, in two woodlands, one polluted, the other relatively unpolluted, that supported Tutt's hypothesis. It was the reciprocal nature of Kettlewell's results in the two woodlands, allied to extensive survey work showing a strong correlation between *carbonaria* frequency and industrial pollutants that made the case so persuasive. Over the next 40 years, various researchers worked with peppered moths to tease apart the fine detail of the case, but none of the new findings seriously undermined Tutt's hypothesis or Kettlewell's evidence in support of it.

The zenith for the case came in 1996, when, reporting work from both sides of the Atlantic that showed similar changes in melanic frequencies were correlated to pollution levels (Grant et al., 1996), *The New York Times* featured the peppered moth on the front page of its science section.

However, since Kettlewell's experiments, the black peppered moth has suffered two types of decline. First, following anti-pollution legislation in the 1950s and thereafter, *carbonaria* began to decrease in frequency, as would be expected from Tutt's theory. Second, the reputation of the case as an example of Darwinian evolution in action has been severely tarnished.

Today, I want to briefly explain the reason for the case's decline in reputation. I will then detail why I chose to undertake a piece of experimental work that has taken me 7 years, before presenting you with a series of observations on the natural resting sites of the peppered moth, and the results of two field experiments. Finally, at the end, I may make a few mild concluding remarks.

The decline in the peppered moth's reputation may be sourced to a book that I wrote in 1998 or, more correctly, a review of it by Jerry Coyne, in *Nature,* in which Coyne concluded, "For the time being we must discard *Biston* as a well-understood example of natural selection in action." I, and others, have dealt with this review previously, showing that the review had little to do with what was said in the book. As Donald Frack put it, "There is essentially no resemblance between Majerus' book and Coyne's review of it. If I hadn't known differently, I would have thought that the review was of some other book."

But the damage had been done. Coyne's review, and an article, titled "Scientists pick holes in Darwin moth theory" by Robert Matthews in the *Sunday Telegraph* (March 1999), began to appear on creationist and anti-evolution Web sites.... I should point out that it was about this time, the year 2000, that I first began to formulate the design of the predation experiment that I will tell you about later. But before we come to that, I must mention one other publication that raised the ante.

In 2002, Judith Hooper, an American journalist and writer, published a book called *Of Moths and Men: Intrigue, Tragedy and the Peppered Moth*, which is, according to the front cover, a riotous story of ambition and deceit. This appalling book is essentially an attack on the peppered moth case, those who have worked on the evolution of industrial melanism, lepidopterists in general, and Kettlewell and Professor E.B. Ford in particular. To cite the front fly-sheet, "*Of Moths and Men* is...a fascinating psychological dissection of the ambitious scientists who will ignore the truth for the sake of fame and recognition."

I do not want to dwell on Hooper's book for long. Various reputable scientists, from both sides of the Atlantic ... have discussed the many flaws in Hooper's book. Coyne, for example, in a review in *Nature* criticizes her "flimsy conspiracy theory," her theme of "ambitious scientists who will ignore the truth for the sake of fame and recognition," by which "she unfairly smears a brilliant naturalist."

... In 2000, while the peppered moth was under initial attack from anti-evolution lobbyists, I conceived two parts of the work that I am going to describe to fill up a major gap in our knowledge of the natural history of the peppered moth: that is, where peppered moths rest in the day. [I also would] check whether various valid criticisms of Kettlewell's experimental protocols could have altered the qualitative validity of his conclusions, by conducting a new field predation experiment.

... The question that I wished to answer with the main predation experiment was: Is differential bird predation sufficient to explain any changes in the frequencies of the *typica* and *carbonaria* forms observed over a period of years. Given previous observations in the Cambridge area and other parts of Britain, I knew that the frequency of *carbonaria* had been declining, and I had no reason to suppose that the decline would not continue. (N.B.: It was not possible to replicate Kettlewell's reciprocal design because in no part of Britain is *carbonaria* frequency increasing.)

The main experiment was designed to take account of all the flaws that had been aimed at Kettlewell's work, that:

1. The densities of moths were too great, and he used too few release sites.
2. Moths were released onto tree trunks, when Kettlewell knew that peppered moths usually rest under lateral branches.
3. Moths were released during the day, and so might not have selected sites that would maximize their crypsis.
4. Kettlewell used mixtures of wild caught and lab-bred moths, which might behave differently.
5. Kettlewell used translocated moths that might have had different behaviors as a result of local adaptation.

My experimental design, which was piloted in 2001, and has already been published (Majerus 2005), allowed me to:

1. Do the experiments in the wild, at low frequency (<10 per hectare per night), and collect back any moths left at each predation run.
2. Release moths in their natural resting positions (initially 103 release sites in a one hectare experimental site).
3. Release moths at dusk, into restricted arenas at their natural resting sites, so that they could take up resting positions at the end of their night flight. Arenas are removed in the forty minutes before sunrise.
4. Use and compare moths that were moth-trap caught males, pheromone-trap caught males, lab-bred males or lab-bred females.
5. Only use moths from Cambridge, within 5 km of the experimental site.

    In addition, I:

6. Released moths at the frequencies that they occurred in the previous year at a site 1.9 km from the experimental site.
7. Ran the experiment during the months that the moth is naturally on the wing.
8. Predation was scored by direct observation or absence after four hours.

This experimental procedure has been adhered to, with the only changes being that six of the release sites were lost due to storm damage, and the experiment, which initially was expected to last for five years, has taken six because of the low frequency of *carbonaria* in 2003, which meant that the number of *carbonaria* being exposed to predation was lower than expected.

The additional experiment that has been done addresses the question of bat predation. This arose directly from Hooper's book, for it reveals Hooper's lack of understanding of Darwinian selection. Hooper (2002, 270) raises the question of bats as predators of peppered moths. She states that "Kettlewell himself admitted that they [bats] probably accounted for 90% of the predation of adult moths."

By e-mail in 2000, she pointed out to me that Kettlewell had "said that this didn't matter because it wasn't selective—ergo, even if only 10% of the predation was by birds hunting by sight, that 10% is what makes the difference and drives evolution." Hooper thought that there were flaws in this argument and asked me about this. By phone I said I agreed with Kettlewell and explained why (Hooper 2002, 270). But not understanding how selection operates, Hooper didn't get it, and concludes, "Can we really be sure that bat predation is not selective . . . ?"

For bats hunting by sonar, it is hard to see how they could be responsible for the changes in *carbonaria* frequency correlated to pollution levels, without some fairly unrealistic logical gymnastics and assumptions. But, I decided to do an experiment anyway: to test whether bats do prey on *typica* and *carbonaria* differentially. The design was simply to release equal numbers of the forms near moth-traps where pipistrelle bats were feeding and watch which moths were eaten.

From the data shown here, it is obvious that there are no significant differences in the predation of the two forms. Across the four runs, 208 *carbonaria* and 211 *typica* were taken. So pipistrelle bats do not show differential selection of *typica* compared to *carbonaria* or vice versa.

During the main predation experiment, I have had occasion to spend time carefully scrutinizing the trunks, branches and twigs of a limited set of trees at the experimental site. During this time I have found 135 peppered moths, resting in what I have no reason to presume are not their freely chosen natural resting sites.

The position of each moth was scored for resting site (trunk, branch, twig); height above ground; on trunks, north or south half; on branches, top or bottom half. Sex and form of each moth was also recorded. In summary, these results show that:

1. The majority (50.4%) of moths rest on lateral branches.
2. Of the moths on lateral branches, the majority (89%) rest on the lower half of the branch.
3. A significant proportion of moths (37%) do rest on tree trunks (so Kettlewell wasn't so wrong in releasing his moths onto tree trunks).
4. Of those that rest on trunks, the majority (86.8%) rest on the north, rather than the south half.
5. A minority of moths (12.6%) rest under or among twigs.
6. There was no significant difference in the resting sites of males and females.
7. There was no significant differences in the resting sites used by *typica*, *carbonaria* or *insularia* (intermediate in color) forms.

While the results may be somewhat biased towards lower parts of the tree, due to sampling technique, I believe that they give the best field evidence that we have to date of where peppered moths spend the day.

... The frequencies of the *typica* and *carbonaria* forms for the seven years 2001–2007 were obtained by moth-trapping in Madingley Wood, 1.9 km from the experimental site. *Insularia* which had a frequency of between about 6% and 10% throughout was excluded from analysis. Thus the frequencies of *carbonaria* here (Fig. 11.1) is the proportion of *typica* + *carbonaria* that were *carbonaria*.

The basic results of the predation experiment are shown (Table 11.1), with the numbers of moths of each form available for predation and the numbers eaten. The bottom line is that a significantly greater proportion of *carbonaria* were eaten than typica. [Proportion of *typica* taken = 0.212; proportion of *carbonaria* taken = 0.292.]

A number of species of bird were observed preying on the moths: These included: robins, hedge sparrows, a lesser-spotted woodpecker, great tits, blue tits, blackbirds, starlings, wrens and magpies.

From the data, the selection coefficient needed to account for the decline in *carbonaria* frequency for each year between 2001 and 2007 can be compared to the

Table 11.1
Numbers of the two forms available for predation and predated (2002–2007).

Year	Numbers available for predators		Numbers eaten	
	typica	*carbonaria*	*typica*	*carbonaria*
2002	706	101	162	31
2003	731	82	204	24
2004	751	53	128	17
2005	763	58	166	18
2006	774	34	145	6
2007	797	14	158	4

Predation on the darker form of the moth was higher than on the lighter form.

selection coefficient observed in the predation experiment for each year 2002–2007 (Table 11.2). Here, were bird predation to be a causative factor of changes in *carbonaria* frequency, compared to *typica*, the observed frequency in one year should be a consequence of the predation the previous year. So, I do not have predation data to account for changes from 2001–2002, and the predation observed in 2007 should be predictive of frequencies in 2008.

However, we can look at two things here. First, the average selection against carbonaria over the period is not very different between that gained from the form

Figure 11.1
Decreasing frequencies of *carbonaria*, 2001–2007. There has been a steady decline in the frequency of the dark form of the moth, *Biston betularia*, during the time of the study.

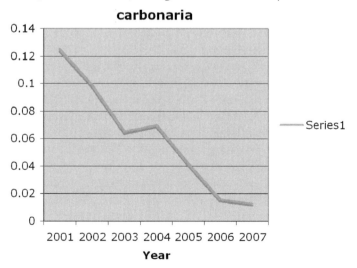

Table 11.2
Comparison of selection coefficients for predation calculated from
observations and predication.

Year	Expected selection against *carb.* based on form frequency differences between years	Observed selection against *carb.* from selection experiment
2001	0.239	Not done
2002	0.337	0.252
2003	−0.096	0.046
2004	0.435	0.469
2005	0.63	0.299
2006	0.13	−0.061
2007	prediction	0.306

The selection coefficient against *carbonaria* calculated from the decline in numbers
over the years compares favorably to the selection coefficient calculated from the
predation experiment.

frequency data and that observed in the predation experiment. Second, the correlation between these for the five years for which we can make the comparison (that's 2002–2006) is rather high. [Average selection against *carbonaria* from form frequency data = 0.286; average selection against *carbonaria* from predation experiment = 0.219. Correlation coefficient for expected compared to observed for the years 2002–2006 = 0.75.]

I conclude that differential bird predation here is a major factor responsible for the decline in *carbonaria* frequency in Cambridge between 2001 and 2007. So Tutt's hypothesis stands, and is once again supported by empirical evidence.

There was another thing that I wanted to do when I started this work: that is, to be able to say with authority whether the peppered moth story should, or should not, be taught in biology class as an example of Darwinian evolution in action.... But the peppered moth story is easy to understand, because it involves things that we are familiar with: vision and predation and birds and moths and pollution and camouflage and lunch and death. That is why the anti-evolution lobby attacks the peppered moth story. They are frightened that too many people will be able to understand.

I said I would have a few mild concluding remarks: Here they are.

... I caught my first butterfly when I was four, and started recording peppered moth forms when I was ten. I am getting old, and have spent my life in scientific enquiry and discovery. And it has been a great life! Until now, for instead of the vision of a world made better by the appliance of science, I see a future of ever-increasing global problems. I probably won't see the worst of what's coming—but I fear for my children, who will face escalating problems of climate change, over-population, pollution, starvation, disease and conflict. And for their children and grandchildren, I have little optimism.

We need to address global problems now, and to do so with any chance of success, we have to base our decisions on scientific facts: and that includes the fact of Darwinian evolution. If the rise and fall of the peppered moth is one of the most visually impacting and easily understood examples of Darwinian evolution in action, it should be taught. It provides after all: The Proof of Evolution.

REFERENCES

Grant, Bruce, D. F. Owen, and C. A. Clarke. 1996. Parallel rise and fall of melanic peppered moths in America and Britain. *Journal of Heredity* 87: 351–357.

Hooper, Judith. 2002. *Of moths and men: Intrigue, tragedy and the peppered moth*. London: Fourth Estate.

Majerus, Michael E. N. 2005. The peppered moth: Decline of a Darwinian disciple. In *Insect Evolutionary Ecology*, ed. M. D. E. Fellowes, G. J. Holloway, and J. Rolff, 367–391. Wallingford; Cambridge MA, CABI.

Rudge, David W. 2005. Did Kettlewell commit fraud? Re-examining the evidence. *Public Understanding of Science* 14 (3): 249–268.

Excerpted from Michael E. N. Majerus, The peppered moth: The proof of Darwinian evolution. Talk delivered to the European Society for Evolutionary Biology in Uppsala, Sweden, August 23, 2007. Written talk and presentation retrieved May 27, 2008, from http:// www.gen.cam.ac.uk/Research/Majerus/Swedentalk220807.pdf. Reprinted with permission of the author.

Texas Board of Education Textbook Hearings: Testimony of Dr. Matt Winkler

Good afternoon—or I guess I should say, good evening, Chairman Miller and Board members. My name is Matt Winkler and I'm the founder and CEO of Ambion, a biotechnology company here in Austin. I'm a scientist by training. I received my Ph.D. from the University of California at Berkeley. I'm also a former University of Texas zoology professor. About fourteen years ago, I started Ambion Inc. to invent and sell kits and products that helped scientists perform biomedical research. Our customers are cancer researchers, urologists, biochemists and other kinds of biologists. We've been very successful. Our products are used by molecular biologists in universities, medical schools, pharmaceutical and biotechnology companies around the world. We grow at over 30 percent a year and in 2003, we'll do almost $40 million in revenue. We currently have about 250 employees here in Austin and another 20 at our European subsidiary in England. The success of my company depends on our ability to recruit the very best scientists. This includes scientists who we recruit outside of Texas and ones that are trained here in Texas. Having high quality biology in science textbooks that are not diluted with creationists' views is important to my ability to recruit first-rate scientists.

The first step in recruiting good scientists is getting them to answer want ads. The state of Kansas shot themselves in the foot by acquiring an international reputation in the scientific community as having an education system that taught watered-down science. I would hate to have to compete to recruit the best scientists with other states if Texas had a reputation for teaching creation science. A second issue is that job candidates are frequently concerned about the quality of the school system that their kids would be attending. When they show up for interviews, they frequently have researched the quality of school systems here in the Austin area. Again, I would not want to have to have them worrying that their children are going to be getting a first-rate scientific education.

A second issue is the ability of the state of Texas to educate first-rate homegrown scientific talent. Again, I would like to see the focus of biology textbooks used in Texas to be on science and not religion. My company depends on being able to hire the very best scientists. This doesn't mean that my employees are not religious or that they do not believe in creation. What it does mean is that they have had a rigorous scientific education. One final issue is that I have three school-age children. I would like to see science textbooks used in Texas get the best quality science education that's available.

Thank you very much.

Excerpted from testimony of Matt Winkler, before the State Board of Education, Texas Education Agency, Public Hearing, September 10, 2004, 347–349. Retrieved June 1, 2008, from http://www.tea.state.tx.us/textbooks/adoptprocess/sept03transcript.pdf

CHAPTER 12

•••••••••••••

Issues Concerning Religion

The creationism/evolution controversy exists in the United States because evolution is incompatible with certain religious views and has profound implications for others. But as discussed in chapter 3, there are many religious perspectives, and many of them embrace evolution—even if we limit ourselves only to Christianity among the many religions practiced in the United States. In this chapter, we will consider ways of relating science and religion. One topic will be to compare literalist and nonliteralist approaches to creation and evolution. We also will consider two issues having to do with the implications of evolution: whether evolution leads to moral decay and societal evils, and at what is perhaps the overriding issue in the creationism/evolution controversy, the question of whether there is an existential loss of purpose or meaning to life if evolution occurred.

•••••••••••••••••••••••••••

MODELS OF SCIENCE AND RELIGION INTERACTION

When Science Meets Religion

In 1990, in the first chapter of *Religion in an Age of Science*, I proposed a fourfold typology as an aid to sorting out the great variety of ways people have related science and religion. . . . In the present volume this typology is used as the organizing structure for every chapter.

1. *Conflict.* Biblical literalists believe that the theory of evolution conflicts with religious faith. Atheistic scientists claim that scientific evidence for evolution is incompatible with any form of theism. The two groups agree in asserting that a person cannot believe in both God and evolution, though they disagree as to which they will accept. For both of them, science and religion are enemies. These two opposing groups get most attention from the media, since a conflict makes a more exciting news story

than the distinctions made by persons between these two extremes who accept both evolution and some form of theism.

2. *Independence.* An alternative view holds that science and religion are strangers who can coexist as long as they keep a safe distance from each other. According to this view, there should be no conflict because science and religion refer to differing domains of life or aspects of reality. Moreover, scientific and religious assertions are two kinds of language that do not compete because they serve contrasting questions. Science asks how things work and deals with objective facts; religion deals with values and ultimate meaning. . . .

3. *Dialogue.* One form of dialogue is a comparison of the methods of the two fields, which may show similarities even when the differences are acknowledged. . . . Both scientists and theologians are engaged as dialogue partners in critical reflection on such topics, while respecting the integrity of each other's fields.

4. *Integration.* A more systematic and extensive kind of partnership between science and religion occurs among those who seek a closer integration of the two disciplines. The long tradition of natural theology has sought in nature a proof (or at least suggestive evidence) of the existence of God. . . . Such an approach I call a *theology of nature* (within a religious tradition) rather than a *natural theology* (arguing from science alone). Alternatively, a philosophical system such as process philosophy can be used to interpret scientific and religious thought within a common conceptual framework.

Excerpted from Ian G. Barbour, *When science meets religion* (San Francisco: HarperSanFrancisco, 2000).

● ●

LITERALIST APPROACHES TO CREATION

Creation: Believe It or Not

Tamper with the Book of Genesis and you undermine the very foundation of Christianity. You cannot treat Genesis I as a fable or a mere poetic saga without severe implications for the rest of Scripture. . . . If Genesis 1 is not accurate, then there's no way to be certain that the rest of Scripture tells the truth.

. . . If you doubt or explain away the Bible's account of the six days of creation, where do you put the reins on your skepticism? Do you start with Genesis 3, which explains the origin of sin, and believe everything from chapter 3 on? Or maybe you don't sign on until sometime after chapter 6, because the Flood is invariably questioned by scientists, too. . . . What about the miracles of the New Testament? Is there any reason to regard *any* of the supernatural elements of biblical history as anything other than poetic symbolism?

After all, the notion that the universe is billions of years old is based on naturalistic presuppositions that (if held consistently) would rule out all miracles. If we're worried about appearing "unscientific" in the eyes of naturalists, we're going to have to reject a lot more than Genesis 1–3.

Once rationalism sets in and you start adapting the Word of God to fit scientific theories based on naturalistic beliefs, there is no end to the process. . . . Why should we doubt the literal sense of Genesis 1–3 unless we are also prepared to deny that Elisha made an axe-head float or that Peter walked on water or that Jesus raised Lazarus from the dead? And what about the greatest miracle of all—the resurrection of Christ? If we're going to shape Scripture to fit the beliefs of naturalistic scientists, why stop at all? Why is one miracle any more difficult to accept than another?

. . . So the question of whether we interpret the Creation account as fact or fiction has huge implications for every aspect of our faith. . . . Frankly, believing in a supernatural, creative God who made everything is the only possible rational explanation for the universe and for life itself. It is also the only basis for believing we have any purpose or destiny.

Excerpted from John MacArthur, *The battle for the beginning: The Bible on creation and the fall of Adam* (Nashville, TN: W. Publishing Group, 2001), 44–45.

One Hand on the Bible

The non-biblical chronology currently accepted in the historical sciences challenges more than just the age of things. The process of developing this chronology has required that a sequence of historical events be developed and accepted. It is the sequence—and not just the timing—of the origin of things that is contrary to the claims of Scripture. So if you believe the Bible to be the ultimate authority of truth, pause and consider how you can square these discrepancies:

1. Genesis 1 claims that the earth came into being before the sun (compare Gen. 1:1 and 1:14–18), and fruit trees before the sea creatures (compare Gen. 1:11–12 and 1:20–21), and flying animals before land animals (compare Gen. 1:20–21 and 1:24–25). The conventional old-age chronology reverses the timing of each of these events.
2. Genesis 1:36 suggests that all the animals before the Fall were vegetarians, whereas old-age chronology suggests there was never a time when all animals were herbivores.
3. Old-age chronology suggests the death, disease, and suffering preceded the appearance of man by hundreds of millions of years. This seems to be counter to numerous biblical claims: for example, God's description of the creation as being "very good" (Gen. 1:31); the strong association between man's sin and animal sacrifice in Romans 5; and the description of Heaven in 1 Corinthians 15 as being free from the curse of death that came with the fall of man.
4. Scripture indicates that man was derived from dust (Gen. 2:7) and not from animals as argued in old-age chronology
5. Scripture indicates that Adam named all the animals (Gen. 2:19–20), something that would not be possible in non-biblical chronology because so many thousands of animals became extinct long before Adam came to be.
6. Eve was created from the side of Adam (Gen. 2:21–22) and not derived from a population of pre-humans as suggested in conventional chronology.
7. According to Genesis 2:10–14, a river flowed out of the Garden of Eden and divided into four rivers, three of which were the Gihon (the ancient name for the Nile), the Hiddekel (the ancient name for the Tigris), and the Euphrates. The rivers currently known by these names are not connected to one another now and in conventional chronology were *never* connected to one another.

8. According to Scripture, there were no thorns or thistles before man's sin (Gen. 3:18), whereas old-age chronology would have these things preceding even man's appearance by millions of years.

9. Old-age chronology would deny the long life spans of pre-Flood and early post-Flood humans as recorded in Genesis 5 and 11.

10. Scripture tells us that all the land animals and birds were represented on the Ark (Gen. 6:17–20), whereas the old-age chronology would suggest that thousands of species were extinct long before humans arrived on the scene and could not possibly have been on the Ark.

11. The Bible claims that the Flood of Noah lasted for more than a year (Gen. 7:11 compared to 8:13) and covered the entire surface of the earth over the highest mountains (Gen. 7:19), whereas old-age chronology suggests that there never was a global flood on this planet.

12. Finally, Genesis 11 tells us that at a time when all the humans on earth were united in language and tower-building, God introduced a number of languages to force them to spread across the Earth's surface. In old-age chronology, however, humans capable of building cities were never all located in one place, and human languages were generated one by one over a long period.

Old-age chronology and Scripture cannot both be true, and they cannot be reconciled. The time argued for in old-age chronology is thousands of times longer than that indicated in biblical chronology. The dates are not just slightly different, so they cannot be easily modified to make them agree. Not only does the sequence of events in the two chronologies differ, but even the nature of the events in the two chronologies are different. It is only with the rejection of old-age chronology in its entirety—not just the age of things but also the actual events of history and their order of occurrence—that the biblical chronology can be accepted. Alternatively, the acceptance of humanly devised old-age chronology or any variation of it would require the rejection or modification of divinely inspired Scripture. Acceptance of Scripture's clearly understood intent requires the rejection of old-age chronology and the acceptance of a radically different biblical chronology.

Excerpted from Kurt Wise, *Faith, form, and time* (Nashville, TN: Broadman and Holman Publishers, 2002), 53–54. Used with permission.

● ●

NONLITERALIST APPROACHES TO CREATION

Comparing Biblical and Scientific Maps of Origin

. . . Now, the biblical accounts of creation in Genesis are different ways of mapping origins than those to which we who have been schooled in science are accustomed. In fact, even the two accounts of creation in Genesis 1 and 2 (the six-day account and the Adam-and-Eve account) have significant differences, reflecting the significant differences between the two cultural traditions in ancient Israel, the agricultural/urban and the shepherd/nomadic. Genesis 1 is a mapping of creation using the imagery, terminology, and perspectives of agricultural/urban Israel; and Genesis 2, of pastoral/nomadic Israel.

The Two Creation Accounts

We immediately recognize this difference in biblical language and usage elsewhere, as among the prophets and psalmists, to depict God's relationship to the world and to humanity. Isaiah, for example, goes back and forth between these two sets of imagery, as in Isaiah 40. On the one hand, Isaiah draws upon agricultural/urban imagery, as he speaks of God surveying the universe ("who has measured the waters in the hollow of his hand, and marked off the heavens with a span," v. 12), or God laying a foundation ("Have you not understood from the foundations of the earth," v. 21). But Isaiah also draws on shepherd/nomadic imagery: "He stretches out the heavens like a curtain, and spreads them like a tent to dwell in" (v. 22), or "He will feed his flock like a shepherd, he will gather the lambs in his arms" (v. 11). Shepherds would speak naturally in terms of tents, curtains, sheep, garden oases, and the simple life of nomads. Farmers and the city-dwellers would speak naturally in terms of foundations, pillars, boundaries, the sedentary life, and cosmic and social order.

What we are given in the first chapters of Genesis are two distinct accounts of creation, the first using the language, imagery, and concerns of the agricultural/urban tradition in Israel, and the second using those of the pastoral/nomadic tradition. This observation helps to explain the inevitable problems in a literal/historical approach to harmonizing the two accounts of creation, as well as, in turn, trying to harmonize both of them with modern scientific accounts of origins. The order of events in the Adam and Eve version in Genesis 2, for example, is quite different from the six-day account with which Genesis begins.

Genesis 1–2:4a	Genesis 2:4b–24
(Water and formless Earth)	(Heavens and Earth presupposed)
Light (day 1)	Water (mist)
Firmament (day 2)	Adam
Earth and vegetation (day 3)	Vegetation
Sun, moon, and stars (day 4)	Rivers
Fish and birds (day 5)	Land animals, birds (no fish)
Land animals, humans (day 6)	Eve

... The attempt to interpret these materials as literal, chronological accounts of origins runs into enormous difficulties internally, well before modern scientific scenarios are introduced. Despite valiant efforts by clever exegetes, the two biblical accounts cannot be reconciled, as long as the assumption is made that they are intended to be read as comparable to a natural history. The clue to the differences is to be found within Israel itself where, broadly speaking, there were two main traditions: the pastoral/nomadic and the agricultural/urban. Genesis 1 has drawn upon the imagery and concerns of the farmers and city-dwellers who inhabited river basins prone to flooding, while Genesis 2 has drawn upon the experiences of shepherds, goat-herders, and camel-drivers who lived on the semiarid fringes of the fertile plains, around and between wells and oases. For the pastoral nomads and desert peoples the fundamental threat to life was dryness and barrenness, whereas for those agricultural and urban peoples in or near flood plains the threat was too much water, and the chaotic possibilities of water. It is also revealing that Genesis 2 does not mention a creation of fish, whereas fish in abundance are prominent in Genesis 1 (fish occupy half of day five, with "swarms of living creatures").

This interpretive approach also helps explain why the two versions of creation present such different—nearly opposite—views of human nature. In Genesis 1 human beings are pictured in the lofty terms of royalty, taking dominion over the earth and subduing it—imagery and values drawn from the very pinnacle of ancient civilizations, which Israel itself achieved in the time of Solomon. In Genesis 2, however, Adam and Eve are pictured as *servants* of the garden, living in a garden oasis: essentially the gardener and his wife. And while Genesis 1 refers to humans as made in "the image and likeness of God," in the continuation of the garden story in Genesis 3 the theme of godlikeness is introduced by the *serpent* who tempts Eve with the promise that by eating of the fruit of the Tree of Knowledge, they would be "like God," knowing good and evil. Celebrants of science and technology beware!

Thus, while Genesis 1 is comfortable with the values of civilization and the fruits of its many achievements and creations, Genesis 2 offers an humble view of humanity, a reminder of the simple life and values of the shepherd ancestors, before farming, and even before shepherding, in an Edenic state of food gathering and tending. In this manner these two views of human nature are counterbalanced. They are not contradictory but complementary. Any celebration of human creation and its achievements is tempered by warnings concerning overweening pride and claims to godlikeness. Our heads at times may be in the clouds, but our feet walk on the Earth and are made of clay.

The two accounts of creation in Genesis are contradictory only if taken as literal history, rather than recognizing that they are operating *analogically*, using the contrasting imagery and concerns of the two main traditions in ancient Israel. The biblical accounts are also not in contradiction with modern scientific accounts either, because (again) the biblical accounts are interpreting origins analogically (albeit using very different sets of analogy), not geologically or biologically. To cite Calvin again: "The Holy Spirit had no intention to teach astronomy, and, in proposing instruction meant to be common to the simplest and most uneducated persons, he made use by Moses and the other prophets of popular language, that none might shelter himself under the pretext of obscurity." Thus it may be said that biblical affirmations of creation are in harmony with the science of any age and culture, not because they have been harmonized by clever argument, but because they have little to do with such concerns.

Excerpted from Conrad Hyers, Comparing biblical and scientific maps of origins, in *Perspectives on an evolving creation*, ed. K. B. Miller, 19–33 (Grand Rapids, MI: Wm. B. Eerdmans, 21–24). Used with permission.

• •

EVOLUTION AND MORAL WEAKNESS/SOCIAL EVIL: CREATIONIST VIEWS

Henry Morris on Evolution

Evolution is at the foundation of communism, fascism, Freudianism, social Darwinism, behaviorism, Kinseyism, materialism, atheism, and in the religious world, modernism and neo-orthodoxy. Jesus said, "A good tree cannot bring forth corrupt fruit." In view of the bitter fruit yielded by the evolutionary system over the past hundred years, a closer look at the nature of the tree itself is well warranted today.

Excerpted from Henry M. Morris, *The twilight of evolution* (Grand Rapids, MI: Baker Book House, 1963, 24).

Darwinism and the Nazi Race Holocaust

Of the many factors that produced the Nazi holocaust and World War II, one of the most important was Darwin's notion that evolutionary progress occurs mainly as a result of the elimination of the weak in the struggle for survival. Although it is no easy task to assess the conflicting motives of Hitler and his supporters, Darwinism-inspired eugenics clearly played a critical role. Darwinism justified and encouraged the Nazi views on both race and war. If the Nazi party had fully embraced and consistently acted on the belief that all humans were descendants of Adam and Eve and equal before the creator God, as taught in both the Old Testament and New Testament Scriptures, the holocaust would never have occurred.

Expunging of the Judeo-Christian doctrine of the divine origin of humans from mainline German (liberal) theology and its schools, and replacing it with Darwinism, openly contributed to the acceptance of Social Darwinism that culminated in the tragedy of the holocaust. Darwin's theory, as modified by Haeckel, Chamberlain and others, clearly contributed to the death of over nine million people in concentration camps, and about 40 million other humans in a war that cost about six trillion dollars. Furthermore, the primary reason that Nazism reached to the extent of the holocaust was the widespread acceptance of Social Darwinism by the scientific and academic community.

Excerpted from Jerry Bergman, Darwinism and the Nazi race Holocaust. *Technical Journal* 13, no. 2 (1999): 101–111.

• •

EVOLUTION AND MORAL WEAKNESS/SOCIAL EVIL: COUNTERCREATIONIST VIEWS

The *Expelled* Controversy: Raising Walls of Division?

Is Evolution Wedded to Atheism?

To start with, a crucial contribution of the film [*Expelled: No Intelligence Allowed*] is its making abundantly clear something that should be but has not always been clear to the public at large: it is not just ID advocates, but also many of the world's leading evolutionists who think Darwinism is completely incompatible with theism or any other tenets of the major religions. Cornell historian of biology and AAAS Fellow William Provine, interviewed in the film, famously asserts that the clear implications of naturalistic evolution are "no gods worth having exist, no life after death exists, no ultimate foundation for ethics exists, no ultimate meaning in life exists" (Provine, 1998). . . . Richard Dawkins, Daniel Dennett, and numerous other prominent interpreters of evolution make similar claims in the public square.

. . . Physicist-priest John Polkinghorne, one of the most esteemed scholars of science and religion featured in the movie, rightly reminds us that "metaphysical claims need to be defended with metaphysical arguments." Dawkins doesn't provide such arguments. And neither does anyone else in the movie.

. . . Now even without argument, it is clear by inspection that atheism must entail evolution: for anyone who rejects the possibility of an intelligence behind the cosmos, there is no viable alternative to some sort of naturalistic evolutionary account of origins. But the reverse—that evolution requires or logically leads to atheism as Stein claims—well, this is not clear without argument. For a film wanting to engage a popular audience, it's not surprising that it raises this issue via personal stories of individuals who (now claim to have) lost some kind of theistic belief upon encountering evolution. But for a film that not only raises the question but ends up endorsing a conclusion, two things seem to be lacking.

First, conspicuously absent are any personal stories on the other side, that could have been drawn from thousands of scientists who simultaneously accept evolution and embrace a vibrant religious faith, many of whom testify that their belief in God has actually been deepened in light of evolutionary science and the grandeur of life's history. This is a regrettable omission, particularly in light of the fact that the film's own promotional materials emphatically claim, "Unlike some other documentary films, *Expelled* doesn't just talk to people representing one side of the story" [reference omitted].

. . . And here is the second lack. I may scandalize my colleagues by suggesting this, but the problem is actually not that Chapman, or Bethell, or Dawkins, is entirely wrong. Some interpretations of Darwinian theory are indeed incompatible with some understandings of divine purpose, and waving the wand of happy imaginings does not make conflicts disappear. The trick is to see where the genuine as opposed to manufactured conflicts are, which ones can be solved by the concessions reason recommends, and which ones cannot be avoided without conceding reason itself. A popular film cannot resolve these issues, but *Expelled*, like Dawkins, doesn't seem to let on that these are issues at all. What appears to be waved off without consideration is even the possibility of mutually enriching commerce between faith and evolution.

"Implicit in most evolutionary theory is either there is no God or he can't have anything to do with the world," the typically very fair-minded journalist Larry Witham asserts in the movie. But this provocative comment could have been used to stimulate rather than settle conversation. Hmm . . . most evolutionary theory? If such implications do exist, but don't exist for all versions, how do we distinguish between the ones that do and don't harbor atheism? How do we know it's "most," and would it make a difference if it were only "some," or even "just a few crackpot extremes"? How could a scientific theory, which just offers an account of how nature operates, ever tell us—even if it's a wrong theory about how the world works—that there is no God beyond the world's workings? Or if there is a God, why would belief that certain features of the world are explainable by natural law, mean that God has "nothing to do" with those features or the law that supports them?

Did Darwin Lead to Hitler?

Without question, *Expelled*'s single most riveting though not necessarily central claim, and the one that has turned out to be a lightning rod for contention, is the assertion that Darwin inspired the Holocaust.

... There are several ways Darwinism (or any idea) could have contributed to the Holocaust. The most modest way is that evolutionary theory could have been used "merely" as a justification for what Nazi social architects wanted to do anyway.... Or, it could actually have contributed to the thinking of some master race theorists, even if such ideas were neither advocated by Darwin himself nor employed by all Nazi thinkers. The historical record amply and indisputably confirms the fact that references to Darwin and to ideological principles attributed to the evolutionary process were frequently employed by the intellectual architects of the Reich, at the very least in this way. That Darwin was used (or abused) in Holocaust thinking seems uncontestable.

But it is also not necessarily very interesting. Darwin has been used in this way for many other social movements very different from fascist eugenics: e.g., racial egalitarianism, feminism, anti-feminism, Marxism, and free enterprise capitalism. Big ideas can be used, or misused, for all manner of big causes, and Darwinism—like the Bible—has been claimed to justify or inspire many. In fact, the Bible and the Christian tradition themselves were used to justify the anti-Semitism of the Holocaust. Martin Luther's fierce denunciation of Jews ("everyone would gladly be rid of them," "we are at fault in not slaying them") [reference omitted] was frequently referred to by Hitler and other influential anti-Semites. Luther was lauded as the "greatest anti-Semite of his time," and the infamous Kristallnacht on the night of November 9–10 [1938], when my own grandfather was taken to a concentration camp, was celebrated with the applauding observation that "on Luther's birthday, the synagogues are burning in Germany" [reference omitted]. Hitler personally claimed:

> My feelings as a Christian point me to my Lord and Savior as a fighter.... How terrific was His fight for the world against the Jewish poison. Today, after two thousand years, with deepest emotion I recognize more profoundly than ever before, the fact that it was for this that He had to shed His blood upon the Cross. [reference omitted]

These words stun me, as they should any follower of Christ. I believe they betray a monstrous distortion of the life and message of Jesus. (And there is considerable evidence that Hitler didn't believe them anyway, but merely used them to manipulate the religious emotions of others.) Either way though, the point is that they did successfully manipulate Jew hatred. The question we should ask—regarding Christian or evolutionary ideas—is did right understanding of such ideas reasonably lead to Nazi racism?

If so, there are two ways this could occur, and *Expelled* features advocates of each interpretation of Darwin's influence. The strongest and most pernicious way would be for Darwinism to "lead" to Hitler by advancing ideas that logically entail it.... The problem with this is that many of the most important aspects of the Hitlerian program have nothing at all to do with Darwin (such as Germanic superiority, Jewish vileness, a racial view of human history). And those ideas that are attributed to Darwin (such as natural selection makes might right in social policy) were actually not advocated but repudiated by Darwin and his immediate colleagues. Nor have ensuing generations of self-professed Darwinians and modern evolutionary biologists been led to conclusions that are remotely similar. Clearly the horrors of Nazism cannot be inevitable outcomes or logical extensions of Darwinian theory.

So another option is that Darwinism did not "lead" to Hitler—the road to the Holocaust is paved with something else—but perhaps it provided some of the necessary gas to get there.... that Darwin was "necessary"—not the whole recipe but a crucial ingredient in the stew, or golden spike in the tracks—and without it we never could have had the evils of the final solution. But there are also serious inadequacies with this seemingly more modest assertion. For one thing, there have been many programs of racial extermination—before and after Darwin—that made no appeal to evolution. So the idea isn't necessary to such evils. And looking specifically at the Holocaust, there are important factual problems with the claim even when applied just to this phenomenon....

Darwin and Hitler: The Idea of a Master Race and Subhuman Jews Does Not Fit Well with Darwin's Theory

The film claims that Darwinism involves a "deprivileging of human life," which was instrumental to the Holocaust. There is absolutely no question that Darwinism, when wedded to atheism, can and for some does lead to this devaluing, and many Darwinians not only recognize but also overtly endorse this (Rachels 1991). On the other hand, many prominent Darwinists, including Richard Dawkins himself, repudiate this and argue that Darwinian theory actually helps illuminate what is most distinctive and precious in humanity.

... But for purposes of argument, what if Darwin does lead to devaluation, at least for some thinkers? Contrary to what the film claims and what it might seem on the face of things, it is actually not the deprivileging or devaluing of human life that was necessary to fuel the Holocaust fires. Rather, it is the selective deprivileging and devaluing of some lives. It is not that humans are claimed to be mere animals with no value, terrible though this would be. It is that some humans are super valuable—Übermenschen—and others are subhuman, toxic pollutants. This is the essence of monstrous notions of "race hygiene" and, in fact, is the core of all genocidal attempts to eliminate groups of people who are viewed as evil or inferior. People are treated inhumanely, when they are viewed as distinctively inhuman or somehow essentially different than ourselves.

This has nothing intrinsically to do with Darwin. It is a tragically archetypal human problem embodied in the self-deluded profession of the Pharisee, "I thank you, Lord, that you have not made me like that other man." And the modern versions of this sentiment, so destructively tied to racism, are themselves pre-Darwinian. The monumental race-based interpretation of human history that inspired all future versions—On the Inequality of Human Races—was written by the nineteenth-century Frenchman Arthur de Gobineau, before Darwin ever published anything about evolution.

Darwin and Hitler: Prominent Anti-Jewish Voices Rejected Darwin

Many of the most prominent advocates of the above ideas knew little about Darwin, or actually repudiated him. So how could Darwinism be necessary for the Holocaust? Gobineau was skeptical of evolution, famously quipping, "I'm not sure if humans came from apes, but we're certainly heading in that direction." Houston Chamberlain, the biologist whose massively influential racial meta-narrative modified Gobineau's ideas into hatred of Jews and elevation of Germans, rejected Darwin outright.

. . . Chamberlain arguably became one of the most expansive master-race theorists in Germany, if not all history. In addition to repudiating Darwinism and rejecting scientific materialism, his views were anchored in a spiritual, explicitly Christocentric understanding of history.

. . . Chamberlain's thinking does not appear to involve mere religious posturing but genuine conviction: "having once seen Jesus Christ—even if it be with half-veiled eyes—we cannot forget Him. . . . [Nothing] can dispel the vision of the Man of Sorrow when once it has been seen." His book was widely discussed throughout Germany, being required reading in civic life. Early in his political career, Hitler visited the nationally prominent ageing anti-Semite several times in his family home. After one such visit, Chamberlain wrote, "Most respected and dear Hitler. . . . That Germany, in the hour of her greatest need, brings forth a Hitler is proof of her vitality. . . . May God protect you!" (reference omitted).

. . . Both Darwin and the Bible were seized upon by anti-Jewish zealots in search of a legitimating ideology. Hatred is notoriously indiscriminate in what it cobbles together to justify itself. Hitler, in particular, evidenced little regard for learning and—as the historical sources cited by recent defenders and critics of *Expelled* acknowledge—he extracted whatever was useful to support his preconceptions, from widely ranging popular, crude sources. . . . In the case of Darwinian and Christian tradition though, there really exist disturbing themes that were (and are) amenable to misuse. However the fundamental ideas of the Holocaust were not just absent from, but contrary to the founders of each tradition. This would seem to represent something considerably weaker than being "necessary for," but rather involves being "amenable to" distortion and employment by Nazism.

REFERENCES

Frankowski, Nathan, director. 2008. *Expelled: No intelligence allowed*. DVD. Salt Lake City, UT: Rocky Mountain Pictures.

Provine, William. 1998. *Evolution: Free will and punishment and meaning in life*. Second Annual Darwin Day Celebration, University of Tennessee Knoxville. February 12, 1998. Retrieved June 14, 2008, from http://eeb.bio.utk.edu/darwin/Archives/1998ProvineAbstract.htm.

Rachels, James. 1991. *Created from animals: The moral implications of Darwinism*. Oxford: Oxford University Press.

Excerpted from Jeffrey P. Schloss, The *Expelled* controversy: Raising walls of division? 2008, http://www.asa3.org/ASA/resources/Schloss200805.pdf. Used with permission.

●●●●●●●●●●●●●●●●●●●●●●●●●●●

EVOLUTION AND PURPOSE (MEANINGFULNESS)

INTRODUCTION

If evolution occurred, and the universe was not specially created in its present form by God, does this imply that there is no ultimate purpose or meaning to the universe? Many conservative Christians believe this to be the case. Other Christians, interpreting Scripture differently, believe that the history of the universe is independent of its purpose, which they believe is determined by God, whether or not we know what it is. Humanists hold the position that there is no cosmic or ultimate meaningfulness to the universe but that individuals may choose to give meaning to their own individual lives.

Both biblical literalists and Christians who are not biblical literalists may be concerned that even if evolution were part of God's plan, and God worked through natural processes, the insertion of such laws between God and creation somehow implies that God is a more distant deity and less involved in human affairs. There is, not surprisingly, a great diversity of opinion within Christian theology on this topic.

●●●●●●●●●●●●●●●●●●●●●●●●●●●

EVOLUTION IMPLIES COSMIC MEANINGLESSNESS

The Church of Darwin

The reason the theory of evolution is so controversial is that it is the main scientific prop for scientific naturalism. Students first learn that "evolution is a fact," and then they gradually learn more and more about what that "fact" means. It means that all living things are the product of mindless material forces such as chemical laws, natural selection, and random variation. So God is totally out of the picture, and humans (like everything else) are the accidental product of a purposeless universe. Do you wonder why a lot of people suspect that these claims go far beyond the available evidence?

Excerpted from Phillip E. Johnson, The Church of Darwin, *Wall Street Journal*, August 16, 1999, A14.

Epilogue: The Prospect Ahead

Whichever cosmological model proves correct, there is not much of comfort in any of this. It is almost irresistible for humans to believe that we have some special relation to the universe, that human life is not just a more-or-less farcical outcome of a chain of accidents reaching back to the first three minutes, but that we were somehow built in from the beginning. As I write this I happen to be in an airplane at 30,000 feet, flying over Wyoming en route home from San Francisco to Boston. Below, the earth looks very soft and comfortable—fluffy clouds here and there, snow turning pink as the sun sets, roads stretching straight across the country from one town to another. It is very hard to realize that all this is just a tiny part of an overwhelmingly hostile universe. It is even harder to realize that this present universe has evolved from an unspeakably unfamiliar early condition, and faces a future extinction of endless cold or intolerable heat. The more the universe seems comprehensible, the more it also seems pointless.

But if there is no solace in the fruits of our research, there is at least some consolation in the research itself. Men and women are not content to comfort themselves with tales of gods and giants, or to confine their thoughts to the daily affairs of life; they also build telescopes and satellites and accelerators, and sit at their desks for endless hours working out the meaning of the data they gather. The effort to understand the universe is one of the very few things that lifts human life a little above the level of farce, and gives it some of the grace of tragedy.

Excerpted from Steven Weinberg, *The first three minutes* (New York: Bantam Books, 1997), 143–144. Used with permission.

● ●

EVOLUTION AND COSMIC MEANING: EVOLUTIONARY VIEWS

Evolution, Tragedy, and Cosmic Purpose

At the end of his important book, *The First Three Minutes*, physicist Steven Weinberg remarks grimly that the more comprehensible the universe has become to modern science, the more "pointless" it all seems. Many other scientists would agree. Alan Lightman has collected several of their reactions to Weinberg's oft-repeated claim. Astronomer Sandra Faber, for example, states that the universe is "completely pointless from a human perspective."

. . . I have been arguing that, from the point of view of Christian theology at least, canvassing nature for evidence of a divine "plan" distracts us from engaging in sufficiently substantive conversation with evolutionary science. Evolutionists have told a story about nature that is extremely difficult to square with the notion of a divine "plan." Nature's carefree discarding of the weak, its tolerating so much struggle and waste during several billion years of life's history on Earth, has made simplistic portraits of a divinely designed universe seem quite unbelievable.

To admit this much, however, is not necessarily to conclude that there is no "point" or purpose to the universe. We must remember that science as such is not equipped, methodologically speaking, to tell us whether there is or is not any "point" to the universe. If scientists undertake nevertheless to hold forth on such matters, they must admit in all candor that their ruminations are not scientific declarations but at best declarations *about* science.

. . . Thus, any respectable argument that evolution makes the universe pointless would have to be erected on grounds other than those that science itself can provide. And yet, even though science cannot decide by itself the question of whether religious hope is less realistic than cosmic pessimism, we must admit that any beliefs we may hold about the universe, whether pessimistic or otherwise, cannot expect to draw serious attention today unless we can at least display their consonance with evolutionary science. We must be able to show that the visions of hope at the heart of the Abrahamic religious traditions provide a coherent metaphysical backdrop for the important discoveries of modern science.

. . . For Christian theology, this would mean seeking to understand the natural world, and especially its evolutionary character, in terms of the outpouring of compassion and the corresponding sense of world renewal associated with the God of Jesus the crucified and risen Christ.

Christians have discerned in the "Christ-event" [the crucifixion] the decisive self-emptying or *kenosis* of God. And at the same time they have experienced in this event a God whose effectiveness takes the form of a power of renewal that opens the world to a fresh and unexpected future. As a Christian theologian, therefore, when I reflect on the relationship of evolutionary science to religion, I am obliged to think of God as both *kenotic love* and *power of the future*. This sense of God as a self-humbling love that opens up a new future for the world took shape in Christian consciousness only in association with the "Christ-event"; and so, as we ponder the implications of such discoveries as those associated with evolution, it would be disingenuous of Christian theologians to suppress the specific features of their own faith community's experience of divine mystery. This means quite simply that in its quest to understand the scientific story of life, Christian theology must ask how evolution might make sense when situated in a universe shaped by God's kenotic compassion and an accompanying promise of new creation.

Excerpted from John F. Haught, *God after Darwin: A theology of evolution* (Boulder, CO: Westview Press, 2000), 105–110. Used with permission.

The Problem of Purpose

The central problem faced by advocates of a theistic interpretation of evolution is not one of reconciling science with a literal reading of Genesis. Nor is the problem one of transformation of one species into another. Nor is it a problem raised by gradual change over long periods of deep time. Rather, it is the problem of purpose.

. . . Why is purpose a problem? Can we not think of nature progressing through evolution just as the design of automobiles is progressing through engineering? After all, we speak of technology "evolving," so is it not appropriate to equate evolution with advance? Evolution makes things better, right?

. . . This cultural interpretation of natural selection leads us to say: yes, evolution is a form of progress. As a doctrine of progress, evolution becomes one more modern Western ideology among others. As a doctrine of progress it is a philosophy, a value system, an ideology, maybe even a materialistic religion.

However, if we seek what is narrowly scientific and treat evolution strictly as a theory to explain biological change, then we must expunge all references to purpose. No such thing as progress can be admitted. This is a principle of scientific research which is dogmatic to today's evolutionary biologists. They appeal to Charles Darwin himself for having set the precedent by relying upon chance variation. Natural selection, suggested Darwin, is not a secular form of divine providence. . . . Natural selection favors the fit, to be sure; but what determines fitness has nothing to do with an overall purpose or direction in nature.

. . . This expunging of purpose from nature gives nightmares to theologians. How can we speak of a creation without purpose? How can we speak of redemption without a goal toward which the creation aspires?

. . . On the one hand, people of faith simply cannot conceive of the natural world without purpose or at least value. To be sure, theologians have no investment in the secular doctrine of progress; nor do they feel obligated to use the language of "evolution" to indicate advancement toward a better and better world. Dropping the idea of progress from the long story of nature is no loss, theologically speaking. Yet, on the other hand, giving up totally on purpose merely to satisfy the scientific method seems like a high price to pay. A purposeless creation would not be a creation at all.

So, alas, what's a theologian to do? One option would be to throw in the towel and become a deist. No purpose within nature is discernible because, according to this option, it simply is not there. The deistic theologian could affirm that God created the initial conditions of law and chance that made evolution possible; and then God left the natural world to run itself ever since. If such deism is unsatisfying, another option would be to mix in a divine plot to the scientist's story of nature. The theologian could rewrite the Epic of Evolution by expanding on the story told by scientists. The theologian could declare that this evolutionary story has had a plot all along. When God created the world in the beginning, according to this option, God placed a potential into the creation which now through evolution is becoming actualized. The problem with this Epic of Evolution approach is that it is a dogmatic superimposition; it does not derive from the science itself. A third option would be to treat creation eschatologically—that is, to locate the world's purpose in God rather than in nature. The problem for theologians as posed by scientists is that no purpose can be seen *within* nature. According to this option, nature's purpose is not inherent within nature itself; rather, its value or direction belongs to the relationship of nature with God. God's redemptive vision becomes the source of the divine declaration that nature is "very good."

Excerpted from Ted Peters and Martinez Hewlett, *Evolution from creation to new creation* (Nashville, TN: Abingdon Press, 2003), 25–27. Used with permission.

Why Are We Here?

Why are we here? That is, what is our role in the body of Life? What contributions can we make, individually and collectively, in the evolutionary process?

Praise God! We now have a way of understanding the role of the human in cosmic evolution that makes sense both scientifically and religiously. Using traditional Christian night language, we might say:

> Our purpose, individually, is to grow in Christ and to support one another in staying true to God's Word and God's will. Collectively, we are here to create Christ-centered institutions that glorify God and embody the values of the Kingdom.

A day language way of saying the same thing might be:

> Our purpose, individually, is to grow in trust, authenticity, responsibility, and service to the Whole, and to support others in doing the same. Collectively, we are here to celebrate and steward what Life has been doing for billions of years and to devise systems of governance and economics that align the self-interest of individuals and groups with the wellbeing of the larger communities of which we are apart.

We are here, as well, to love as broadly and as deeply as we possibly can—knowing that we cannot do this without the support of the entire community of Life. Our purpose is to consciously further evolution in ways that serve everyone and everything, not just ourselves. This is our calling. This is our Great work. Indeed, this is our destiny!

Excerpted from Michael Dowd, *Thank God for evolution! How the marriage of science and religion will transform your life and our world* (Tulsa, OK: Council Oak Books, 2007), 273–274.

Humanist Manifesto III

Humanism is a progressive philosophy of life that, without supernaturalism, affirms our ability and responsibility to lead ethical lives of personal fulfillment that aspire to the greater good of humanity.

The lifestance of Humanism—guided by reason, inspired by compassion, and informed by experience—encourages us to live life well and fully. It evolved through the ages and continues to develop through the efforts of thoughtful people who recognize that values and ideals, however carefully wrought, are subject to change as our knowledge and understandings advance.

This document is part of an ongoing effort to manifest in clear and positive terms the conceptual boundaries of Humanism, not what we must believe but a consensus of what we do believe. It is in this sense that we affirm the following:

> Knowledge of the world is derived by observation, experimentation, and rational analysis. Humanists find that science is the best method for determining this knowledge as well as for solving problems and developing beneficial technologies. We also recognize the value of new departures in thought, the arts, and inner experience-each subject to analysis by critical intelligence.
>
> Humans are an integral part of nature, the result of unguided evolutionary change. Humanists recognize nature as self-existing. We accept our life as all and enough, distinguishing things as they are from things as we might wish or imagine them to be. We welcome the challenges of the future, and are drawn to and undaunted by the yet to be known.
>
> Ethical values are derived from human need and interest as tested by experience. Humanists ground values in human welfare shaped by human circumstances, interests, and concerns and extended to the global ecosystem and beyond. We are committed to treating each person as having inherent worth and dignity, and to making informed choices in a context of freedom consonant with responsibility.
>
> Life's fulfillment emerges from individual participation in the service of humane ideals. We aim for our fullest possible development and animate our lives with a deep sense of purpose, finding wonder and awe in the joys and beauties of human existence, its

challenges and tragedies, and even in the inevitability and finality of death. Humanists rely on the rich heritage of human culture and the lifestance of Humanism to provide comfort in times of want and encouragement in times of plenty.

Humans are social by nature and find meaning in relationships. Humanists long for and strive toward a world of mutual care and concern, free of cruelty and its consequences, where differences are resolved cooperatively without resorting to violence. The joining of individuality with interdependence enriches our lives, encourages us to enrich the lives of others, and inspires hope of attaining peace, justice, and opportunity for all.

Working to benefit society maximizes individual happiness. Progressive cultures have worked to free humanity from the brutalities of mere survival and to reduce suffering, improve society, and develop global community. We seek to minimize the inequities of circumstance and ability, and we support a just distribution of nature's resources and the fruits of human effort so that as many as possible can enjoy a good life.

Humanists are concerned for the well being of all, are committed to diversity, and respect those of differing yet humane views. We work to uphold the equal enjoyment of human rights and civil liberties in an open, secular society and maintain it is a civic duty to participate in the democratic process and a planetary duty to protect nature's integrity, diversity, and beauty in a secure, sustainable manner.

Thus engaged in the flow of life, we aspire to this vision with the informed conviction that humanity has the ability to progress toward its highest ideals. The responsibility for our lives and the kind of world in which we live is ours and ours alone.

American Humanist Association, *The Humanist Manifesto III*, http://www.americanhumanist. org/3/HumandItsAspirations.php.

CHAPTER 13

• • • • • • • • • • • • • • •

The Nature of Science

INTRODUCTION

The creationism/evolution controversy concerns issues of religion, politics, culture—and, of course, science, an institution with considerable influence and authority in American society. Both antievolutionists and the defenders of evolution seek the imprimatur of science to promote their respective positions. Both sides seek to define science and determine its application, with factions differing strikingly on what they consider good or acceptable scientific practice.

This chapter will consider three nature-of-science themes that recur in the literature of the creationism/evolution controversy. The first concerns the use of two terms discussed in chapter 1, fact and theory. Evolution, we hear from antievolutionists, is only a theory and should not be presented as fact. "Of course evolution is a theory," retort the evolutionary scientists. Theories are much more important than facts! Antievolutionists respond that even if evolution is a theory in the scientific sense (of explanation), it isn't a very good theory, and isn't supported by the evidence.

The second topic in this chapter is whether there should be different rules for how science works, depending on the type of question being asked. Antievolutionists say yes, and encourage distinguishing "origins science" from "operation science"; evolutionary scientists, while drawing a somewhat similar distinction between historical sciences and experimental sciences, maintain that both are scientifically legitimate and are not fundamentally different.

This distinction is related to the third topic of this chapter, the relationship between methodological and philosophical naturalism. As discussed in chapter 3, philosophical naturalism is the idea that material causes (matter, energy, and their interaction) are the only realities in the universe: there is no God or gods, nor any supernatural forces. Methodological naturalism is a habit of scientists of restricting explanations to natural causes. Creationists, especially ID proponents, contend that this is too limiting; science would produce "truer" answers if scientists were allowed to invoke supernatural causation. A question posed is whether methodological naturalism implies or entails philosophical naturalism; if one decides to use only natural explanations in one's science, does this imply a philosophical conclusion that there *can be* no divine involvement in the universe? There are contrasting views.

• •

FACT VERSUS THEORY: CREATIONISTS VIEWS
Theory, Not Fact, Resolutions and Legislation

A classic example of evolution as theory (meaning "guess" or "hunch") is found in the disclaimer that the Alabama Board of Education ordered placed in biology textbooks from about 1996 until 2002. The disclaimer in its full form is presented in chapter 10 as Figure 10.1. The first two sentences, however, exemplify the idea that theories are somehow suspect and not as reliable as facts. Other legislation or resolutions from other states similarly illustrate nonscientific usage of these terms.

The Alabama Disclaimer, 1995

This textbook discusses evolution, a controversial theory some scientists present as a scientific explanation for the origin of living things, such as plants, animals, and humans. No one was present when life first appeared on earth. Therefore, any statement about life's origins should be considered as theory, not fact. . . .

Columbia County, Georgia, Resolution, 1996 (failed)

The teaching of science should distinguish between theory and fact. Scientific hypotheses which cannot be proven or replicated, such as the theory of evolution, must always be taught as theories and not fact. . . .

Mississippi HB No. 1397, 2003 (failed)

BE IT ENACTED BY THE LEGISLATURE OF THE STATE OF MISSISSIPPI: SECTION 1. Beginning with the 2004–2005 school year, the State Board of Education shall require that any textbook that includes the teaching of evolution in its contents shall have the following language inserted on the inside front cover of those textbooks: "The word 'theory' has many meanings: systematically organized knowledge, abstract reasoning, a speculative idea or plan, or a systematic statement of principles. Scientific theories are based on both observations of the natural world and assumptions about the natural world. They are always subject to change in view of new and confirmed observations.

"This textbook discusses evolution, a controversial theory some scientists present as a scientific explanation for the origin of living things. No one was present when life first appeared on earth. Therefore, any statement about life's origins should be considered a theory. . . ." [The legislation cites the remainder of the Alabama disclaimer.]

Tennessee SB 3229, 1996 (failed)

BE IT ENACTED BY THE GENERAL ASSEMBLY OF THE STATE OF TEN-
NESSEE: SECTION 1. Tennessee Code Annotated, Title 49, Chapter 6, part 10, is
amended by adding the following new section thereto:

. . . No teacher or administrator in a local education agency shall teach the theory
of evolution except as a scientific theory. Any teacher or administrator teaching such
theory as fact commits insubordination, as defined in Section 49-5-501(s)(6), and
shall be dismissed or suspended as provided in Section 49-5-511. . . .

Darwin's Leap of Faith

In essence, the theory of evolution should never have been accepted as a legitimate
scientific theory. True science uses logical inference from empirical observations to
arrive at truth (Bird 1991, 2:14). What the proper definition of science and description
of the scientific method will indicate is that, while scientists who study nature may
utilize the scientific method, evolutionary theory *itself* is not ultimately good science
because evolution has few, if any, "demonstrated truths" or "observed facts." It does
not arrive at truth inductively, i.e., the supporting of a general truth (evolution) by
observing particular cases that exist (Bird 1991, 1:15–16). A strictly limited change
within species can be demonstrated, such as crossbreeding among some plants and
animals, but this has nothing to do with evolution as commonly understood, nor,
when examined critically, can the mechanisms involved explain how evolution might
occur.

Let's further explain why we do not think the term theory is appropriately applied
to evolution. As noted, in science, the term has a more profound meaning than in
common usage. In science, the phrase, "It's *just* a theory" is inappropriate. A good
scientific theory explains a great deal of scientific knowledge, including "both laws
and the facts dependent on scientific laws" (Broad and Wade 1982).

> "Theory"—to a scientist—is a concept firmly grounded in and based upon facts, contrary
> to the popular opinion that it is a hazy notion of undocumented hypothesis. Theories
> do not become facts; they explain facts. A theory must be verifiable. If evidence is found
> that contradicts the stated theory, the theory must be modified or discarded. (Matsumura
> 1995: 119)

Evolution is not verifiable; it explains few facts, and it contradicts several scientific
laws. Further, there is a great deal of undeniable evidence against it. So, scientifically,
it must be modified or discarded. It is inappropriate to apply the term *scientific theory* to
evolution. Applied to evolution, the term actually fits the popular idea much better,
e.g., "a hazy notion or undocumented hypothesis."

REFERENCES

Bird, W. R. 1991. *The Origin of Species revisited: The theories of evolution and of abrupt appearance.*
2 vols. Cambridge, ON: Thomas Nelson.

Broad, William, and Nicholas Wade. 1982. *Betrayers of the truth: Fraud and deceit in the halls of science*. New York: Simon & Schuster.

Matsumura, M., ed. 1995. *Voices for evolution*. Berkeley, CA: National Center for Science Education.

Excerpted from John Ankerberg and John Weldon, *Darwin's leap of faith* (Eugene, OR: Harvest House, 1998), 52–54.

• •

FACT VERSUS THEORY: RESPONSES TO CREATIONIST VIEWS

Speech of Hon. Rush D. Holt, Ph.D.

[Mr. HOLT]: Mr. Speaker, I rise today to address my colleagues regarding H.R. 1, No Child Left Behind. Although we passed this important legislation last week, I must express my reservations about certain language included in the conference report:

> The conferees recognize that a quality science education should prepare students to distinguish the data and testable theories of science from the religious or philosophical claims that are made in the name of science. Where topics are taught that may generate controversy (such as biological evolution), the curriculum should help students to understand the full range of scientific views that exist, why such topics may generate controversy, and how scientific discoveries can profoundly affect society.

Outside of the scientific community, the word *theory* is used to refer to a speculation or guess that is based on limited information or knowledge. Among scientists, however, a theory is not a speculation or guess, but a logical explanation of a collection of experimental data. Thus, the theory of evolution is not controversial among scientists. It is an experimentally tested theory that is accepted by an overwhelming majority of scientists, both in the life sciences and the physical sciences. The implication in this language that there are other scientific alternatives to evolution represents a veiled attempt to introduce creationism—and, thus, religion—into our schools. Why else would the language be included at all? In fact, this objectionable language was written by proponents of an idea known as "intelligent design." This concept, which could also be called "stealth creationism," suggests that the only plausible explanation for complex life forms is design by an intelligent agent. This concept is religion masquerading as science. Scientific concepts can be tested; intelligent design can never be tested. This is not science, and it should not be taught in our public schools.

Mr. Speaker, I am a religious person. I take my religion seriously and feel it deeply. My point here is not to attack or diminish religion in any way. My point is to make clear that religion is not science and science is not religion. The language in this bill can result in diminishing both science and religion.

Excerpted from Rush Holt, Conference report on H.R. 1, No Child Left Behind Act of 2001, December 13, 2001. *The Congressional Record*, December 20 (Extensions), E2365. See also http://frwebgate.access.gpo.gov/cgi-bin/getpage.cgi, retrieved September 30, 2003.

What Is a "Fact" and What Is a "Theory"?

A fact is a confirmed observation. For example, it is a confirmed observation that every tetrapod known has at some stage of its life, a humerus, a radius and ulna, and a distal cluster of bones corresponding to carpals, metacarpals and phalanges. The general public (and even some scientists) use the word *fact* to imply capital T *Truth*: unchanging agreement. In science, facts, like theories, may change: it was once a fact (for about 10 years) that *Homo sapiens* had 48 chromosomes. But other observations were confirmed and explanations found for the erroneous observations, and now we know that there are 46. In general, though, in science we treat facts as statements we don't need to test and question anymore, but rather can use as givens to build more complex understandings.

A theory, in science, is a logical construct of facts and hypotheses that attempts to explain a natural phenomenon. It is an explanation, not a guess or hunch that one can casually disregard. Theory formation—explanation—is the goal of science, and nothing we do is more important. A scientist joked that we should applaud the Tennessee law punishing teachers for teaching evolution as a "fact rather than a theory" because "everyone knows that theories are more important than facts!" Theories explain facts, but the general public doesn't know that.

Concerning evolution, then, what's a fact and what's a theory? One hears from many scientists, "Evolution is FACT!!!" The meaning here is that evolution, the "what happened," is so well supported that we don't argue about it, anymore than we argue about heliocentrism versus geocentrism. We accept that change through time happened, and go on to try to explain how. What we mean and what is heard is often different, however. What the public often hears when scientists say "Evolution is FACT!" is that we treat evolution as unchallengeable dogma, which it isn't.

We must learn to present evolution not as "a fact" in this dogmatic sense, but "matter of factly," as we would present heliocentrism and gravitation. Most people consider heliocentrism and gravitation as "facts," but they are not "facts" in my definition of "confirmed observations." Instead, they are powerful inferences from many observations, which are not in themselves questioned, but used to build more detailed understandings.

From the standpoint of philosophy of science, the "facts of evolution" are things like the anatomical structural homologies such as the tetrapod forelimb, or the biochemical homologies of cross species protein and DNA comparisons, or the biogeographical distribution of plants and animals. The "facts of evolution" are observations, confirmed over and over, such as the presence and/or absence of particular fossils in particular strata of the geological column (one never finds mammals in the Devonian, for example). From these confirmed observations we develop an explanation, an inference, that what explains all of these facts is that species have had histories, and that descent with modification has taken place. Evolution is thus a theory, and one of the most powerful theories in science.

We may also speak of "theories" (plural) of evolution, in the sense of the explanations for how descent with modification has taken place. It is conceptually sound to separate evolution as something that did or did not happen from explanations about how, or how fast, or which species are related to which. . . .

Indeed, teachers have to be sure that students know what theories are and why they are important. Students also must—this is crucial—learn as part of their science instruction that our explanations change with new data or better ways of looking at things. Anti-evolutionists make the statement that "evolution isn't science because you guys are always changing your minds about stuff." This is not a criticism. That's the way a vigorous science works.

Excerpted from Eugenie C. Scott, Dealing with antievolutionism, in *Learning from the Fossil Record*, ed. J. Scotchmoor and F. K. McKinney (Pittsburgh: Paleontological Society, 1996) pp. 21–22. Used with permission.

• •

ORIGINS SCIENCE AND OPERATION SCIENCE: CREATIONIST VIEW

Summary of Thaxton, Bradley, and Olsen

My intent was to present a selection ("Operation Science and the God Hypothesis") from one of the founding documents of the intelligent design movement, *The Mystery of Life's Origin* (Charles B. Thaxton, Walter L. Bradley, Roger L. Olsen, New York: Philosophical Library, 1984), but the authors denied permission for the excerpt. Readers are encouraged to consult pages 202–205 of that book, or a book by N. L. Geisler and J. K. Anderson expressing the same ideas from a Young-Earth Creationist perspective (*Origin Science: A Proposal for the Creation-Evolution Controversy*. Grand Rapids, MI: Baker Book House, 1987).

Thaxton and colleagues begin by defining science as having the ability (1) to explain what has been observed; (2) to explain what has not yet been observed; and (3) to be tested by further experimentation and modification. They claim that this approach works only if there are recurring events to test theories against, and since these natural events recur during the operation of the universe, this kind of science is known as "operation science." Appealing to God in operation science is illegitimate, because "by definition God's supernatural action would be willed at His pleasure and not in a recurring manner" (202). They argue against "God of the gap" explanations (see chapter 6), where God's hand is invoked to fill a gap in our scientific knowledge, and note how Newton made this error in explaining the rotation of planets around the sun.

Origin science, on the other hand, refers to attempts to understand singular events such as the origin of life. The authors claim that theories about such events cannot be falsified because they occur only once. They contend that because of the differences between origin science and operation science, it is legitimate to allow divine explanations in the former if not the latter.

There are significant and far-ranging consequences in the failure to perceive the legitimate distinction between origin science and operation science. Without the distinction we inevitably lump origin and operation questions together as if answers to both are sought in the same manner and can be equally known. Then, following the accepted practice of omitting appeals to divine action in recurrent nature, we extend it to origin questions too. The blurring of these two categories partially explains the widely held view that a divine origin of life must not be admitted into the *scientific* discussion, lest it undermine the motive to inquire and thus imperil the scientific enterprise.

...The perception of a threat to scientific inquiry and the possible end of science are legitimate concerns. But we question whether the God-hypothesis in origin science would necessarily have this disastrous effect.... In our view, as long as one acknowledges and abides by the above distinction between origin science and operation science, there is no *necessary* reason that Special Creation would have the disastrous effects predicted for it. One must be careful, however, to follow the tradition of early modern scientists and *disallow* any divine intervention in operation science. (204–205)

● ●

ORIGINS SCIENCE AND OPERATION SCIENCE: OPPOSING CREATIONIST VIEW

How Does Biology Explain the Living World?

When a biologist tries to answer a question about a unique occurrence such as "Why are there no hummingbirds in the Old World?" or "Where did the species *Homo sapiens* originate?" he cannot rely on universal laws. The biologist has to study all the known facts relating to the particular problem, infer all sorts of consequences from the reconstructed constellations of factors, and then attempt to construct a scenario that would explain the observed facts of this particular case. In other words, he constructs a historical narrative.

Because this approach is so fundamentally different from the causal-law explanations, the classical philosophers of science—coming from logic, mathematics, or the physical sciences—considered it quite inadmissible. However, recent authors have vigorously refuted the narrowness of the classical view and have shown not only that the historical-narrative approach is valid but also that it is perhaps the only scientifically and philosophically valid approach in the explanation of unique occurrences.

It is, of course, never possible to prove categorically that a historical narrative is "true." The more complex a system is with which a given science works, the more interactions there are within the system, and these interactions very often cannot be determined by observation but can only be inferred. The nature of such inference is likely to depend on the background and the previous experience of the interpreter; and therefore, not surprisingly, controversies over the "best" explanation frequently occur. Yet every narrative is open to falsification and can be tested again and again.

For instance, the demise of the dinosaurs was once attributed to the occurrence of a devastating disease to which they were particularly vulnerable, or to a drastic change of climate caused by geological events. Neither assumption was supported by credible evidence, however, and both ran into other difficulties. Yet, when in 1980 the asteroid theory was proposed by Walter Alvarez and, particularly, after the presumed impact crater was discovered in Yucatan, all previous theories were abandoned, since the new facts fit the scenario so well.

Among the sciences in which historical narratives play an important role are cosmogony (the study of the origin of the universe), geology, paleontology, phylogeny, biogeography, and other parts of evolutionary biology. All these fields are characterized by unique phenomena. Every living species is unique and so is, genetically speaking, every individual. But uniqueness is not limited to the world of life. Each of the nine planets of the solar system is unique. On earth, every river system and every mountain range has unique characteristics.

Unique phenomena have long frustrated the philosopher. Hume noted that "science cannot say anything satisfactory about the cause of any genuinely singular phenomenon." He was correct if he had in mind that unique events cannot be fully explained by causal laws. However, if we enlarge the methodology of science to include historical narratives, we can often explain unique events rather satisfactorily, and sometimes even make testable predictions.

The reason why historical narratives have explanatory value is that earlier events in a historical sequence usually make a causal contribution to later events. For instance, the extinction of the dinosaurs at the end of the Cretaceous vacated a large number of ecological niches and thus set the stage for the spectacular radiation of the mammals during the Paleocene and Eocene, owing to their invasion of these vacant niches. The most important objective of a historical narrative is to discover causal factors that contributed to the occurrence of later events in a historical sequence. The establishment of historical narratives does not in the least mean the abandonment of causality, arrived at strictly empirically.

Excerpted from Ernst Mayr, *This is biology: The science of the living world* (Cambridge, MA: Belknap Press, 1998), 64–66.

Creationism, Ideology, and Science

One largely old-earth creationist proposal is that there are two different kinds of science: "operation science," and "origins science" (Thaxton et al., 1984; Geisler and Anderson, 1987). A distinction is made between phenomena which occur "with regularity" and those which occur "singularly." Regularly occurring phenomena can be studied in the fashion most of us associate with normal science, or "operation science." But one-time phenomena, such the Big Bang, and other evolutionary events comprise what creationists call "origins science."

Of course there are differences in the study of repeatable events vs. non-repeatable ones, but mainstream philosophers of science agree that phenomena of historical sciences like geology, paleontology, and astronomy can be studied scientifically, and even experimentally. Mount St. Helens erupted as a singular event, but this does not prevent there being a science of volcanoes. Similarly, even if bears and dogs split from a common ancestor only once, we can still evaluate the hypothesis that bears and dogs are closely related against empirical evidence (from fossils, comparative anatomy, biochemistry, etc.). We can also learn about the processes that influence evolution by looking at the evidence for other such splits. There are many ways to scientifically study events of this type.

Creationists add an additional factor to this bimodal division of the scientific world, which I believe sheds light on why the division was invented in the first place: it allows the intrusion of the supernatural into scientific explanation. Geisler proposes that to accompany the two kinds of science, there are two kinds of causation: primary causes and secondary causes. Operation science relies properly on secondary causes, but origins science is allowed to invoke primary causes. Thaxton et al. refer to primary cause more bluntly as the "God hypothesis," and agree that in operation science, "the appeal to God is quite illegitimate, since by definition God's supernatural action would be willed at His pleasure and not in a recurring manner" (Thaxton et al., 1984: 203). But when dealing with "origins science," it is not only permissible, but essential, to allow recourse to supernatural causation (i.e., miracles).

Few would argue with not resorting to miracles in operation science, but proponents of this artificial division do not make a solid case for resorting to miracles in origin science. Arguably, non-recurrent events may be more difficult and challenging to study than repeated events, but that in itself is insufficient to require resorting to the supernatural.

REFERENCES

Geisler, Norman L., and J. Kerby Anderson. 1987. *Origin science: A proposal for the creation–evolution controversy.* Grand Rapids, MI: Baker Book House.
Thaxton, Charles B., Walter L. Bradley, and Roger L. Olsen. 1984. *The mystery of life's origin: Reassessing current theories.* New York: Philosophical Library.

Excerpted from Eugenie C. Scott. Creationism, ideology, and science, in *The Flight from Science and Reason*, ed. P. Gross, N. Levitt and M. W. Lewis (New York: New York Academy of Sciences, 1996), 505–522; 354; 514.

● ●

PHILOSOPHICAL AND METHODOLOGICAL NATURALISM: CREATIONIST VIEWS

Position Paper on Darwinism

1.0 The important issue is not the relationship of science and creationism, but the relationship of science and materialist philosophy.

...1.2... The question I raise is not whether science should be forced to share the stage with some Biblically based rival known as creationism, but whether we ought to be distinguishing between the doctrines of scientific materialist philosophy and the conclusions that can legitimately be drawn from the empirical research methods employed in the natural sciences.

1.3 Scientific materialism (or naturalism...) is the philosophical doctrine that everything real has a material basis, that the path to objective knowledge (as distinguished from subjective belief) is exclusively through the methods of investigation accepted by the natural sciences, and that teleological conceptions of nature ("we are here for a purpose") are invalid. To a scientific materialist there can be no "ghost in the machine," no non-material intelligence which created the first life or guided its development into complex form, and no reality which is in principle inaccessible to scientific investigation, i.e., supernatural.

1.4 The metaphysical assumptions of scientific materialism are not themselves established by scientific investigation, but rather are held a priori as unchallengeable and usually unexamined components of the "scientific" world view. . . . The naturalistic evolution of life from prebiotic chemicals and its subsequent naturalistic evolution into complexity and humanity is assumed as a matter of first principle, and the only question open to investigation is how this naturalistic process occurred.

1.5 The question is whether this refusal to consider any but naturalistic explanations has led to distortions in the interpretation of empirical evidence, and especially to claims of knowledge with respect to matters about which natural science is in fact profoundly ignorant. . . .

2.0 The continued dominance of neo-Darwinism is the most important example of distortion and overconfidence resulting from the influence of scientific materialist philosophy upon the interpretation of the empirical evidence. . . .

3.0 The refusal (or inability) of the scientific establishment to acknowledge that Darwinism is in serious evidential difficulties and probably false as a general theory is due to the influence of scientific materialist philosophy and certain arbitrary modes of thought that have become associated with the scientific method.

3.1 Science requires a paradigm or organizing set of principles and Darwinism has fulfilled this function for more than a century. It is the grand organizing theoretical principle for biology—a statement which does not imply that it is true.

3.2 Once established as orthodox, a paradigm customarily is not discarded until it can be replaced with a new and better paradigm which is acceptable to the scientific community. Disconfirming evidence (anomalies) can always be classified as "unsolved problems," and the situation remains satisfactory for researchers because even an inadequate paradigm can generate an agenda for research.

3.3 To be acceptable a paradigm must conform to the philosophical tenets of scientific materialism. For example, the hypothesis that biological complexity is the product of some preexisting creative intelligence or vital force is not acceptable to scientific materialists. They do not fairly consider this hypothesis and then reject it as contrary to the evidence; rather they disregard it as inherently ineligible for consideration.

3.4 Given the above premises, something very much like Darwinism simply must be accepted as a matter of logical deduction, regardless of the state of the evidence. Random mutation and natural selection must be credited with shaping biological complexity, because nothing else could have been available to do the job. . . . Because the escape from Darwinism seems to lead nowhere, Darwinism for scientific materialists is inescapable. . . .

5.0 The important debate is not between "evolutionists" and "creationists," but between Darwinists (i.e., scientific materialists) and persons who believe that purely naturalistic or materialistic processes may not be adequate to account for the origin and development of life.

5.1 Once separated from its materialistic-mechanistic basis in Darwinism, "evolution" is too vague a concept to be either true or false. If I am told that the phyla of the Cambrian explosion evolved in some non-Darwinian sense from preexisting bacteria or algae, I do not know what the claim adds to the simple factual statement that the prokaryotes came first. It conveys no information about how the new forms came into existence, and the "evolution" in question could be something as metaphysical as the evolution of an idea in the mind of God.

5.2 Similarly, whether "creation" occurred over a greater or lesser period of time, or whether new forms were developed from older ones rather than from scratch, is not fundamental. The truly fundamental question is whether the natural world is the product of a preexisting intelligence, and whether we exist for a purpose which we did not invent ourselves. If Darwinists have not been overstating their case, they have disproved the theistic alternative, or at least made consideration of it superfluous....

6.0 Whatever may be its utility as a paradigm within the restrictive conventions of scientific materialism, Darwinism has continually been presented to the public as the factual basis for a comprehensive world view that excludes theism as a possibility. A few representative quotations will suffice to make the point:

6.1 George Gaylord Simpson: "Although many details remain to be worked out, it is already evident that all the objective phenomena of the history of life can be explained by purely naturalistic or, in a proper sense of the sometimes abused word, materialistic factors. They are readily explicable on the basis of differential reproduction in populations (the main factor in the modern conception of natural selection) and of the mainly random interplay of the known processes of heredity.... Man is the result of a purposeless and natural process that did not have him in mind." (*The Meaning of Evolution*)...

6.3 Richard Dawkins: "Darwin made it possible to be an intellectually fulfilled atheist." (*The Blind Watchmaker*)...

7.0 Whether the materialist-mechanist program has succeeded as the Darwinists have so vehemently claimed is a legitimate subject for intellectual exploration. Scientists rightly fight to protect their freedom from dogmas that others would impose upon them. They should also be willing to consider fairly the possibility that they have been seduced by a dogma which they found too attractive to resist....

Excerpted from Phillip E. Johnson, Position paper on Darwinism, 2003. Apologetics.org. Retrieved September 30, 2003, from http://www.apologetics.org/articles/positionpaper.html. Used with permission.

● ●

PHILOSOPHICAL AND METHODOLOGICAL NATURALISM: RESPONSE TO CREATIONIST VIEWS

Methodological Naturalism and Evidence

But is the methodological rule itself dogmatic?...Does science put forward the methodological principle not to appeal to supernatural powers or divine agency simply on authority?...Certainly not. There is a simple and sound rationale for the principle based upon the requirements of scientific evidence.

Empirical testing relies fundamentally upon use of the lawful regularities of nature that science has been able to discover and sometimes codify in natural laws. For example, telescopic observations implicitly depend upon the laws governing optical phenomena. If we could not rely upon these laws—if, for example, even when under the same conditions, telescopes occasionally magnified properly and at other occasions produced various distortions dependent, say, upon the whims of some supernatural entity—we could not trust telescopic observations as evidence. The same problem would apply to any type of observational data. Lawful regularity is at the very heart of the naturalistic world view and to say that some power is supernatural is, by definition, to say that it can violate natural laws. So, when Johnson argues that science should allow in supernatural powers and intelligences he is in effect saying that it should allow beings that are above the law (a rather strange position for a lawyer to take). But without the constraint of lawful regularity, inductive evidential inference cannot get off the ground. Controlled, repeatable experimentation, for example, which Johnson explicitly endorses in his video *Darwinism on Trial*, would not be possible without the methodological assumption that supernatural entities do not intervene to negate lawful natural regularities.

Of course science is based upon a philosophical system, but not one that is extravagant speculation. Science operates by empirical principles of observational testing; hypotheses must be confirmed or disconfirmed by reference to empirical data. One supports a hypothesis by showing consequences obtained that would follow if what is hypothesized were to be so in fact. Darwin spent most of the *Origin of Species* applying this procedure, demonstrating how a wide variety of biological phenomena could have been produced by (and thus explained by) the simple causal processes of the theory. Supernatural theories, on the other hand, can give no guidance about what follows or does not follow from their supernatural components. For instance, nothing definite can be said about the processes that would connect a given effect with the will of the supernatural agent—God may simply say the word and zap anything into or out of existence. Furthermore, in any situation, any pattern (or lack of pattern) of data is compatible with the general hypothesis of a supernatural agent unconstrained by natural law. Because of this feature, supernatural hypotheses remain immune from disconfirmation. Johnson's form of creationism is particularly guilty on this count. Creation Science does include supernatural views at its core that are not testable and it was rightly dismissed as not being scientific because of these in the Arkansas court case, but it at least was candid about a few specific nonsupernatural claims that are open to disconfirmation (and indeed that have been disconfirmed), such as that the earth is less than 10,000 years old and that many geological and paleontological features were caused by a universal flood (the Noachian Deluge). Johnson, however, does not provide any creationist claim beyond his generic one that "God creates for some purpose," and as a purely supernatural hypothesis this is not open to empirical test. Science assumes Methodological Naturalism because to do otherwise would be to abandon its empirical evidential touchstone (pp. 88–89).

Excerpted from Robert T. Pennock, *Tower of Babel: The evidence against the new creationism* (Cambridge, MA: Bradford Book/MIT Press, 1999), 88–89. Used with permission.

Creationist Strategy 1: The Nature of Science

First, what does Johnson mean by "naturalism"? Johnson begins his book, *Reason in the Balance*, by characterizing naturalism as a metaphysical assumption: "the doctrine that nature is all there is," and he goes on to claim that natural science is "based on naturalism." The only content that Johnson gives to the doctrine that "nature is all there is" is atheism. Naturalists, he says, are those who "assume that

God exists only as an idea in the minds of religious believers." And: "If naturalism is true, then humankind created God—not the other way around." If naturalism is the assumption that God exists only as an idea in the mind of religious believers, then it is pretty clear that natural science is *not* based on naturalism. It is absurd to suppose that science, which is totally silent on the question of God, is based on the assumption that God exists only as an idea in the mind of religious believers. So, this metaphysical naturalism—whether it is a bias or not—is not an underpinning of science.

But there is another kind of naturalism, *methodological* naturalism, that Johnson discusses in his appendix. "A methodological naturalist," he says, "defines science as the search for the best naturalistic theories," where a naturalistic theory abjures supernatural causes. This amounts to saying that methodological naturalism is the view that no naturalistic explanation can appeal to God or to any supernatural phenomena. In this sense of methodological naturalism, I would agree with Johnson, science is committed to naturalism. But is methodological naturalism just a bias in science? Is it a bias to exclude supernatural explanations from science?

Surely not. If the methodological naturalism that is a hallmark of science were a mere bias, the explosion of scientific knowledge from the 16th and 17th centuries on would be totally inexplicable. The proof of the pudding is in the eating, and the sciences are unparalleled as generators of knowledge of the natural world. It makes little sense to say that scientists have misunderstood their own enterprise, that they should count as scientific explanations those that appeal to a supernatural being. It makes little sense to rebuke such a successful practice for having the character that it has. What counts as a scientific explanation is determined by science. So, taking methodological naturalism as the view that no explanation that appeals to a Creator or an Intelligent Designer is a scientific explanation, Johnson's charge that methodological naturalism is a *bias* in science is off the mark.

... Scientific explanations—explanations put forward on the basis of scientific consideration—are fully naturalistic, and have no place for appeal to a supernatural agent. It does not follow from this, however, that *all* correct explanations are scientific explanations. We must distinguish between scientific claims—claims made from *within* science—and claims made *about* science. One important claim about science (one that I reject) is that science is the arbiter of all knowable truth, that there is nothing to be known beyond what science delivers. Call this claim "scientism."

(Scientism) Science is the arbiter of all knowable truth.

If scientism were correct, then from the commitment of science to methodological naturalism, it would follow that all correct explanations (not just scientific explanations) are naturalistic. That stance would rule out, a priori, any explanation that appealed to God. This, I think, would be a bias. But this does not follow from the methodological naturalism of science; it follows only with the addition of the metaphysical, extra-scientific thesis of scientism. Scientism is like a closure principle—"and that's all there is." If we reject scientism, as I think that we should, then from the fact that all scientific explanations are naturalistic, it does not follow that all legitimate explanations are naturalistic. So, exclusion of God from the science classroom is not necessarily exclusion of God elsewhere—for example, where we are trying to give a metaphysical account of why there is something rather than nothing at all. This latter question—Why is there anything rather than nothing at all?—is not a scientific question and will not be susceptible to a scientific explanation. But unless we are scientistic, we may think that there is some explanation—albeit not a scientific one. Again, however, questions not susceptible of scientific answers do not belong in a science class.

. . . To sum up: Science is not committed to the nonexistence of God, as it would be if it were based on metaphysical naturalism. Science is committed to naturalistic explanations. Science does not count any explanation that appeals to God or to supernatural phenomena as a scientific explanation (thus, it is committed to methodological naturalism). But methodological naturalism is no bias: it is in the nature of science. And unless one conjoins methodological naturalism with scientism, nothing at all follows about the nonexistence of God. So, methodological naturalism (but not scientism) is part of science, and given the success of science, it is idle to charge that science should be something other than what it is.

Excerpted from Lynn Rudder Baker. God and science in the public schools. *Philosophic Exchange* 30 (2000): 53–69 (from 57–59). Used with permission.

The Game of Science

Science, fundamentally, is a game. It is a game with one overriding and defining rule:

> Rule No. 1: Let us see how far and to what extent we can explain the behavior of the physical and material universe in terms of purely physical and material causes, without invoking the supernatural.

Operational science takes no position about the existence or non-existence of the supernatural; it only requires that this factor is not to be invoked in scientific explanations. Calling down special-purpose miracles as explanations constitutes a form of intellectual "cheating." A chess player is perfectly capable of removing his opponent's king physically from the board and smashing it in the midst of a tournament. But this would not make him a chess champion, because the rules had not been followed. A runner may be tempted to take a short-cut across the infield of an oval track in order to cross the finish line ahead of his faster colleague. But he refrains from doing so, as this would not constitute "winning" under the rules of the sport.

Similarly, a scientist also can say to himself, "I believe that *Homo sapiens* was placed on this planet by a special act of divine creation, separate and apart from the rest of living creatures." While this can be a genuinely held private belief, it can never be advanced as a scientific explanation, because once again it violates the rules of the game. If that situation were true, and if *H. sap.* were indeed the result of a special miracle, then, in view of Rule No. 1, above, the only proper scientific assessment would be: "Science has no explanation." The problem with any such statement is that we know from past experience that it probably should have been qualified: "Science has no explanation—yet." As people who have grown up amid the current scientific revolution know full well, last year's miracle is this year's technology.

The vital importance of excluding miracles and divine intervention from the game of science, as is advocated even today by the creationist movement, is that allowing such factors to be invoked as explanations discourages the search for other and more systematic causes. Two centuries ago, if Benjamin Franklin and his contemporaries had been content to regard vitreous and resinous forms of static electricity only as expressions of divine humor, we would be unlikely to have the science of electromagnetism today. A century later, a passive belief that God made all the molecules "after their own kind" would have stunted the infant science of chemistry. And a contemporary who believes devoutly that there are no connections between branches of living organisms is unlikely ever to discover such connections as do exist. The most insidious evil of supernatural creationism is that it stifles curiosity and therefore blunts the intellect.

There are those who demand, in a bizarre misapplication of courtroom standards, that the claims of modern science either be proven beyond a shadow of a doubt at this present moment, or else be given up entirely. Such people do not understand the structure of science as a game. We do not say, "Science absolutely and categorically denies the existence and intervention of the supernatural." Instead, as good game players, we say, "So far, so good. We haven't needed special miracles yet." The particular glory of science is that such an attitude has been so successful, over the past four centuries, in explaining so much of the world around us. A good maxim is: *If it isn't broke, don't fix it*. The game of rational science has been enormously successful. We change the rules of that game at our peril.

To be sure, many areas exist where we as scientists do not yet know all the answers. But these problem areas change from one generation to another, and that which might have seemed miraculous (to some) a generation ago now is seen to be perfectly explicable by natural causes. In hindsight we would have felt foolish had we written off those areas as the result of miracles fifty years ago; and we would be ill-advised to set ourselves up for ridicule by those who will follow us fifty years from now. It is a reasonable prediction that the attitude of future generations toward twentieth-century "scientific creationism" (an inherent oxymoron according to Rule No. 1, above) will be one of ridicule.

It would augur well, for both science and religion, if creationists and evolutionary biologists would realize jointly that the question of the existence or the nonexistence of a Deity is irrelevant to the study of biological evolution. Both the die-hard atheist and the theistic evolutionist can function as modern biologists with absolute integrity. The people who are entirely beyond the pale intellectually are those who can be characterized as short-Earth creationists and Biblical literalists—those who maintain that it all happened in 6 standard 24-hour days, with the celestial equivalent of a wave of a magic wind. A clear line of demarcation must be drawn between such people and evolutionists of either theistic or nontheistic inclination. Some creationist rhetoricians would like to draw the line between nontheistic and theistic evolutionists and to lump the second group (which probably includes the majority of nonscientists) together with the 6-day [Y]oung-Earth modern "Know-Nothings." We absolutely must not let them get away with such a tactic.

Science is not a closed body of dogma; it is a continuing process of enquiry. A dry and querulous legalism that tends to inhibit or close off that process is antithetical to science. The cartoonist Sidney Harris once published a cartoon depicting two scientists in consultation before a blackboard filled with equations—obviously some kind of proof in the making. One scientist points to a particular equation and proclaims confidently, "And at this point a miracle occurs!" Real scientists don't talk that way—not because some of them don't believe in miracles, sometime, somewhere—but *because invoking miracles and special creation violates the rules of the game of science and inhibits its progress.* People who do not understand that concept can never be real scientists, and should not be allowed to misrepresent science to young people from whom the ranks of the next generation of scientists will be drawn.

Excerpted from Richard E. Dickerson, The game of science, *Journal of Molecular Evolution* 34 (1992): 277–279.

CHAPTER 14

• • • • • • • • • • • • • • •

Evolution and Creationism in the Media and Public Opinion

INTRODUCTION

The creationism/evolution controversy has been much in the media, partly because of public interest, but also because of increased anti-evolutionary activity at the state and local level. In addition, courtroom dramas such as *Kitzmiller v. Dover* and *Selman v. Cobb County* have attracted national, and even international, attention from television and radio, as well as newspapers and magazines.

The coverage of this controversy is uneven; as discussed by Mooney and Nisbet, science and education reporters generally do a more satisfactory job than general news or political reporters. Television has its own stumbling blocks to competent coverage; television coverage tends to be visual and emotional, skimping on complexity and nuance—and there is plenty of complexity and nuance surrounding the creationism/evolution controversy. Mooney and Nisbet suggest ways that media can more accurately cover the controversy, while Rosenhouse and Branch encourage scientists to be more effective communicators on this topic. Selections from these articles appear in the "Coverage of the Controversy" section of this chapter.

A continuing feature of media coverage of the creationism/evolution controversy is the presentation of public opinion on these issues, as assessed by survey research. Every year there are a number of polls seeking to describe the views of Americans about creationism, evolution, or religion. Survey research is difficult work, and it is surprisingly difficult to produce accurate data. Not everyone is aware of the problems that beset survey research, so I'll take a moment to discuss some of the challenges.

As in almost all survey research, pollsters cannot question everyone (all scientists, all teachers, all American adults); they can survey only a sample of the population they seek to describe. How this sample is chosen is critical to whether the results are valid: one would not want to generalize the opinions of females over the age of sixty who write science books (however important those opinions are) to the entire American population. Such a sample (probably rather few people, actually) would not accurately reflect the opinions of the general American population on hardly anything.

Therefore, readers should carefully scrutinize sampling practices as the first step in evaluating the validity of a survey. Voluntary Internet polls, where people self-select and therefore likely have stronger opinions about the questions being asked, are not reliable. There is a built-in bias in such polls, and the results are almost always less representative of the whole population than polls where the sample is chosen according to established polling techniques. It is also often possible for people to respond multiple times or to recruit like-minded people to swamp the polls, making them even less reliable.

To make valid generalizations about the attitudes or knowledge of the population, then, a pollster would want to be certain to include a representative sample. The more specific the population (scientists at elite institutions, for example, rather than all scientists), the easier it is to draw a sample. The broader the population (as all adult Americans), the more the pollster has to work to be sure that the sample includes representatives of all the relevant subgroups. To conduct a valid survey of adult Americans on their views about the creationism/evolution controversy, a pollster would need to draw a sample that takes into consideration sex, age, race, economic status, and education, and because religion is involved, the religious beliefs of the respondents as well. To be able to generalize to Americans as a whole, the responses of smaller subsegments of a sample might have to be weighted (i.e., counted more than once) according to formulae determined by demographics of American society. Sampling is perhaps the most important single component determining a survey's validity.

As important as drawing a valid sample, a pollster needs to pay considerable attention to how questions are worded. The wording of questions regarding acceptance of evolution or creationism is quite sensitive to how the question is asked—and even which questions are asked early in the questionnaire versus later. For example, Bishop (2007) shows that if a people are primed to think about religion by first being asked about their belief in God, they tend to report a lower acceptance of evolution. Similarly, if the Bible is specifically mentioned in the question, people report less acceptance of evolution. When a question on acceptance of evolution is asked in a more neutral fashion—calling on only scientific rather than religious opinions—more respondents answer that they accept evolution.

Of course, surveys are not conducted in a vacuum: questions have to take into account events occurring previous to or near the time of the survey, previous understanding of the topic by the respondents, and even cultural factors, or inaccurate information might be collected or inaccurate inferences drawn from the data. In a survey conducted in Ohio by the Zogby polling organization for the intelligent design–promoting Discovery Institute, for example, results suggested that Ohio citizens were very favorable toward the teaching of ID (Zogby 2002). A whopping 78 percent of respondents agreed that, "When Darwin's theory of evolution is taught in school, students should also be able to learn about scientific evidence that points to an intelligent design of life." Yet the University of Cincinnati's Ohio Poll conducted within a few months of the DI poll demonstrated that most Ohioans were uninformed about intelligent design. The Ohio poll asked, "Do you happen to know anything about the concept of 'intelligent design'?" and 84 percent of citizens answered no (14 percent said yes and 2 percent were not sure). The largest category showing any knowledge of ID were college graduates, but only 28 percent of them answered yes (Hoffman 2002).

If such a high percentage of Ohioans avowed their ignorance of ID, it is unlikely that the 78 percent figure from the DI poll reflects a sincere enthusiasm for the teaching of intelligent design. The high frequencies might instead reflect the well-known American propensity for fairness (discussed in the introduction to this book) as well as the relative unpopularity of evolution.

The American population is composed of different groups that overlap one another, among them adults, children, males, females, college students, scientists, nonscientists, K–12 science teachers, people of faith, and nonbelievers. It is of interest how attitudes about creationism and evolution in the United States compare to those in other countries, and how different groups of Americans compare with one another. The readings in the "Surveying Opinions" section present some results of surveys about American attitudes toward evolution and creationism.

LEGAL CASES

Kitzmiller v. Dover Area School District 400 F. Supp. 2d 707 (M.D. Pa. 2005)
Selman v. Cobb County School District, 390 F. Supp. 2d 1286 (N.D. Ga. 2005).

REFERENCES

Bishop, G. 2007. Polls apart on human origins. *Reports of the National Center for Science Education* 27 (5–6): 35–41.
Hoffman, C. 2002. Majority of Ohio science professors and public agree: "Intelligent Design" mostly about religion. University of Cincinnati press release. Retrieved April 26, 2008, from http://www.uc.edu/news/idpoll.htm.
Zogby International. 2002. Results from Ohio poll on Darwin's theory of evolution, public schools. Retrieved on October 15, 2008 from http://www.sciohio.org/OhioZogbyPoll.pdf.

• •

COVERING THE CONTROVERSY

Undoing Darwin

On March 14, 2005, the *Washington Post*'s Peter Slevin wrote a front-page story on the battle that is "intensifying across the nation" over the teaching of evolution in public-school science classes. Slevin's lengthy piece took a detailed look at the lobbying, fund-raising, and communications tactics being deployed at the state and local level to undermine evolution. The article placed a particular emphasis on the burgeoning "intelligent design" movement, centered at Seattle's Discovery Institute, whose proponents claim that living things, in all their organized complexity, simply could not have arisen from a mindless and directionless process such as the one so famously described in 1859 by Charles Darwin in his classic, *The Origin of Species*.

Yet Slevin's article conspicuously failed to provide any background information on the theory of evolution, or why it's considered a bedrock of modern scientific knowledge among both scientists who believe in God and those who don't. Indeed, the few defenders of evolution quoted by Slevin were attached to advocacy groups, not research universities; most of the article's focus, meanwhile, was on anti-evolutionists and their strategies. Of the piece's thirty-eight paragraphs, twenty-one were devoted to this "strategy" framing—an emphasis that, not surprisingly, rankled the *Post*'s science reporters. "How is it that *The Washington Post* can run a feature-length A1 story about the battle over the facts of evolution and not devote a single paragraph to what the evidence is for the scientific view of evolution?" protested an internal memo from the paper's science desk that was copied to Michael Getler, the *Post*'s ombudsman. "We do our readers a grave disservice by not telling them. By turning this into a story of dueling talking heads, we add credence to the idea that this is simply a battle of beliefs." Though he called Slevin's piece "lengthy, smart, and very revealing," Getler assigned Slevin a grade of "incomplete" for his work.

Slevin's incomplete article probably foreshadows what we can expect as evolution continues its climb up the news agenda, driven by a rising number of newsworthy events.

As evolution, driven by such events, shifts out of scientific realms and into political and legal ones, it ceases to be covered by context-oriented science reporters and is instead bounced to political pages, opinion pages, and television news. And all these venues, in their various ways, tend to deemphasize the strong scientific case in favor of evolution and instead lend credence to the notion that a growing "controversy" exists over evolutionary science. This notion may be politically convenient, but it is false.

We reached our conclusions about press coverage after systematically reading through seventeen months of evolution stories in *The New York Times* and *The Washington Post*; daily papers in the local areas embroiled in the evolution debate (including both papers covering Dover, Pennsylvania, the *Atlanta Journal-Constitution*, and the Topeka, Kansas, *Capital-Journal*); and relevant broadcast and cable television news transcripts. Across this coverage, a clear pattern emerges when evolution is an issue: from reporting on newly discovered fossil records of feathered dinosaurs and three-foot humanoids to the latest ideas of theorists such as Richard Dawkins, science writers generally characterize evolution in terms that accurately reflect its firm acceptance in the scientific community. Political reporters, generalists, and TV news reporters and anchors, however, rarely provide their audiences with any real context about basic evolutionary science. Worse, they often provide a springboard for anti-evolutionist criticism of that science, allotting ample quotes and sound bites to Darwin's critics in a quest to achieve "balance." The science is only further distorted on the opinion pages of local newspapers [a discussion of the science of evolution deleted here]. . . .

If attacks on evolution aren't anything new in America, neither is the tendency of U.S. journalists to lend undue credibility to theological attacks that masquerade as being "scientific" in nature. During the early 1980s, for example, the mega-evolution trial *McLean v. Arkansas* pitted defenders of evolutionary science against so-called "scientific creationists." Today, few take the claims of these scientific creationists very seriously. At the time, however, proponents of creation science were treated quite seriously indeed by the national media, which had parachuted in for the trial.

As media scholars have noted, reporters generally "balanced" the scientific-sounding claims of the scientific creationists against the arguments of evolutionary scientists. They also noted that religion and public-affairs reporters, rather than science writers, were generally assigned to cover the trial.

Now, history is repeating itself: intelligent-design proponents, whose movement is a descendant of the creation science movement of yore, are enjoying precisely the same kind of favorable media coverage in the run-up to another major evolution trial. This cyclical phenomenon carries with it an important lesson about the nature of political reporting when applied to scientific issues. In strategy-driven political coverage, reporters typically tout the claims of competing political camps without comment or knowledgeable analysis, leaving readers to fend for themselves. . . .

Political reporting in newspapers is just part of the problem. Television news reporting often makes the situation even worse, even in the most sophisticated of venues. Consider, for example, a March 28 report on *The NewsHour* with Jim Lehrer, in which the correspondent Jeffrey Brown characterized evolution's new opponents as follows: "Intelligent design's proponents carefully distinguish themselves from creation scientists. They use only the language of science, and avoid speaking of God as the ultimate designer." Brown appears oblivious to the scientific-sounding arguments employed by earlier creationists. Moreover, references to God and religion aren't particularly difficult to find among ID defenders, if you know where to look. The pro-ID Discovery Institute's strategic Wedge Document, exposed on the Internet years ago and well known to those who follow the evolution issue, baldly stated the hope that intelligent design would "reverse the stifling dominance of the materialist worldview, and . . . replace it with a science consonant with Christian and theistic convictions." . . .

Even the best TV news reporters may be hard-pressed to cover evolution thoroughly and accurately on a medium that relies so heavily upon images, sound bites, drama, and conflict to keep audiences locked in. These are serious obstacles to conveying scientific complexity. And with its heavy emphasis on talk and debate, cable news is even worse. The adversarial format of most cable news talk shows inherently favors ID's attacks on evolution by making false journalistic "balance" nearly inescapable. . . .

Besides citing the overwhelming scientific consensus in support of evolution, journalists can also contextualize the claims of ID proponents by applying clear legal precedents. Instead of ritually likening the contemporary intelligent-design debate to the historic Scopes "monkey trial" of 1925, journalists should ask the same questions about ID that more recent court decisions (especially the *McLean v. Arkansas* case) have leveled at previous challenges to evolution: First, is ID religiously motivated and does it feature religious content? In other words, would it violate the separation of church and state if covered in a public school setting? Second, does ID meet the criteria of a scientific theory, and is there strong peer-reviewed evidence in support of it? In short, to better cover evolution, journalists don't merely have to think more like scientists (or science writers). As the evolution issue inevitably shifts into a legal context, they must think more like skeptical jurists.

And as evolution becomes politicized in state after state through trials and school board maneuverings, it rises to prominence on the opinion pages as well as in news stories. Here, competing arguments about evolution and intelligent design tend to be paired against one another in letters to the editor and sometimes in rival guest op-eds, providing a challenge to editors who want to give voice to alternative ideas yet provide an accurate sense of the state of scientific consensus. The mission of the opinion pages and a faithfulness to scientific accuracy can easily come into conflict.

In our study of media coverage of recent evolution controversies, we homed in on local opinion pages, both because they represent a venue where it's easy to keep score of how the issue is being defined and because we suspected they would reflect a public that is largely misinformed about the scientific basis for the theory of evolution yet itching to fight about it. That's especially so since many opinion-page editors see their role not as gatekeepers of scientific content, but rather as enablers of debate within pluralistic communities—even over matters of science that are usually adjudicated in peer-reviewed journals. Both editorial-page editors of the York papers, for example, emphasized that they try to run every letter they receive that's "fit to print" (essentially meaning that it isn't too lengthy or outright false or libelous). . . .

Rather stunningly, we found that the heated political debate in Dover, Pennsylvania, produced a massive response: 168 letters, op-eds, columns, and editorials appearing in the *York Daily Record* alone over the seventeen-month period analyzed (plus ninety-eight in the *York Dispatch*). A slight plurality of opinion articles at the *Dispatch* (40.9 percent) and the *Daily Record* (45.3 percent) implicitly or explicitly favored teaching ID and/or "creation science" in some form in public schools, while 39.8 percent and 36.3 percent of opinion articles at those two papers favored teaching only evolution. On the question of scientific evidence, more than a third of opinion articles at the two papers contended or suggested that ID and/or "creation science" had scientific support.

In short, an entirely lopsided debate within the scientific community was transformed into an evenly divided one in the popular arena, as local editorial-page editors printed every letter they received that they deemed "fit." At the *York Dispatch* this populism was partly counterbalanced by an editorial voice that took a firm stand in favor of teaching evolution and termed intelligent design the "same old creationist wine in new bottles." The *York Daily Record*, however, was considerably more sheepish in its editorial stance. The paper generally sought to minimize controversy and seemed more willing to criticize Dover school board members who resigned over the decision to introduce intelligent design into the curriculum (asking why they didn't stay and fight) than to rebuke those board members who were responsible for attacking evolution in the first place. . . .

Interestingly, however, not all local opinion pages fit the mold of the York papers. Given the turmoil in Cobb County, Georgia, over the introduction of anti-evolutionist textbook disclaimers, the *Atlanta Journal-Constitution* also covered the debate heavily on its opinion pages. But the paper took a very firm stand on the issue, with the editorial-page editor, Cynthia Tucker, declaring in one pro-evolution column that "our science infrastructure is under attack from religious extremists." Tucker, along with the deputy editorial-page editor, Jay Bookman, also warned repeatedly of the severe negative economic consequences and national ridicule that anti-evolutionism might bring on the community. . . .

Yet despite the strong stance of the *Journal-Constitution* editorial staff, the editors also actively worked to include at least some balance in perspectives, inviting guest op-eds that countered the strongly pro-evolution editorial position of the paper. Roughly

30 percent of the letters and op-eds to the paper featured pro-ID and/or creationist views. . . .

At two elite national papers, *The New York Times* and *The Washington Post*, the opinion pages sided heavily with evolution. But even there a false sense of scientific controversy was arguably abetted when *The New York Times* allowed Michael Behe, the prominent ID proponent, to write a full-length op-ed explaining why his is a "scientific" critique of evolution. And when *USA Today* took a strong stand for evolution on its editorial page on August 8 ('INTELLIGENT DESIGN' SMACKS OF CREATIONISM BY ANOTHER NAME), the paper, using its point-counterpoint editorial format, ran an anti-evolution piece with it (EVOLUTION LACKS FOSSIL LINK), written by a state senator from Utah, D. Chris Buttars. It was filled with stark misinformation, such as the following sentence: "There is zero scientific fossil evidence that demonstrates organic evolutionary linkage between primates and man." . . .

At the end of August, the *Times* weighed in with a three-part series on the evolution "controversy," drawing from its deep well of expertise. On Sunday, August 21, reporter Jodi Wilgoren provided background on the history, funding, and tactics of the Discovery Institute. On Monday, science writer Kenneth Chang tackled the science, giving considerable space to an explanation of evolutionary theory. Cornelia Dean broke new ground on Tuesday with a piece about how scientists, including devout Christian scientists, view religion.

The series was nuanced and comprehensive, and will likely boost even higher the profile of evolution in the news. Still, the unintended consequence may be that increased media attention only helps proponents present intelligent design as a contest between scientific theories rather than what it actually is—a sophisticated religious challenge to an overwhelming scientific consensus. As the Discovery Institute's vice president, Jay Richards, put it on *Larry King Live* the day of the final *Times* story: "We think teachers should be free to talk about intelligent design, and frankly, I don't think that it can be suppressed. It's now very much a public discussion, evidenced by the fact that you're talking about it on your show tonight."

Without a doubt, then, political reporting, television news, and opinion pages are all generally fanning the flames of a "controversy" over evolution. Not surprisingly, in light of this coverage, we simultaneously find that the public is deeply confused about evolution. . . .

At the very least, the flaws in the journalistic presentation of evolution by political reporters, TV news, and op-ed pages aren't clarifying the issues. Perhaps journalists should consider that unlike other social controversies—over abortion or gay marriage, for instance—the evolution debate is not solely a matter of subjective morality or political opinion. Rather, a definitive standard has been set by the scientific community on the science of evolution, and can easily be used to evaluate competing claims. Scientific societies, including the National Academy of Sciences and the American Association for the Advancement of Science, have taken strong stances affirming that evolution is the bedrock of modern biology. In such a situation, journalistic coverage that helps fan the flames of a nonexistent scientific controversy (and misrepresents what's actually known) simply isn't appropriate.

So what is a good editor to do about the very real collision between a scientific consensus and a pseudo-scientific movement that opposes the basis of that consensus?

At the very least, newspaper editors should think twice about assigning reporters who are fresh to the evolution issue and allowing them to default to the typical strategy frame, carefully balancing "both sides" of the issue in order to file a story on time and get around sorting through the legitimacy of the competing claims. As journalism programs across the country systematically review their curriculums and training methods, the evolution "controversy" provides strong evidence in support of the contention that specialization in journalism education can benefit not only public understanding, but also the integrity of the media. For example, at Ohio State, beyond basic skill training in reporting and editing, students focusing on public-affairs journalism are required to take an introductory course in scientific reasoning. Students can then specialize further by taking advanced courses covering the relationships between science, the media, and society. They are also encouraged to minor in a science-related field.

With training in covering science-related policy disputes on issues ranging from intelligent design to stem-cell research to climate change, journalists are better equipped to make solid independent judgments about credibility, and then pass these interpretations on to readers. The intelligent-design debate is one among a growing number of controversies in which technical complexity, with disputes over "facts," data, and expertise, has altered the political battleground. The traditional generalist correspondent will be hard-pressed to cover these topics in any other format than the strategy frame, balancing arguments while narrowly focusing on the implications for who's ahead and who's behind in the contest to decide policy. If news editors fail to recognize the growing demand for journalists with specialized expertise and backgrounds who can get beyond this form of writing, the news media risk losing their ability to serve as important watchdogs over society's institutions.

When it comes to opinion pages, meanwhile, there's certainly more room for dissent because of the nature of the forum—but that doesn't mean editorial-page editors can't act as responsible gatekeepers. Unlike the timidity of the *York Daily Record* and *The Topeka Capital-Journal*, *The York Dispatch* and *The Atlanta Journal-Constitution* serve as examples of how papers can inform their readers about authoritative scientific opinion without stifling the voices of anti-evolutionists.

One thing, above all, is clear: a full-fledged national debate has been reawakened over an issue that once seemed settled. This new fight may not simmer down again until the U.S. Supreme Court is forced (for the third time) to weigh in. In these circumstances, the media have a profound responsibility—to the public, and to knowledge itself.

Excerpted from Chris Mooney and Matthew C. Nisbet. Undoing Darwin. *Columbia Journalism Review* 44, no. 3 (2005): 30–39. Used with permission from the authors.

Media Coverage of Intelligent Design: Conclusions

We have only scratched the surface of this topic. Regional newspapers vary widely in their coverage of, and respect for, evolution. Partisan magazines, especially conservative ones, have a great many things to say on the subject. Television coverage on the major networks is substantially more sedate than it is on cable. And the extent and quality of coverage of these issues is influenced by a host of journalistic, social, political, political, economic, and religious factors too numerous and too broad to discuss thoroughly in this article. But, even on the basis of such a limited and preliminary discussion, there are clear morals to extract for the scientific community.

Antievolutionists have a very attractive message to market. They do not tell journalists that they want a certain myopic religious viewpoint presented as legitimate science. Instead, they talk about presenting both sides, being open-minded, opposing censorship, and presenting all the evidence. The only way for the evolutionist to counteract this is to show that creationism's scientific pretensions are nonsense. That is precisely what cannot be done in a brief newspaper article or television appearance.

Scientists therefore need to become more savvy in their dealings with the media. Toward that end, we offer the following suggestions.

In any encounter between scientists and the media on the subject of creationism, declare first and foremost that the specific scientific assertions of ID proponents are false. State unambiguously that evolutionary theory is perfectly capable in principle of explaining the formation of complex biological systems, and, indeed, has done so in practice many times.

Avoid arguing simply that ID is unscientific because of its reliance on the supernatural, or that present-day mysteries may eventually yield to scientific explanations. Both of these assertions are certainly correct, but they play into the hands of ID proponents. The former fosters the impression, which ID proponents are keen to convey, that defenders of evolution are merely ruling ID unscientific by definitional fiat, while the latter seems to concede that there are vast explanatory holes in modern evolutionary theory.

Invest time in preparation. Read the books and articles produced by Young-Earth Creationists and ID proponents. Scientifically knowledgeable readers may find this a frustrating and aggravating experience, but it has to be done in order to respond. Also read material on the historical, religious, philosophical, educational, and legal issues associated with the dispute (a good starting place is Scott 2005).

Watch your language, with respect to both terminology and tone. Don't assume that your readers and listeners understand that a theory is more than a hunch or a guess, for example. Similarly, speak of accepting rather than believing in evolution, since the latter will strike many as expressing a statement of faith rather than a judgment based on the evidence. As for tone, the manner in which you deliver your message can be as important as the content of your message. Try to sound calm, informed, and knowledgeable—especially in public appearances and on radio or television.

Expect the religion card to be played. Creationists—abetted by a handful of scientists, to be sure—have convinced a large segment of the public that evolution is intrinsically atheistic. Whether or not you are religious yourself, be prepared to point out that evolution is accepted simply on the basis of the overwhelming evidence in its favor by scientists of all faiths, and that quite a few religious denominations have made their theological peace with evolution.

Be ready, too, to rebut the inevitable appeal to fairness. Perhaps the most powerful argument in the creationist repertoire is the idea of giving students "both views" and leaving it up to them to decide. What is truly unfair, of course, is to cheat students of an adequate science education by telling them anything other than the truth: Evolution is at the core of modern biology.

Look for opportunities to become a spokesperson. You might begin small, by submitting letters to the editor of your local newspapers applauding, criticizing, or expanding on recent articles on evolution–creationism issues. More ambitiously, inquire about the possibility of submitting an op-ed piece supporting evolution education. You can ask your university press office to list you as an expert on the topic. You can also cultivate reporters on your own: Drop a friendly note to reporters who write on evolution–creationism issues, commenting on their stories and offering your help when they next do a story on the topic.

With the rise of blogs (short for "Web log") as cheap but influential media sources, consider speaking out on the Internet. Among the scientists who use blogs as platforms to defend the teaching of evolution are P. Z. Myers (University of Minnesota at Morris; http://scienceblogs.com/pharyngula), John M. Lynch (Arizona State University; http://scienceblogs.com/strangerfruit), and the first author of this essay [http://scienceblogs.com/evolutionblog]. Such blogging can have profound effects. For example, when an article arguing for ID was published, under suspicious circumstances, in a legitimate scientific journal, a detailed critique quickly appeared on the collaborative blog *The Panda's Thumb* (www.pandasthumb.org, to which Myers, Lynch, and the first author contribute). This critique was subsequently cited in news stories in *The Scientist* and *Nature*.

For most scientists, it is natural to be circumspect when discussing complex scientific issues. That approach is totally ineffective in dealing with the media. What seems like sober reflection in an academic setting comes off as weakness when printed in a newspaper or stated on television. Proponents of ID are effective precisely because they spend so much time thinking about public relations. Scientists need to do likewise. Anticipating a storm of controversy over *On the Origin of Species*, Thomas Henry Huxley wrote, "I am sharpening up my beak and claws in readiness." Scientists today, too, should be taking themselves to the grindstone.

REFERENCE

Scott, Eugenie C. 2005. *Evolution vs. creationism: An introduction.* Berkeley: University of California Press.

Excerpted from Jason Rosenhouse and Glenn Branch. 2006. Media coverage of "intelligent design". *BioScience* 56, no. 3 (2006): 247–252, 251–252. Copyright © American Institute of Biological Sciences. Used with permission.

● ●

SURVEYING OPINIONS

Public Acceptance of Evolution

Beginning in 1985, national samples of U.S. adults have been asked whether the statement, "Human beings, as we know them, developed from earlier species of animals," is true or false, or whether the respondent is not sure or does not know. We compared the results of these surveys with survey data from nine European countries in 2002, surveys in 32 European countries in 2005, and a national survey in Japan in 2000. Over the past 20 years, the percentage of U.S. adults accepting the idea of evolution has declined from 45% to 40% and the percentage of adults overtly rejecting evolution declined from 48% to 39%. The percentage of adults who were not sure about evolution increased from 7% in 1985 to 21% in 2005. After 20 years of public debate, the public appears to be divided evenly in terms of accepting or rejecting evolution, with about one in five adults still undecided or unaware of the issue. This pattern is consistent with a number of sporadic national newspaper surveys reported in recent years.

A dichotomous true-false question format tends to exaggerate the strength of both positions. In 1993 and 2003, national samples of American adults were asked about the same statement but were offered the choice of saying that the statement was "definitely true, probably true, probably false, definitely false," or that they did not know or were uncertain. About a third of American adults firmly rejected evolution, and only 14% of adults thought that evolution is "definitely true." Treating the "probably" and "not sure" categories as varying degrees of uncertainty, ~55% of American adults have held a tentative view about evolution for the last decade.

This pattern is different from that seen in Europe and Japan. Looking first at the simpler true-false question, our analysis found that significantly (at the 0.01 to 0.05 level by difference of proportions) more adults in Japan and 32 European countries accepted the concept of evolution than did American adults (see Figure 14.1). Only Turkish adults were less likely to accept the concept of evolution than American adults. In Iceland, Denmark, Sweden, and France, 80% or more of adults accepted the concept of evolution, as did 78% of Japanese adults.

A cross-national study of the United States and nine European nations in 2002–2003 used the expanded version of the question. The results confirm that a significantly lower proportion of American adults believe that evolution is absolutely true than adults in nine European countries [see Fig. S-1 in the Supporting Online Material (SOM), here, Figure 14.2]. A third of American adults indicated that evolution is "absolutely false"; the proportion of European adults who thought that evolution was absolutely false ranged from 7% in Denmark, France, and Great Britain to 15% in the Netherlands. Regardless of the form of the question, one in three American adults firmly rejects the concept of evolution, a significantly higher proportion than found in any western European country. How can we account for this pattern of American reservations about the concept of evolution in the context of broad acceptance in Europe and Japan?

First, the structure and beliefs of American fundamentalism historically differ from those of mainstream Protestantism in both the United States and Europe. The biblical literalist focus of fundamentalism in the United States sees Genesis as a true and accurate account of the creation of human life that supersedes any scientific finding or interpretation. In contrast, mainstream Protestant faiths in Europe (and their U.S. counterparts) have viewed Genesis as metaphorical and—like the Catholic Church—have not seen a major contradiction between their faith and the work of Darwin and other scientists.

Figure 14.1: Public Acceptance of Evolution in 34 Countries, 2005.
The United States has the second-lowest acceptance rate of evolution
among the countries surveyed.

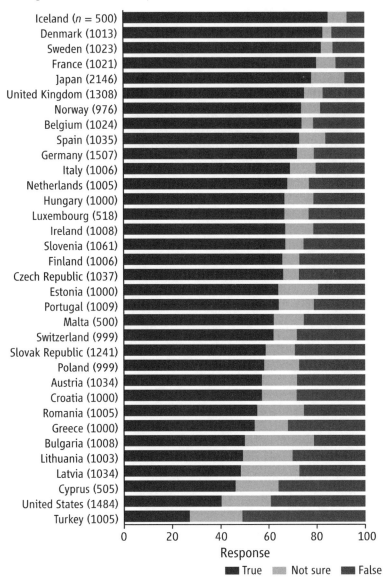

To test this hypothesis empirically, a two-group structural equation model (SEM) was constructed using data from the United States and nine European countries (see statistical analyses in the Supporting Online Materials, available at http://www.sciencemag.org/cgi/data/313/5788/765/DC1/1). The SEM allows an examination of the relation between several variables simultaneously on one or more outcome variables. In this model, 10 independent variables—age, gender, education, genetic literacy, religious belief, attitude toward life, attitude toward science and technology

Figure 14.2: Public Acceptance of Evolution in Ten Countries, 2002–2003.
When the question on acceptance of evolution is expanded from "true" or "false" to
a five-part question, the differences in American opinions become even more striking.
Far more Americans consider evolution "definitely false" than in nine other developed
countries. The percentage of "unsure" also increases.

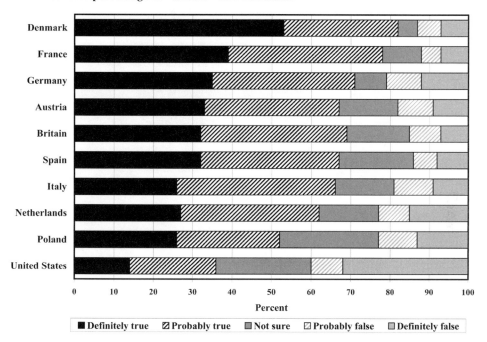

(S&T), belief in S&T, reservations about S&T, and political ideology—were used to
predict attitude toward evolution. The total effect of fundamentalist religious beliefs
on attitude toward evolution (using a standardized metric) was nearly twice as much
in the United States as in the nine European countries (path coefficients of −0.42 and
−0.24, respectively), which indicates that individuals who hold a strong belief in a
personal God and who pray frequently were significantly less likely to view evolution
as probably or definitely true than adults with less conservative religious views [see
Fig. S-2, from the SOM statistical analyses, here, Figure 14.3].

Second, the evolution issue has been politicized and incorporated into the current
partisan division in the United States in a manner never seen in Europe or Japan. In
the second half of the 20th century, the conservative wing of the Republican Party
has adopted creationism as a part of a platform designed to consolidate their support in
southern and Midwestern states—the "red" states. In the 1990s, the state Republican
platforms in seven states included explicit demands for the teaching of "creation
science." There is no major political party in Europe or Japan that uses opposition to
evolution as a part of its political platform.

The same SEM model discussed above offers empirical support for this conclusion.
In the United States, the abortion issue has been politicized and has become a key
wedge issue that differentiates conservatives and liberals. In the SEM, individuals who

Figure 14.3

Analysis of factors influencing acceptance or rejection of evolution show a different pattern in the United States and in European nations. Religious conservatism, for example, exists in both the United States and in Europe, but is almost twice as likely to predict rejection of evolution in the United States as in Europe. Attitudes towards abortion similarly differ greatly, with a pro-life attitude being much more positively associated with rejection of evolution in the United States than in Europe.

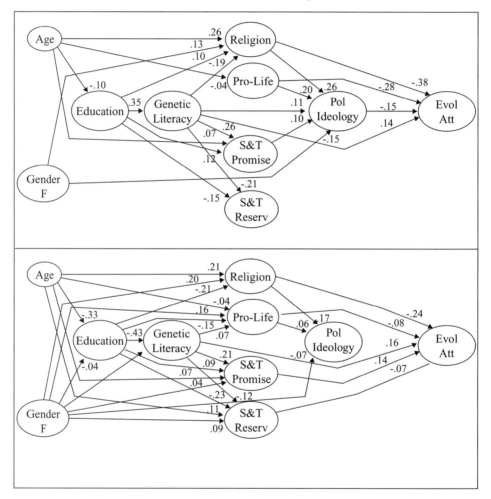

held strong pro-life beliefs were significantly more likely to reject evolution than individuals with pro-choice views. The total effect of pro-life attitudes on the acceptance of evolution was much greater in the United States than in the nine European countries (–0.31 and –0.09, respectively) [see Figure S-2 from the SOM statistical analyses, here, Figure 14.3]. The same model also documents the linkage of religious conservative beliefs and a conservative partisan view in the United States. The path coefficient for the relation between fundamentalist religious views and self-identification as a conservative was 0.26 in the United States and 0.17 in the nine European countries.

The path coefficient between pro-life views and self-identification as a conservative was 0.20 in the United States and 0.06 in the nine European countries. Because the two-group SEM computes path coefficients on a common metric, these results are directly comparable and the impact of fundamentalist religious beliefs and pro-life attitudes may be seen as additive . . .

These results should be troubling for science educators at all levels. Basic concepts of evolution should be taught in middle school, high school, and college life sciences courses and the growing number of adults who are uncertain about these ideas suggests that current science instruction is not effective. Because of the rapidly emerging nature of biomedical science, most adults will find it necessary to learn about these new concepts through informal learning opportunities. The level of adult awareness of genetic concepts (a median score of 4 on a 0-to-10 scale) suggests that many adults are not well informed about these matters. The results of the SEM indicate that genetic literacy is one important component that predicts adult acceptance of evolution.

The politicization of science in the name of religion and political partisanship is not new to the United States, but transformation of traditional geographically and economically based political parties into religiously oriented ideological coalitions marks the beginning of a new era for science policy. The broad public acceptance of the benefits of science and technology in the second half of the 20th century allowed science to develop a nonpartisan identification that largely protected it from overt partisanship. That era appears to have closed. (references deleted)

Excerpted from Jon D. Miller, Eugenie C. Scott, and Shinji Okamoto, Public acceptance of evolution. *Science* 313 (2006): 775–776. Supporting online materials available at http://www.sciencemag.org/cgi/data/313/5788/765/DC1/1. Used with permission.

Prevailing Wisdom

A nationwide study of 1,000 likely U.S. voters has revealed new information about public perceptions and attitudes toward evolution and creationism. Teaching evolution is the most favored public science school curriculum, with 54 percent of those surveyed strongly or somewhat favoring it, compared to 28 percent for intelligent design, which was declared "a religious view, a mere re-labeling of creationism" by a U.S. Federal judge in a landmark December 2005 decision. The survey revealed that only 22 percent of the U.S. voting public was not sure whether evolution should be taught in public schools.

The respondents found both privacy and taxpayer-spending issues to be persuasive reasons for keeping religion out of public school science classrooms: public schools should not impose one religious viewpoint on students (56 percent of respondents) and tax dollars should not support religion in science class (54 percent).

The study also explored how the public perceives science's contribution to society (figure omitted). Developing new medicines and curing diseases was the contribution that most respondents said was important, 63 percent.

Figure 14.4
Knowledge about basic science among American voters is spotty, at best.

Scientific Knowledge (Percent responding accurately)

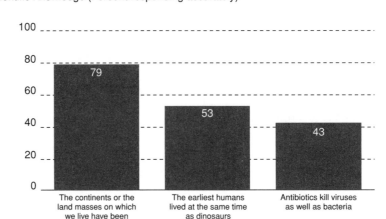

*Percent responding accurately

Other contributions to society that were thought of as important are improving the general quality of life (39 percent), identifying the best ways to protect our environment and natural resources (32 percent), and developing new technologies to protect our national security (25 percent).

Many respondents viewed science education as an important component of a larger, general education [figure omitted]. According to an overwhelming majority of respondents, science instruction should promote how to draw conclusions from evidence (80 percent) and to learn how to think critically (78 percent). According to the survey, more people can agree that critical thought and college preparation are the purpose of science education rather than scientific literacy or practical applications of science to the real world.

The study also shows the voting population has distinct ideas about religion and public life. A majority said they go to church at least once a month, and most respondents agree that there should be a "high wall of separation" between church and state, with 59 percent agreeing and 43 percent strongly agreeing. Nearly all respondents, 89 percent, agree that parents today are not taking enough responsibility for teaching their children moral values. And the majority of respondents, 57 percent, oppose efforts by the government to legislate moral values, while only 35 percent favor it.

Another goal of the study was to take a brief inventory of U.S. voters' knowledge of basic science information (Figure 14.4). It was found to be spotty, at best.

Respondents were given three statements. They had to answer whether they agree with, disagree with or were unsure about each of them: The continents or the land-masses on which we live have been moving for millions of years and will continue to move in the future; The earliest humans lived at the same time as the dinosaurs; Antibiotics kill viruses as well as bacteria. Seventy-nine percent correctly agreed with

Figure 14.5
Only 23 percent of respondents could correctly answer all three general science questions; 30 percent could answer two, and 39 percent could correctly answer only one of the three questions.

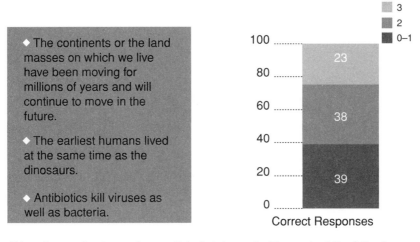

"Now, I am going to read you a list of statements. For each of the following please tell me if you agree or disagree with the statement. If you neither agree nor disagree with the statement, or don't know, please say so."

the first statement, 53 percent correctly disagreed with the second statement, and 43 percent correctly disagreed with the third statement. As illustrated in Figure 14.5, 39 percent of respondents gave less [sic] than two correct answers, 38 percent gave two correct answers and only 23 percent gave correct answers to all three.

Which authority should inform the voting public about evolution and creationism? The respondents said they'd be most interested in hearing from scientists (77 percent somewhat or very interested) and science teachers (76 percent somewhat or very interested), with members of the clergy a strong third (62 percent somewhat or very interested) (Figure 14.6).

The study was performed by Greenberg Quinlan Rosner Research and Mercury Public Affairs in August 2006 and surveyed 1,000 U.S. citizens likely to vote. It was commissioned by more than 30 scientific organizations, including the American Institute of Physics, American Physical Society, the Federation of Associated Societies for Experimental Biology, American Chemical Society, American Institute of Biological Sciences, Consortium of Social Science Associations, and the Society for Developmental Biology.

Excerpted from Ben Stein, Prevailing wisdom. *Interactions* 38, no. 1 (2008): 16–18. Reprinted with permission from the American Institute of Physics.

Religion among Academic Scientists: Distinctions, Disciplines, and Demographics

Using data we collected as part of the Religion among Academic Scientists study (RAAS), which includes a survey of natural and social scientists, we examine how

Figure 14.6
Respondents would prefer to hear about science from scientists and science
teachers rather than school board members or celebrities.

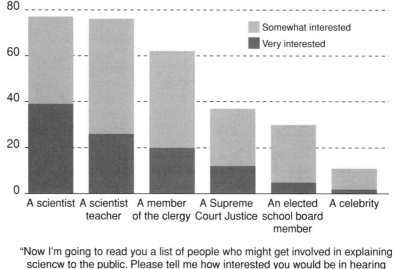

"Now I'm going to read you a list of people who might get involved in explaining
sciencw to the public. Please tell me how interested you would be in hearing
from each person about Evolution, Creationism, and Intelligent Design"

scientists differ from the general public, from each other, and some of the sources of
these differences. We start by showing broad religious differences between academic
scientists and the American population, which reveal that academic scientists are
much less religious than the general public according to traditional indicators of
religiosity. We then turn to difference within the academy. Since most of the scholarly
literature on faculty attitudes towards religiosity addresses the field-specific differences
between natural and social scientists, we ask how these two groups compare to one
another according to traditional measures of religiosity and find that only a small
amount of variance is explained by broad field-specific differences between natural
and social scientists or even specific interdisciplinary differences. Instead, particular
demographic factors, such as age, marital status, and presence of children in the
household, seem to explain some of the religious differences among academic scientists
(although our data do not allow direct analysis of causal direction). Most important,
respondents who were raised in religious homes, especially those raised in homes where
religion was important are most likely to be religious at present.

Differences between Scientists and the General Population

Comparing scientists at elite institutions to the general population, the psychologist
James Leuba did surveys in 1916 and 1934 on the attitudes of American scientists
towards Christian belief, which he defined as participation in Christian worship and
a Christian theology of life after death. Leuba discovered that scientists were less
likely than the general population to believe in God. In particular, the most successful
scientists were the least likely to be religiously involved. An implication of Leuba's

findings was that religion ought to completely capitulate to science in order to remain an influence on American society, a sentiment that very much reflects the twentieth-century modernist project

. . . Although American society did not experience the widespread decline in traditional forms of religiosity that Leuba predicted, later research did support Leuba's assertions about the differences between scientists and the general population. . . . Although there are exceptions, the general tenor of older research on the religious beliefs of scientists, however, supports the perception there is a conflict between the principles of religion and those of science, such that those who pursue science tend to abandon religion, either because of an inherent conflict between knowledge claims or because scientific education exerts a secularizing force.

Data and Methods

The data we examine were part of a broader study of religion, spirituality, and ethics among academics in seven different natural and social science disciplines at twenty-one elite U.S. research universities. Faculty members included in the study were randomly selected from seven natural and social science disciplines at universities that appeared on the University of Florida's annual report of the "Top American Research Universities." The University of Florida ranked elite institutions according to nine different measures, which included: total research funding, federal research funding, endowment assets, annual giving, number of national academy members, faculty awards, doctorates granted, postdoctoral appointees, and median SAT scores for undergraduates. . . .

During a seven-week period from May through June 2005, the study's PI [Principal Investigator] randomly selected 2,198 faculty members in the disciplines of physics, chemistry, biology, sociology, economics, political science, and psychology from the universities in the sample. Although faculty were randomly selected, oversampling occurred in the smaller fields and undersampling in the larger fields. For example, a little more than 62 percent of all sociologists in the sampling frame were selected, while only 29 percent of physicists and biologists were selected, reflecting the greater numerical presence of physicists and biologists at these universities when compared to sociologists. In analyses where discipline was not controlled for, we used data weights to correct for the over/under sampling. Table 1 describes the sample and weighting in greater detail. [Table omitted.]

An initial contact letter was written by the PI and contained a fifteen-dollar cash preincentive (i.e., the PI and research team sent fifteen dollars in cash to each of the potential respondents regardless of whether they decided to participate in the survey). Each respondent received a unique ID with which to log into a Web site and complete the survey. After five reminder e-mails the research firm commissioned to field the survey, Schulman, Ronca, and Bucuvalas Inc. (SRBI), called respondents up to a total of twenty times, requesting participation over the phone or Web. Six and a half percent of the respondents completed the survey on the phone and 93.5 percent completed the Web-based survey. Overall, this combination of methods resulted in a relatively high response rate of 75 percent or 1,646 respondents, ranging from a 68 percent rate for psychologists to a 78 percent rate for biologists. This is a high

Table 14.1
Religious Self-Identification of Natural and Social Scientists Compared
to the General Population

Current affiliation	Scientists	U.S. population
Evangelicals/fundamentalist	1.5	13.6
Mainline Protestant	2.9	9.5
Liberal Protestant	10.8	9.9
Other/no Protestant identification	1.5	21.4
Traditional Catholic	.7	6.9
Moderate Catholic	1.7	7.4
Liberal Catholic	6.2	7.0
Other/no Catholic identification	.1	3.9
Jewish	15.3	1.8
Buddhist	1.8	.3
Hindu	1.0	.2
Muslim	.5	.5
Eastern Orthodox	.8	.4
Other Eastern	.2	.1
Other	3.0	3.0
None	51.8	14.2
Affiliation as a child		
Protestant	39.0	54
Catholic	22.6	31
Jewish	18.5	2.2
Other	6.5	4.6
None	13.4	8.3

response rate for a survey of faculty. For example, even the highly successful Carnegie Commission study of faculty resulted in only a 59.8 percent rate.

Findings

Distinctions: Scientists and the General Public

Table 4 compares the general U.S. population, according to questions from the GSS, to the population of academic scientists who participated in the RAAS study. While nearly 14 percent of the U.S. population self-describes as "evangelical" or "fundamentalist," less than 2 percent of the RAAS population identifies with either of these combined labels. The only traditional religious identity category where the RAAS population has a much higher proportion of religious adherents than the general population is among those who identify as Jewish. While out of all participants in the GSS, a little less than two percent of the respondents identify as Jewish, about 15 percent of the academic scientists identify as Jewish. Based on our other findings it is likely that many of those who identify as Jewish in the RAAS survey would also identify as reformed or liberal Jews rather than with the conservative or Orthodox traditions.

... Most salient, about 52 percent of the scientists see themselves as having no religious affiliation when compared to about 14 percent of the entire GSS population. These results seem to confirm other research, which shows that a much smaller proportion of academics identify with traditional religious identity categories when compared to the general U.S. population.

Demographics: Gender, Generation, and Family Effect

... The academy is not a homogenous institution, nor is its heterogeneity captured purely by disciplinary differences. Since gender, age, family characteristics, and immigrant generation are all important predictors of religiosity in the general population, we begin by examining how these factors work among this population of academic scientists.

We find some surprising results when looking at the role of gender and age. Although a variety of scholarly reasons are provided for gender differences in religious belief and commitment, there is almost universal agreement that women tend to be more religious than men. When examining differences in religiosity between men and women in the sciences, however, we find that gender is not a significant predictor for any of our religion measures. That is, academic scientists who are women, when compared to men, are not significantly more likely to believe in God, believe that religion provides truth, or to regularly attend a house of worship.

Similarly, although data from the GSS reveal that older individuals express higher levels of religious belief and practice when compared to younger individuals, we do not find this relationship among scientists. The age predictors have no significant effect on whether scientists say there is little truth in religion. We do note, however, that the middle age cohorts are more likely than the youngest scientists (eighteen to thirty-five) to say that they do not believe in God. A similar pattern is seen with the attendance measure, with those fifty-six to sixty-five more likely than the youngest group to report not attending religious services over the past year.

Family status is also a significant predictor of religiosity in the general population, with married individuals who have children more likely to attend a house of worship than those who are childless and unmarried. Indeed, we find that, unlike the gender and age predictors, most of the family characteristics generate the same effect among scientists as they do among the general population. When partnership status is added to the model, the "cohabiting without being married" is a significant predictor of views toward religion. Those who cohabit are more likely than married scientists to believe there is little truth in religion. None of the partner status categories, including cohabitation, however, are significant predictors of religious attendance or having a belief in God. We also find that scientists with more children are less likely to believe there is little truth in religion and less likely to go a year without attending religious services.

Recent research on religiosity among the foreign-born based on the New Immigrant Survey Pilot finds that a lower percentage of immigrants, when compared to those in the general population, practice some form of religion. Similar to findings from the New Immigrant Survey, analysis of the RAAS data reveals that foreign- born scientists are more likely to say that there is little truth in religion and less likely to attend religious services. Being foreign-born had no significant impact on the odds of believing in God, however. These findings are particularly interesting in light of the high number of foreign-born in the data set.

Childhood Religiosity

. . . Measures of childhood religiosity, the final set of predictors seen in Table 4, provide greater potential for understanding differences between academic scientists and the general population. We added two measures of childhood religiosity or upbringing. The first is a set of predictors representing the tradition in which the respondent was raised and the second is the scientists' reported "importance of religion in your family while you were growing up." The latter is measured by four responses ranging from "very important" to "not at all important." We find that scientists raised as Protestants are more likely to retain religious beliefs and practices than those raised without a religious affiliation. Similarly, those who say that religion was important in their family when growing up are less likely to say that they currently do not see truth in religion, do not believe in God, and do not attend religious services.

Another way to examine the impact of religious upbringing is through predicted probabilities. For instance, consider two sociologists who are male, in the eighteen to thirty-five range, born in the United States, have no children, and are currently married. One was raised some form of Protestant and religion was "very important" while growing up. The other was raised as a religious "none" and religion was "not at all important" while growing up. The former has a predicted probability of 14 percent for saying that he does not believe in God. This compares to a 54 percent chance of the latter saying he does not believe, a striking difference. Such differences do not offer conclusive evidence about the causes of disproportionate self-selection of scientists from certain religious backgrounds into the scientific disciplines. They do offer potential theoretical pathways for explaining the differences in religiosity between scientists and the general population.

These findings become even more important when reexamining Table 4, which compares the religious upbringing of the general population with that of scientists. We find that, overall, when compared to the general population, a larger proportion of scientists were raised in liberal Jewish or non-religious homes. When one considers that scientists come disproportionately from non-religious or religiously liberal backgrounds, the distinctions between the general population and the scientific community make more sense. These data reveal that at least some part of the difference in religiosity between scientists and the general population is likely due simply to religious upbringing rather than scientific training or institutional pressure to be irreligious.

Discussion and Conclusions

It is an assumption of much scholarly work that the religious beliefs of scientists are a function of their commitment to science. The findings presented here show that indeed academics in the natural and social sciences at elite research universities are less religious than many of those in the general public, at least according to traditional indicators of religiosity. Assuming, however, that becoming a scientist necessarily leads to loss of religious commitments is untenable when we take into account the differential selection of scientists from certain religious backgrounds. Our results indicate that people from certain backgrounds (the non-religious, for example) disproportionately self-select into scientific professions. In contrast, being raised a Protestant and in a home where religion was very important, for example, leads to a greater likelihood that a scientist will remain relatively religious.

We found that the oft-discussed distinction between the natural and social science fields was inconsistent and weak. [Note that data are not presented in this excerpt; students should see original article for supporting data and analysis.] Common predictors of religiosity in the general population operated quite differently among academic scientists. Although gender was a significant predictor of religiosity in the general population, with women more likely than men to be religious, in the RAAS population gender was not a significant predictor of religiosity. Although such assessments are made cautiously, given that the GSS data and the RAAS data were collected at different points in time and with different methodologies (face-to-face interview [GSS] versus Web and phone based [RAAS]), these analyses make a strong case that future research ought to examine the possible self-selection effect operating among women in the natural and social sciences, with women who go into the natural or social science disciplines less likely to be religious when compared to women in the general population.

. . . Finding that the strongest predictor of religious adherence among this group was childhood religiosity recasts previous theories about lack of religiosity among academic scientists in a new light. The idea that scientists simply drop their religious identities upon professional training, whether due to an inherent conflict between science and faith or institutional pressure, is not strongly supported by these data. If this was the case, then religious upbringing would have little effect on religion among scientists, with even those scientists who were raised in religious homes losing religion once they entered the academy or received scientific training. Instead, as shown by our results in Table 5, [not presented in this excerpt] religious socialization and heritage remains the strongest predictor of present religiosity among this population of scientists.

. . . Of course, this obviously raises the question of why scientists are self-selected from nonreligious households or households where religion does not play a major role. This is a question that will need further exploration beyond the data presented here. We will, however, provide some possible routes of inquiry. In some cases this selection effect may indeed be due to the tension between the religious tenets of some groups (e.g., those that advocate young earth creationism) and the theories and methods of particular sciences. On the other hand, some of the selection effect may simply be due to differential emphasis on education and/or differential resources. These could be mediating factors between religious background and likelihood of becoming a scientist or independent from religious background. The possibility of such mediating factors reveals that the story is unlikely to be the simple one of "religion is contradictory to science and hence religious individuals do not go into science." Finding that scientists who are raised religious often stay relatively religious casts doubt on this simple cause and effect scenario.

Another possibility is that religious individuals might select into science graduate programs equally but that the graduate programs and scientific environments themselves have strong anti-religious messages and reward structures, either passive or active, such that some abandon their faith in the process and others leave programs. To study the previous we would need data not just on faculty at elite institutions but a data collection including a broader set of individuals in the academic sciences (graduate students, postdoctoral fellows, researchers, in addition to faculty) as well as the ability to follow these same individuals over time.

The population of academic scientists surveyed for the RAAS study comes from a selection of the most elite research universities in the United States. It may be that academic scientists at elite research universities are significantly different than those at other kinds of research universities in ways that would influence religiosity. One possibility is that there is a kind of pressure towards ir-religiosity as a marker of legitimacy, which is found at elite institutions but not at other kinds of institutions. If future research finds evidence that faculty who received undergraduate degrees from religiously-based colleges, for example, are discriminated against in the hiring process more at elite universities when compared to less elite institutions, this would support a theory that takes into account institutional differences in elite status as a correlate of faculty religiosity.

We expect that students, at least undergraduate students at the universities studied, will be significantly more religious than faculty at these schools. Research on education in the general population reveals that a four-year college or university degree is becoming ever more common and that overall the religious are as educated as the general population, although some forms of religion tend to impede educational advancement.

These findings also have implications for the current social issues related to public discussion about the connection between religion and science that motivated these analyses. In the wake of recent public events about teaching intelligent design and evolution in public schools, we see the necessity of increasing dialogue between scientists and the general public as important and timely. . . . Our findings also reveal there is clearly a sizeable minority of academic scientists who are committed religious adherents and thus potentially crucial commentators in the context of an American public trying to find a way to meaningfully connect religion and science. That the scientists in this population are from elite universities makes them all the more poised to productively contribute to significant dialogue about what distinguishes scientific and religious claims. [References omitted.]

Excerpted from Elaine Howard Ecklund and Christopher P. Scheitle, Religion among academic scientists: Distinctions, disciplines, and demographics. *Social Problems* 54, no. 2 (2007): 289–307 (from 290–304). Used with permission.

Evolution and Creationism in America's Classrooms: A National Portrait

Teaching Evolution: Law, Policy, and Practice

Community pressures place significant stress on teachers as they try to teach evolution, stresses that can lead them to de-emphasize, downplay, or ignore the topic (Griffith and Brem 2004). This is particularly true of the many teachers who lack a full understanding of evolution, or at least confidence in their knowledge of it. Such a lack of confidence can lead teachers to avoid confrontations with students, parents, and the wider community. They may, for example, not stress evolution as the classes' organizing principle, or may avoid effective hands-on activity to teach it, or not ask students to apply natural selection to real-life situations (Donnelly and Boone 2007). There are many reasons to believe that scientists are winning in the courts, but losing in the classroom. This is partially due to the occasional explicit teaching of creationism and ID, but most especially because of inconsistent emphasis in the minimal rigor in the teaching of evolution.

Studies of science teachers seem to confirm these fears by suggesting "that instruction in evolutionary biology at the high school level has been absent, cursory or fraught with misinformation" (Rutledge and Mitchell 2002). But we are wary of this conclusion. Most of the previous studies are now dated; the recent ones each examine a single state, and many states (most notably California, New York, and all of New England) have never been studied [references omitted]. Collectively, the studies employ incomparable measures, and some of them sacrifice scientific sample survey methods in favor of higher cooperation rates (such as surveys of teachers attending conventions and professional meetings). As a result, we lack a systematic and coherent account of how instruction varies from teacher to teacher across the nation as a whole. To remedy this, we provide a statistical portrait of evolution and creationism in America's classrooms, from which we draw conclusions about the unevenness of how evolutionary biology is taught and some of the causes of that variation.

The National Survey of High School Biology Teachers

We advance this long tradition of surveying teachers with reports from the first nationally representative survey of teachers concerning the teaching of evolution. The survey permits a statistically valid and current portrait of US science teachers that complements US and international surveys of the general public on evolution and scientific literacy and on evolution in the classroom (references omitted). Between March 5 and May 1, 2007, 939 teachers participated in the study, either by mail or by completing an identical questionnaire online. Our overall response rate of 48 percent yielded a sample that may be generalized to the population of all public school teachers who taught a high school level biology course in the 2006–2007 academic year, with all percentage estimates reported in this essay's tables and figures having a margin of error of no more than 3.2 percent at the 95 percent confidence level. Detailed discussion of the methods of the survey and assessments of non-response can be found in Text S1 (data not presented). Our results confirm wide variance in classroom instruction and indicate a clear need to focus not only on state and federal policy decisions, but on the everyday instruction in American classrooms.

Evolution in the classroom

How much time should be spent on evolution in the typical high school biology class? There is no clear answer to this question. Neither the strongest nor the weakest state standards specify a precise amount of time that should be spent on any particular topic. As we noted above, there are three widely circulated documents that serve as guidelines at the national level, but these, too, refrain from offering directions on the amount of time that should be spent on evolution relative to other topics. In general, these national reports and state standards offer ideas for the content of high school science, biology, and life science classes, but not the curriculum; in other words, they enumerate and elaborate on outcomes—what students should learn—but not on any particular ordering or allocation of time for each subject.

It is clear, however, that all three of these reports expect and recommend a substantial investment in evolutionary biology and evolution-related topics.

... We followed most previous studies in asking teachers to think about how they allocate time over the course of the school year. We went a step further in also asking whether evolution serves as a unifying theme for the content of the course. Over the entire year of high school biology we found substantial variation among America's high school teachers [Table omitted]. Not surprisingly, we found that those who take most seriously the advice of NSES [the National Science Education Standards] to make evolution a unifying theme spent the most time on evolution. Overall, teachers devoted an average of 13.7 hours to general evolutionary processes (including human evolution), with 59 percent allocating between three and 15 hours of class time (table omitted). Only 2 percent excluded evolution entirely. But significantly fewer teachers covered human evolution, which is not included as an NSES benchmark. Of teachers surveyed, 17 percent did not cover human evolution at all in their biology class, while a majority of teachers (60 percent) spent between one and five hours of class time on it.

Those teachers who stressed evolution by making it the unifying theme of their course spent more time on it. Overall, only 23 percent strongly agreed that evolution served as the unifying theme for their biology or life sciences course [table omitted]; and these teachers devoted 18.5 hours to evolution, 50 percent more class time than other teachers. When we asked whether an excellent biology course could exist without mentioning Darwin or evolutionary theory at all, 13 percent of teachers agreed or strongly agreed that such a course could exist.

Creationism in the classroom

We also asked teachers whether they spent classroom time on creationism or intelligent design. We found that 25 percent of teachers indicated that they devoted at least one or two classroom hours to creationism or intelligent design. However, those numbers can be misleading because while some teachers may cover creationism to expose students to an alternative to evolutionary theory, others may bring up creationism in order to criticize it or in response to student inquiries. Questions that simply ask about time devoted to creationism, therefore, will overstate support for creationism or intelligent design by counting both those who teach creationism as a serious subject and those holding it up for criticism or ridicule. We asked a series of supplemental questions that provided some additional insight into the character of creationism in the classroom. Of the 25 percent of teachers who devoted time to creationism or intelligent design, nearly half agreed or strongly agreed that they teach creationism as a "valid scientific alternative to Darwinian explanations for the origin of species." Nearly the same number agreed or strongly agreed that when they teach creationism or intelligent design they emphasize that "many reputable scientists view these as valid alternatives to Darwinian Theory." [Data not presented.]

Figure 14.7
Teachers have a lower percentage of rejection of evolution than the general public, although there remains substantial sympathy for Young[-Earth] Creationism.

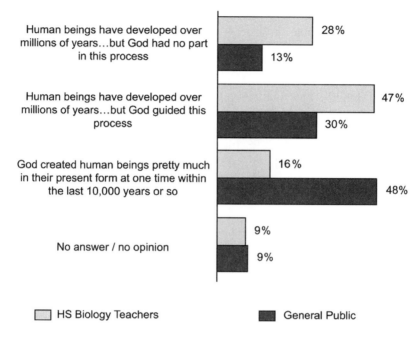

Human beings have developed over millions of years...but God had no part in this process — 28% / 13%

Human beings have developed over millions of years...but God guided this process — 47% / 30%

God created human beings pretty much in their present form at one time within the last 10,000 years or so — 16% / 48%

No answer / no opinion — 9% / 9%

☐ HS Biology Teachers   ■ General Public

On the other hand, many teachers devoted time to creationism either to emphasize that religious theories have no place in the science classroom or to challenge the legitimacy of these alternatives. Of those who spent time on the subject, 32 percent agreed or strongly agreed that when they teach creationism they emphasize that almost all scientists reject it as a valid account of the origin of species, and 40 percent agreed or strongly agreed that when they teach creationism they acknowledge it as a valid religious perspective, but one that is inappropriate for a science class.

Explaining differences and teachers['] emphasis

Why do some teachers spend so much more time on evolution than others? Our data weigh heavily against one possible explanation: differences in state standards. We find that nearly 90 percent of cross-teacher variation is within states (eta-square from a one-way analysis of variance by state is 0.11) as opposed to between states. As an upper limit, then, state standards cannot account for more than 11 percent of the variance.

However, our data lend support to potential explanations: teachers' personal beliefs about evolution and the number of college-level science classes.

Our teachers were each asked a question about their own personal beliefs about human origins. This question is identical to a question that major polling organizations have asked members of the general public since 1981. Figure 14.7 compares the results of our sample of teachers surveyed during March and April of 2007 with the results of a public opinion poll conducted for *Newsweek* on March 28–29 of 2007. Among the biology teachers, 16 percent believe that human beings were created by God in their present form at one time within the last ten thousand years (and an additional 9 percent decline to answer). Although this is a far smaller proportion than found among the general public (48 percent), our data demonstrate substantial sympathy for the "young [E]arth" creationist position among nearly one in six members of the science teaching profession. The teachers who choose the "young earth" creationist position the voted 35 percent fewer class hours to evolution than all other teachers. [Data not presented.]

Teacher qualifications

The No Child Left Behind Act requires that all teachers of core subjects be "highly qualified." Definitions of "highly qualified" vary by state, but most include demonstrated competence in the teacher's teaching assignments. Our data suggest that high school teachers who completed the largest number of college-level credits in biology and life science classes and whose coursework included at least one class in evolutionary biology devote substantially more class time to evolution and teachers with fewer credit hours [data not presented]. The best prepared teachers devote 60 percent more time to evolution than the least prepared.

Evolution in the Classroom? It's about the Teachers

Our survey of biology teachers is the first nationally representative, scientific sample survey to examine evolution and creationism in the classroom. Three different survey questions all suggest that between 12 percent and 16 percent of the nation's biology teachers are creationist in orientation. Roughly [one-sixth] of all teachers profess a "young [E]arth" personal belief and about one in eight reported that they teach creationism or intelligent design in a positive light. The number of hours devoted to these alternative theories is typically low—but this nevertheless must surely convey to students that these theories should be accorded respect as scientific perspectives.

The majority of teachers, however, see evolution as central and essential to high school biology courses. Yet the amount of time devoted to evolutionary biology varies substantially from teacher to teacher, and a majority either avoid human evolution altogether or devote only one or two class periods to the topic. We showed that some of these differences were due to personal beliefs about human origins. However, an equally important factor is the science education the teacher received while in college. Additional variance is likely to be rooted in pressures—subtle or otherwise—emerging from parents and community leaders in each school's community, in combination with teacher's confidence in their ability to deal with such pressures given their knowledge of evolution, as well as their personal beliefs.

These findings strongly suggest that victory in the courts is not enough for the scientific community to ensure that evolution is included in high school science courses. Nor is success in persuading states to adopt rigorous content standards consistent with recommendations of the National Academy of Sciences and other scientific organizations. Scientists concerned about the quality of evolution instruction might have a bigger impact in the classroom by focusing on the certification standards for high school biology teachers. Our study suggests that requiring all teachers to complete a course in evolutionary biology would have a substantial impact on the emphasis on evolution and its centrality in high school biology courses. In the long run, the impact of such change would have a more far-reaching effects of the victories in courts and in state governments.

REFERENCES

Donnelly L., and W. J. Boone. 2007. Biology teachers' attitudes toward and use of Indiana's evolution standards. *Journal of Research in Science Teaching* 44: 236–257.

Griffith J., and S. Brem. 2004. Teaching evolutionary biology: Pressures, stress, and coping. *Journal of Research in Science Teaching* 41: 791–809.

Rutledge, M. L. and M. A. Mitchell. 2002. Knowledge structure, acceptance and teaching of evolution. *American Biology Teacher* 64: 21–27.

Excerpted from Michael B. Berkman, Julianna Sandell Pacheco, and Eric Plutzer, Evolution and creationism in America's classrooms: A national portrait. *PLoS Biology* 6, no. 5 (2008): e124 foiz;10.1371/journal.pb io.0060124 (from 1–4).

References for Further
Exploration

I hope that you have found the creationism/evolution controversy interesting enough to want to learn more about it than is possible in this small book. Or you may wish to write a research paper on one of the many topics touched on in the previous chapters. This section is for you! But be sure to first check the references listed at the end of chapters 1–14; they will not be repeated here.

The topics covered in this section parallel the content of the preceding fourteen chapters. Both Web sites and printed material (books and articles) can be found, as well as occasional references to video sources. The effort has been made to provide sources that are more easily available in libraries or on the Internet. There are occasional references to scientific journals; these are most easily found in a college or university library. Public libraries may have access to online versions of these journals, but this access will vary from library to library.

First a disclaimer: neither I nor the publisher vouches for the accuracy of the information in the following books, articles, videotapes, and Web sites. Because positions taken in these resources may be diametrically opposed to one another, not all of them can be correct and accurate; readers are encouraged to read all sources critically. Similarly, references to a person's or an organization's published work should not be taken as agreement with or endorsement of any of the positions taken by that person or organization.

Most of the books selected were in print or about to be published as of April 2008. All of the Web pages were current as of April 2008: although these Web pages were chosen in part because they are likely to be stable, the Internet is notoriously changeable, and neither I nor the publisher can guarantee that any particular Web site will continue to exist.

Abbreviations used:
YEC—young-Earth creationism
OEC—old-Earth creationism
IDC—intelligent design creationism
NCSE—National Center for Science Education   www.ncseweb.org
NAS—National Academy of Sciences   www.nationalacademies.org
AAAS—American Association for the   www.aaas.org
    Advancement of Science
ICR—Institute for Creation Research (YEC)   www.icr.org
AIG—Answers in Genesis (YEC)   www.answersingenesis.org
CRS—Creation Research Society (YEC)   www.creationresearch.org
CMI—Creation Ministries International (YEC)   www.creationontheweb.org
RTB—Reasons to Believe (OEC)   www.reasons.org
ARN—Access Research Network (IDC)   www.arn.org
CSC—Center for Science and Culture (IDC)   www.discovery.org/csc

CHAPTER 1: SCIENCE: TRUTH WITHOUT CERTAINTY

John A. Moore's *Science as a Way of Knowing* (Cambridge, MA: Harvard University Press, 1999) discusses science as a way of knowing in the context of the history of modern biology; Elliott Sober's *Philosophy of Biology*, 2nd ed. (Boulder, CO: Westview Press, 2000), is probably the most accessible textbook on the philosophy of biology. There are many books that discuss the nature of scientific reasoning; a popular introductory text is Ronald N. Giere's *Understanding Scientific Reasoning*, 5th ed. (Belmont, CA: Wadsworth, 2005). The AAAS publication *Science for All Americans* (New York: Oxford University Press, 1989) is intended to promote scientific literacy in general; chapter 1 deals with the nature of science. A book for teachers from the NAS, *Teaching about Evolution and the Nature of Science* (Washington, D.C.: National Academy Press, 1998; available at http://www.nap.edu/html/evolution98), discusses the nature of science in the particular context of evolution; chapters 3 and 5—"Evolution and the Nature of Science" and "Frequently Asked Questions about Evolution and the Nature of Science"—are particularly relevant. The book *Science, Evolution, and Creationism* (Washington, D.C.: National Academy Press, 2008; available at http://www.nap.edu/sec), also from the NAS, and chapters 2–4 of Philip Kitcher's classic *Abusing Science: The Case against Creationism* (Cambridge, MA: MIT Press, 1982) are also helpful.

CHAPTER 2: EVOLUTION

A brief treatment of astronomical evolution is Andrew Fraknoi, George Greenstein, Bruce Partridge, and John Percy's "An Ancient Universe: How Astronomers Know the Vast Scale of Cosmic Time," *The Universe in the Classroom* 56 (2001): 1–23 (available at http://www.astrosociety.org/education/publications/tnl/56/). Richard Fortey's *Earth: An Intimate History* (New York: Harper Perennial, 2005) is a highly readable introduction to geology in general; G. Brent Dalrymple's *Ancient Earth, Ancient Skies* (Stanford, CA: Stanford University Press, 2004) explains how scientists ascertain the age of the earth. For origin-of-life research, see Christopher

Wills and Jeffrey Bada's *The Spark of Life* (New York: Oxford University Press, 2000); and Jeffrey L. Bada and Antonio Lazcano's "Prebiotic Soup: Revisiting the Miller Experiment," *Science* 300, no. 5620 (2003): 745–746. Richard Fortey's *Life: A Natural History of the First Four Billion Years of Life on Earth* (New York: Knopf, 1998) and Colin Tudge's *The Variety of Life: A Survey and a Celebration of All the Creatures That Have Ever Lived* (Oxford: Oxford University Press, 2000) are two impressive attempts to survey the history of life. On the Web, the Tree of Life Web Project (http://tolweb.org/tree/phylogeny.html) endeavors to chart the phylogenetic tree; for paleontology and evolutionary biology, see the American Museum of Natural History's Division of Paleontology (http://paleo.amnh.org/fossil/FRC.frontdoor) and the University of California Museum of Paleontology (http://www.ucmp.berkeley. edu).

For both evolution and its mechanisms, the video series *Evolution*, which aired on PBS in 2001, is excellent, as are its companion volume, Carl Zimmer's *Evolution: The Triumph of an Idea*, rev. ed. (New York: Harper Perennial, 2006) and its companion Web site (http://www.pbs.org/wgbh/evolution/). Michael Ruse's *The Evolution Wars: A Guide to the Debates* (New Brunswick, NJ: Rutgers University Press, 2001) discusses controversies in (and about) evolution throughout its history. The Hawaiian honeycreeper is a fine example of adaptive radiation; see H. Douglas Pratt's *The Hawaiian Honeycreepers* (Oxford: Oxford University Press, 2004). A classic telling of the most famous adaptive radiation story, the Galápagos Islands finches, is Jonathan Weiner's Pulitzer Prize–winning *The Beak of the Finch: A Story of Evolution in Our Time* (New York: Knopf, 1994). A readable and interesting book on how the human body reflects evolution is Neil Shubin's *Your Inner Fish: A Journey into the 3.5-Billion-Year History of the Human Body* (New York: Pantheon Books, 2008).

For human evolution, Donald Johanson and Blake Edgar's *From Lucy to Language*, rev. ed. (New York: Simon & Schuster, 2006), is not only useful but also beautifully illustrated, and the Becoming Human Web site (http://www.becominghuman.org) is worth a visit. The science writer Carl Zimmer writes eloquently about evolution on his blog, http://blogs.discovermagazine.com/loom/.

CHAPTER 3: BELIEFS: RELIGION, CREATIONISM, AND NATURALISM

A recent introductory text on science and religion is Alister E. McGrath's *Science and Religion: An Introduction* (Oxford: Blackwell, 1998). For a scholarly treatment of the creation/evolution continuum, see Eugenie C. Scott, "Antievolutionism and Creationism in the United States," *Annual Review of Anthropology* 26 (1997): 263–289; for a briefer but more accessible treatment, see Eugenie C. Scott, "The Creation–Evolution Continuum," *Skeptic* 10, no. 4 (2000): 50–54. Ample resources promoting the YEC position are available on the Web sites of AIG, CMI, the CRS, and the ICR; on the ICR Web sites, see especially the online version of the *Impact* pamphlets. *Scientific Creationism*, edited by Henry M. Morris (San Diego: Creation-Life Publishers, 1974) is a central YEC text; Henry M. Morris and Gary E. Parker's *What Is Creation Science?* rev. ed. (Green Forest, AR: Master Books, 1987), is a polished presentation.

The TalkOrigins Archive Web site (www.talkorigins.org) contains many refutations of YEC positions; a good place to begin there is Mark Isaak's "An Index to Creationist Claims" (http://www.talkorigins.org/indexcc/index.html), a version of which was published as *The Counter-Creationism Handbook* (Berkeley: University of California Press, 2006). The most significant OEC organization is RTB; Hugh Ross's *Creation as Science: A Testable Model Approach to End the Creation/Evolution Wars* (Colorado Springs, CO: NavPress, 2006) is a representative book by its president. The major IDC organization is CSC, which provides a lot of material on its Web site. *Signs of Intelligence: Understanding Intelligent Design*, edited by William A. Dembski and James M. Kushiner (Grand Rapids, MI: Brazos Press, 2001), is a collection of essays reflecting religious views associated with IDC. *Perspectives on an Evolving Creation*, edited by Keith B. Miller (Grand Rapids, MI: Eerdmans, 2003) is a collection of essays mostly by Protestant theistic evolutionists; John F. Haught's *Responses to 101 Questions on God and Evolution* (Mahwah, NJ: Paulist Press, 2001), from a Catholic theistic evolutionist, is in a convenient question-and-answer format. Richard Dawkins's *The Blind Watchmaker* (New York: W. W. Norton, 1987) is often regarded as a paradigm of atheistic evolutionism; it is joined by his more recent *The God Delusion* (New York: Houghton Mifflin, 2006).

CHAPTER 4: BEFORE DARWIN TO THE TWENTIETH CENTURY

A classic history of the pre-Darwinian revolution in geology is C. C. Gillispie's *Genesis and Geology* (Cambridge, MA: Harvard University Press, 1996). The best history of the theory of evolution is Peter J. Bowler's *Evolution: The History of an Idea*, 3rd ed. (Berkeley: University of California Press, 2003), but Michael Ruse's *The Darwinian Revolution: Science Red in Tooth and Claw*, 2nd ed. (Chicago: University of Chicago Press, 1999) and *The Evolution Wars: A Guide to the Debates* (New Brunswick, NJ: Rutgers University Press, 2001) are perhaps a better read. A brief introduction in cartoon form to Darwin's life and times is Jonathan Miller and Borin Van Loon's *Darwin for Beginners* (New York: Pantheon Books, 1982). Among the many excellent biographies of Darwin, Adrian Desmond and James Moore's *Darwin: The Life of a Tormented Evolutionist* (New York: Warner Books, 1991) is perhaps the most readable, and Janet Browne's *Charles Darwin: Voyaging* (New York: Knopf, 1995) and *Charles Darwin: The Power of Place* (New York: Knopf, 2002) are perhaps the most complete, but his autobiography (*The Autobiography of Charles Darwin*, edited by Nora Barlow [New York: W. W. Norton, 1958]) ought not to be skipped. On the Web, the companion to the 2001 *Evolution* series (http://www.pbs.org/wgbh/evolution) is useful. *On the Origin of Species* is essential reading, preferably in a facsimile of the first edition such as *On the Origin of Species: A Facsimile of the First Edition* (Cambridge, MA: Harvard University Press, 1966); it is also worthwhile to sample a range of Darwin's work either through an anthology such as Mark Ridley's *The Darwin Reader*, 2nd ed. (New York: W. W. Norton, 1996) or Philip Appleman's *Darwin*, 3rd ed. (New York: W. W. Norton, 2000), or on the Web at the Writings of Charles Darwin Online (http://darwin-online.org.uk/). For the initial responses to *Origin* in the United States, see Jon H. Roberts's *Darwinism and the Divine in America: Protestant Intellectuals and Organic Evolution, 1859–1900* (Madison: University of Wisconsin Press, 1988), David

N. Livingstone's *Darwin's Forgotten Defenders: The Encounter between Evangelical Theology and Evolutionary Thought* (Grand Rapids, MI: Eerdmans, 1987), and Ronald L. Numbers's *Darwinism Comes to America* (Cambridge, MA: Harvard University Press, 1998). A definitive treatment of the idea of design is Michael Ruse's *Darwin and Design: Does Evolution Have a Purpose?* (Cambridge, MA: Harvard University Press, 2003).

CHAPTER 5: ELIMINATING EVOLUTION, INVENTING CREATION SCIENCE

For fundamentalism in general, see George M. Marsden's *Understanding Fundamentalism and Evangelicalism* (Grand Rapids, MI: Eerdmans, 1991). By far the best book on the history of creationism is Ronald L. Numbers's *The Creationists: From Creation Science to Intelligent Design*, expanded ed. (Cambridge, MA: Harvard University Press, 2006). The definitive book on the Scopes trial, by Edward J. Larson, is the Pulitzer Prize–winning *Summer for the Gods: The Scopes Trial and America's Continuing Debate over Science and Religion* (New York: Basic Books, 1997); two useful Web sites on the trial are *Tennessee vs. John Scopes*: The "Monkey Trial" (http://www.law.umkc.edu/faculty/projects/ftrials/scopes/scopes.htm), administered by a law professor at the University of Missouri, in Kansas City, and American Experience: Monkey Trial (http://www.pbs.org/wgbh/amex/monkeytrial), which complements a documentary about the trial. The AIG, CMI, CRS, and ICR all maintain Web sites expounding YEC. For the legal history of the creationism/evolution controversy, which largely drove the development of creationism in the period under discussion, see Edward J. Larson's *Trial and Error: The American Controversy over Creation and Evolution*, 3rd ed. (New York: Oxford University Press, 2003). A useful but out-of-print collection of papers about the *McLean* trial is *Creationism, Science, and the Law*, edited by Marcel Chotkowski La Follette (Cambridge, MA: MIT Press, 1983). Transcripts of the opinions in *Epperson v. Arkansas* and *McLean v. Arkansas*, together with links to related material, are available on the Web at the Talk Origins Archive (www.talkorigins.org); look for "legal decisions" in the index.

CHAPTER 6: NEOCREATIONISM

A transcript of the opinion in *Edwards v. Aguillard*, together with links to related material, are available on the Web at the Talk Origins Archive (www.talkorigins.org); look for "legal decisions" in the index. Edward J. Larson's *Trial and Error: The American Controversy over Creation and Evolution*, 3rd ed. (New York: Oxford University Press, 2003) is excellent on the legal issues. Although Wendell Bird's *The Origin of Species Revised* (New York: Philosophical Library, 1987/1989, 2 vols.) is out of print, a number of articles are available online at the ICR Web site. The Foundation for Thought and Ethics is on the Web at http://www.fteonline.com/. Various articles by Michael Behe and William Dembski are available from CSC and ARN; for criticism, see Kenneth R. Miller's *Finding Darwin's God*, rev. ed. (New York: Harper Perennial, 2007), and *Only a Theory: The Battle for America's Soul* (New York: Penguin Group, 2008), as well as various articles in Robert T. Pennock's collection *Intelligent*

Design Creationism and Its Critics (Cambridge, MA: MIT Press, 2001), and Behe's Empty Box on the Web (http://www.simonyi.ox.ac.uk/dawkins/WorldOfDawkins-archive/Catalano/box/behe.shtml). The latter has not been updated for several years, yet its criticisms of ID are still valid, perhaps sadly for the ID movement. Two other good Web sites are Talk Design (http://www.talkdesign.org), and the Talk Origins Archive (http://www.talkorigins.org). A series of point-counterpoint articles between IDC proponents Michael Behe and William Dembski and their critics appeared in the *Boston Review* (http://www.bostonreview.net/evolution.html). For articles by Phillip Johnson espousing the cultural renewal aspect of intelligent design, browse through his articles available from CSC and ARN; Robert T. Pennock's *Tower of Babel: The Evidence against the New Creationism* (Cambridge, MA: MIT Press, 1999) perceptively criticizes Johnson. Larry Witham's *Where Darwin Meets the Bible: Creationists and Evolutionists in America* (New York: Oxford University Press, 2002) attempts to give the history in a neutral fashion; Barbara Forrest and Paul R. Gross's *Creationism's Trojan Horse: The Wedge of Intelligent Design*, rev. ed. (New York: Oxford University Press, 2007), is a scathing attack.

CHAPTER 7: TESTING INTELLIGENT DESIGN AND EVIDENCE AGAINST EVOLUTION IN THE COURTS

No fewer than four books about the *Kitzmiller v. Dover* trial are available—Matthew Chapman's *Forty Days and Forty Nights: Darwin, Intelligent Design, God, Oxycontin and Other Oddities on Trial in Pennsylvania* (New York: HarperCollins, 2007), Edward Humes's *Monkey Girl: Evolution, Education, Religion, and the Battle for America's Soul* (New York: Ecco, 2007), Lauri Lebo's *The Devil in Dover: An Insider's Story of Dogma v. Darwin in Small-Town America* (New York: New Press, 2008), and Gordy Slack's *The Battle over the Meaning of Everything: Evolution, Intelligent Design, and a School Board in Dover, PA* (San Francisco: Jossey-Bass, 2007)—to say nothing of the two-hour *Nova* documentary *Judgment Day: Intelligent Design on Trial* (WGBH/*Nova*, 2007, 112 minutes; see http://www.pbs.org/wgbh/nova/id). Transcripts, exhibits, and the decision are available at the NCSE's Web site (http:www.ncseweb.org/creationism/legal/creationism-law). The book at the center of the case was by Percival W. Davis and Dean H. Kenyon, *Of Pandas and People*, 2nd ed. (Dallas, TX: Haughton, 1993); a collection of critiques and analyses is available from NCSE (http://www.ncseweb.org/creationism/analysis/critique-pandas-people). Information about *Selman v. Cobb County* and *Hurst v. Newman* is available from NCSE.

CHAPTER 8: COSMOLOGY, ASTRONOMY, GEOLOGY

For YEC perspectives on these topics, consult *Scientific Creationism*, edited by Henry M. Morris (San Diego: Creation-Life Publishers, 1974), Henry M. Morris and Gary E. Parker's *What Is Creation Science?* rev. ed. (Green Forest, AR: Master Books, 1987), and various articles available from AIG, CMI, CRS, and ICR; it is instructive to compare these with OEC productions such as Hugh Ross's *Creation and Time* (Colorado Springs, CO: NavPress 1994) and various articles available from Reasons to Believe (www.reasons.org) that are closer to mainstream science. Victor J. Stenger's *Has Science Found God?* (Amherst, NY: Prometheus Books, 2003) is skeptical about the OEC interpretations of cosmology. There is no book that specifically aims to refute YEC views about cosmology and astronomy, although Philip Plait's *Bad Astronomy* (New York: Wiley, 2002) devotes a chapter to doing so; Arthur N. Strahler's *Science and Earth History* (Amherst, NY: Prometheus Books, 1999) does an admirable job of refuting YEC views not only about geology but about other disciplines as well. Mark Isaak's "An Index to Creationist Claims" (http://www.talkorigins.org/indexcc/index.html), a version of which was published as *The Counter-Creationism Handbook* (Berkeley: University of California Press, 2006), lists many YEC and OEC cosmological arguments, as well as the scientific response. G. Brent Dalrymple's *Ancient Earth, Ancient Skies* (Stanford, CA: Stanford University Press, 2004) explains how scientists ascertain the age of Earth. Andrew Fraknoi, George Greenstein, Bruce Partridge, and John Percy's "An Ancient Universe: How Astronomers Know the Vast Scale of Cosmic Time," *The Universe in the Classroom* 56 (2001): 1–23 (available at http://www.astrosociety.org/education/publications/tnl/56/) is good on astronomy, and on the Web, the Talk Origins Archive (http://www.talkorigins.org) is a reliable source of accessible refutations of creationism claims about cosmology, astronomy, and geology.

CHAPTER 9: PATTERNS AND PROCESSES OF BIOLOGICAL EVOLUTION

For creationist views on biology, consult Duane T. Gish's *Evolution: The Fossils Still Say No!* (San Diego: ICR, 1985), Percival W. Davis and Dean H. Kenyon, *Of Pandas and People*, 2nd ed. (Dallas, TX: Haughton, 1993); and (for human evolution in particular) Marvin L. Lubenow's *Bones of Contention: A Creationist Assessment of Human Fossils*, rev. ed. (Grand Rapids, MI: Baker Book House, 2004); on the Web, there are various articles available from AIG, CMI, CRS, and ICR. Many of the readings for chapters 3 and 6 would provide useful background for understanding these views. For refutations from the point of view of mainstream science, Tim M. Berra's *Evolution and the Myth of Creationism* (Stanford, CA: Stanford University Press, 1990) and Philip Kitcher's *Living with Darwin* (New York: Oxford University Press, 2006) are basic and readable introductions. Philip Kitcher's *Abusing Science: The Case against Creationism* (Cambridge, MA: MIT Press, 1982) and Douglas J. Futuyma's *Science on Trial: The Case for Evolution*, rev. ed. (Sunderland, MA: Sinauer, 1995) are detailed refutations of YEC; *Why Intelligent Design Fails*, edited by Matt Young and Taner Edis (New Brunswick, NJ: Rutgers University Press, 2007) and Sahotra Sarkar's *Doubting Darwin?* (Malden, MA: Blackwell 2007) provide detailed refutations of IDC. On the Web, the TalkOrigins Archive (http://www.talkorigins.org) is a reliable source of accessible refutations of creationism claims about biology; the section on fossil hominids is especially good for human evolution. Many of the readings for chapters 1, 2, and 4 provide useful background for understanding these refutations.

CHAPTER 10: LEGAL ISSUES

The definitive legal history of the creationism/evolution controversy is Edward J. Larson's *Trial and Error: The American Controversy over Creation and Evolution*, 3rd ed. (New York: Oxford University Press, 2003). Randy Moore's *Evolution in the*

Courtroom (Santa Barbara, CA: ABC/Clio, 2001) is a legal history by a professor of biology who served as the editor of *The American Biology Teacher*. For recent developments, see the readings suggested for chapter 7. As always, the TalkOrigins Archive (http://www.talkorigins.org) has useful information and links; look in the index under "legal decisions." Although Wendell Bird's *The Origin of Species Revised* (New York: Philosophical Library, 1987/1989, 2 vols.) is out of print, a number of Bird's articles addressing the legal issues are available online from ICR. Francis J. Beckwith's *Law, Darwinism and Public Education* (Lanham, MD: Rowman & Littlefield, 2003) argues that it would not be unconstitutional to teach IDC in the public schools; a number of articles by Beckwith, David DeWolf, and their colleagues addressing the legal issues are available from CSC. For opposing views, see Jay Wexler, "Of Pandas, People, and the First Amendment," *Stanford Law Review* 49 (1997): 439–470; Matthew Brauer, Steven Gey, and Barbara Forrest, "Is It Science Yet? Intelligent Design Creationism and the Constitution," *Washington University Law Review* 83, no. 1 (2005): 1–149; and the decision in *Kitzmiller v. Dover*. The Santorum language is discussed by Glenn Branch and Eugenie C. Scott in "The Anti-Evolution Law That Wasn't" (*The American Biology Teacher* 65, no. 3 [2003]: 165–166).

CHAPTER 11: EDUCATIONAL ISSUES

For the creationist point of view, consult Percival W. Davis and Dean H. Kenyon, *Of Pandas and People*, 2nd ed. (Dallas, TX: Haughton, 1993) as well as a number of articles available online from CSC (http://www.discovery.org/csc/) and ICR. Jonathan Wells's *Icons of Evolution: Science or Myth?* (Washington, D.C.: Regnery, 2000) and articles by Wells that are available online at the CSC Web site (http://www.discovery.org/csc/) criticize biology textbooks; extensive responses are available online at NCSE (http://www.ncseweb.org/creationism/analysis/icons-evolution) and the TalkOrigins Archive (http://www.talkorigins.org). Statements supporting the teaching of evolution and opposing the teaching of creationism are collected in *Voices for Evolution*, 3rd ed., edited by Carrie Sager (Berkeley, CA: NCSE, 2008); section 9 of Robert T. Pennock's collection *Intelligent Design Creationism and Its Critics* (Cambridge, MA: MIT Press, 2001) contains a debate between two philosophers on the topic. For teachers, Brian J. Alters and Sandra M. Alters's *Defending Evolution: A Guide to the Evolution/Creation Controversy* (Sudbury, MA: Jones and Bartlett, 2001) is a useful discussion of the challenges of teaching evolution at the high school level; on the Web, the Evolution and the Nature of Science Institutes (http://www.indiana.edu/~ensiweb/home.html) encourages teachers to teach evolutionary thinking in the context of a more complete understanding of modern scientific thinking. Amy J. Binder's *Contentious Curricula: Afrocentrism and Creationism in American Public Schools* (Princeton, NJ: Princeton University Press, 2002) gives a sociologist's perspective on attempts to introduce creationism into the public schools. The NAS book for teachers, *Teaching about Evolution and the Nature of Science* (Washington, DC: National Academy Press, 1998; available at http://www.nap.edu/html/evolution98/), also deals with issues discussed in this chapter. A new journal published by Springer, *Evolution: Education and Outreach* (http://www.springer.com/life+sci/journal/12052), focuses on improving evolution education.

CHAPTER 12: ISSUES CONCERNING RELIGION

Dialogue between scientists interested in religion and theologians interested in science has expanded greatly since the mid-1990s, and so has the literature. Not all of this literature concerns the creationism/evolution issue, but there is an abundance nonetheless. A good place to begin to explore the science and religion movement is Robert John Russell's "Theology and Science: Current Issues and Further Directions" (Berkeley, CA: Center for Theology and the Natural Sciences, 2000; available at http://ctns.org/russell_article.html); then see the Center for Theology and the Natural Sciences, which publishes the journal *Science and Theology*, on the Web at http://ctns.org. The CTNS site links to Counterbalance (http://counterbalance.org/), which has video clips from a number of conferences featuring scientists and theologians. The AAAS sponsors the Web site Dialogue on Science, Ethics, and Religion (http://www.aaas.org/spp/dser/evolution/), which posts science and religion articles and other references. Ian G. Barbour's treatments of science and religion—*Religion and Science: Historical and Contemporary Issues*, rev. ed. (San Francisco: HarperSanFrancisco, 1997) and *Nature, Human Nature, and God* (Minneapolis: Fortress Press, 2002)—are classic and reflected in most subsequent scholarship. The Catholic theologian John F. Haught deals specifically with evolution and Christian theology in a string of books: *God after Darwin: A Theology of Evolution* (Boulder, CO: Westview Press, 2000); *Responses to 101 Questions on God and Evolution* (Mahwah, NJ: Paulist Press, 2001); *Deeper Than Darwin* (Boulder, CO: Westview Press, 2003). *Perspectives on an Evolving Creation*, edited by Keith B. Miller (Grand Rapids, MI: Eerdmans, 2003), is a collection of essays mostly by Protestant theistic evolutionists. The philosopher Michael Ruse argues in the affirmative to the answer he poses in the title of his book *Can a Darwinian Be a Christian? The Relationship between Science and Religion* (Cambridge: Cambridge University Press, 2001). Scientists who discuss their embrace of both their faith and evolution include Darrel R. Falk, in his *Coming to Peace with Science* (Downers Grove, IL: InterVarsity Press, 2003); Kenneth R. Miller, in his *Finding Darwin's God*, rev. ed. (New York: Harper Perennial, 2007) and *Only a Theory: Evolution and the Battle for America's Soul* (New York: Viking); and Francis S. Collins, in his *The Language of God* (New York: Free Press, 2006).

CHAPTER 13: THE NATURE OF SCIENCE

In addition to the readings for chapter 1, Robert T. Pennock and Michael Ruse's collection *But Is It Science? The Philosophical Question in the Creation/Evolution Controversy* (Amherst, NY: Prometheus, 2008) is a useful collection of papers on the philosophical issues. Although the popular idea that evolution is "just a theory" is widespread, it is more a cultural perspective than a position advocated; for critical discussions, see Stephen Jay Gould's 1981 essay "Evolution as Fact and Theory" (reprinted in his *Hen's Teeth and Horse's Toes*, New York: W. W. Norton, 1994; available at http://www.stephenjaygould.org/library/gould_fact-and-theory.html); T. Ryan Gregory's "Evolution as Fact, Theory, and Path," *Evolution: Education and Outreach* 1, no. 1 (2008): 46–52 (available at http://www.springerlink.com/content/21p11486w0582205/fulltext.pdf); and material available online at the TalkOrigins

Archive (http://www.talkorigins.org) and the Web site for the 2001 *Evolution* series (http://www.pbs.org/wgbh/evolution). Percival W. Davis and Dean H. Kenyon, *Of Pandas and People*, 2nd ed. (Dallas, TX: Haughton, 1993) contains a discussion of evolution as fact and as theory from the point of view of IDC. The operations-origins science distinction is frequently invoked in articles online at AIG's Web site. For discussions of methodological and philosophical naturalism, see sections 2 and 5 of Robert T. Pennock's collection *Intelligent Design Creationism and Its Critics* (Cambridge, MA: MIT Press, 2001). Various articles by IDC proponents on naturalism are available online at ARN and CSC; look especially for articles by Phillip Johnson, Stephen C. Meyer, and Paul Nelson. Robert T. Pennock's *Tower of Babel: The Evidence against the New Creationism* (Cambridge, MA: MIT Press, 1999) is a good response.

CHAPTER 14: EVOLUTION AND CREATIONISM IN THE MEDIA AND PUBLIC OPINION

For useful discussions of public opinion polling with respect to evolution, see George Bishop, "The Religious Worldview and American Beliefs about Human Origins," *Public Perspective* 9, no. 5 (1998): 39–44; George Bishop, "'Intelligent Design': Illusions of an Informed Public," *Public Perspective* 14, no. 3 (2003): 40–42; Matthew C. Nisbet and Erik C. Nisbet, "Evolution and Intelligent Design: Understanding Public Opinion," *Geotimes* 50, no. 8 (2005): 28–33; and George Bishop, "Polls Apart on Human Origins," *Public Opinion Pros* 2006 August; available at http://www.publicopinionpros.com/features/2006/aug/bishop.asp. For polling on the views of scientists, see Edward J. Larson and Larry Witham, "Scientists and Religion in America," *Scientific American*, September 1999, 88–93, or the appendix to Larry Witham's *Where Darwin Meets the Bible: Creationists and Evolutionists in America* (New York: Oxford University Press, 2002), but C. Mackenzie Brown advises caution in interpreting those results in "The Conflict between Religion and Science in Light of the Patterns of Religious Belief among Scientists," *Zygon* 38, no. 3 (2003): 603–632. For advice to scientists, see Glenn Branch, "The Battle over Evolution: How Geoscientists Can Help," *The Sedimentary Record* 3, no. 3 (2005): 4–8 (available at http://www.sepm.org/sedrecord/SR%203-3.pdf), and "Evolution and Its Discontents: A Role for Scientists in Science Education," by a coalition of scientific societies, in *The FASEB Journal* 22 (2008): 1–4 (available at http://www.fasebj.org/cgi/content/full/22/1/1). A central theme of Randy Olson's comic documentary *Flock of Dodos: The Evolution–Intelligent Design Circus* (Prairie Starfish Productions, 2006, 84 minutes; see http://www.flockofdodos.com) is that creationism prospers because scientists are ineffective communicators.

Name Index

Subject Index

Abington v. Schempp, 112
Abrupt appearance theory, 120–21
Academic freedom: invoked by creationist teachers, 154, 220, 244; antievolution legislation and policies invoking, 114, 159–61, 220, 226, 228, 232, 244–45
Access Research Network, 110
Adaptation, 37–39, 199–200
Adaptive radiation, 40–43
Alpha Omega Institute, 110
American Atheists, 73
American Center for Law and Justice, 154
American Civil Liberties Union, 99–100, 114, 116, 147
American Humanist Association, 73
Americans United for Separation of Church and State, 147, 156
Answers in Genesis, 108–10
Anthropic Principle, 168; defense of, 183–85; criticism of, 185–86

Bible: and creationism, 57, 63–70; and geocentrism, 57, 65–66; literalist and inerrantist interpretations of, 74, 87–88, 94, 98, 170–71, 270–72; non-literalist and modernist interpretations of, 74, 87–88, 91, 98; and science, 55, 74, 81–82; symbolism in, 60–61, 272–74. *See also* Genesis.
Bible-Science Association, 65, 110
Big Bang, 24

Biola University, 98
Biological Sciences Curriculum Study, 104
Butler Act, 99–100

Cambrian Explosion, 31–32, 193–94; creationist view of, 194–96; evolutionary biologists' view of, 197–99
Cells: first, 26–27, 29; eukaryotic, 29, 31
Center for Scientific Creation, 110
Center for Renewal of Science and Culture. *See* Discovery Institute.
Center for Science and Culture. *See* Discovery Institute.
Chance and evolution, 37–38, 88, 200–1, 202–3
Cladistics. *See* Systematics.
Common ancestry. *See* Descent with modification.
Common descent. *See* Descent with modification.
Complex specified information, 127–28
Council for Secular Humanism, 73
County of Allegheny v. ACLU, 222
Creation Ministries International, 109
Creation Moments, 110
Creation Research Society, 65, 106, 107, 110
Creation-Science Association of Mid-America, 110
Creation Science Fellowship, 110
Creation Science Research Center, 112

About the Author

EUGENIE C. SCOTT is Executive Director of the National Center for Science Education, the leading advocacy group for the teaching of evolution in the United States. She has written extensively on the evolution vs. creationism controversy in scholarly and popular venues, and she has won numerous awards for her work from scholarly organizations, including the Public Service Award from the National Science Board. She is a recent past-president of the American Association of Physical Anthropologists.